Examining the Role of Environmental Change on Emerging Infectious Diseases and Pandemics

Maha Bouzid
University of East Anglia, UK

A volume in the Advances in Human Services and Public Health (AHSPH) Book Series

IGI GLOBAL
DISSEMINATOR OF KNOWLEDGE
www.igi-global.com

Published in the United States of America by
IGI Global
Information Science Reference (an imprint of IGI Global)
701 E. Chocolate Avenue
Hershey PA 17033
Tel: 717-533-8845
Fax: 717-533-8661
E-mail: cust@igi-global.com
Web site: http://www.igi-global.com

Library of Congress Cataloging-in-Publication Data

Names: Bouzid, Maha, 1978- editor.
Title: Examining the role of environmental change on emerging infectious diseases and pandemics / Maha Bouzid, editor.
Description: Hershey PA : Information Science Reference, [2017] | Includes bibliographical references and index.
Identifiers: LCCN 2016017817| ISBN 9781522505532 (hardcover) | ISBN 9781522505549 (ebook)
Subjects: | MESH: Communicable Diseases, Emerging--epidemiology | Communicable Diseases, Emerging--transmission | Climate Change | Pandemics | Public Health | Environment
Classification: LCC RA643 | NLM WA 110 | DDC 362.1969--dc23 LC record available at https://lccn.loc.gov/2016017817

This book is published in the IGI Global book series Advances in Human Services and Public Health (AHSPH) (ISSN: Pending; eISSN: Pending)

British Cataloguing in Publication Data
A Cataloguing in Publication record for this book is available from the British Library.

Advances in Human Services and Public Health (AHSPH) Book Series

ISSN: Pending
EISSN: Pending

Mission

The well-being of the general public should be a primary concern for any modern civilization. Ongoing research in the field of human services and public healthcare is necessary to evaluate, manage, and respond to the health and social needs of the global population.

The **Advances in Human Services and Public Health (AHSPH)** book series aims to publish high-quality reference publications focused on the latest methodologies, tools, issues, and strategies for managing the health and social welfare of the public. The AHSPH book series will be especially relevant for healthcare professionals, policy makers, government officials, and students seeking the latest research in this field.

Coverage

- Domestic Violence
- Poverty
- Medicare and Medicaid
- Assistance Programs
- Social Work
- Public Welfare
- Health Policy
- Social Welfare Policy
- Healthcare Reform
- Public Funding

IGI Global is currently accepting manuscripts for publication within this series. To submit a proposal for a volume in this series, please contact our Acquisition Editors at Acquisitions@igi-global.com or visit: http://www.igi-global.com/publish/.

The Advances in Human Services and Public Health (AHSPH) Book Series (ISSN Pending) is published by IGI Global, 701 E. Chocolate Avenue, Hershey, PA 17033-1240, USA, www.igi-global.com. This series is composed of titles available for purchase individually; each title is edited to be contextually exclusive from any other title within the series. For pricing and ordering information please visit http://www.igi-global.com/book-series/advances-human-services-public-health/102256. Postmaster: Send all address changes to above address.

Titles in this Series

For a list of additional titles in this series, please visit: www.igi-global.com

Public Health Genomics and International Wealth Creation
E. William Ebomoyi (Chicago State University, USA)
Information Science Reference • copyright 2015 • 463pp • H/C (ISBN: 9781466685598) • US $235.00 (our price)

Transforming Public Health in Developing Nations
Mohamud Sheikh (The School of Public Health and Community Medicine, The University of New South Wales, Sydney Australia) Aziza Mahamoud (Sidra Medical and Research Center, Qatar) and Mowafa Househ (King Saud Bin Abdul Aziz University for Health Sciences, College Of Public Health and Health Informatics, Saudi Arabia)
Information Science Reference • copyright 2015 • 416pp • H/C (ISBN: 9781466687028) • US $215.00 (our price)

Innovative Collaborative Practice and Reflection in Patient Education
Jennifer Lynne Bird (Florida Atlantic University, USA)
Medical Information Science Reference • copyright 2015 • 308pp • H/C (ISBN: 9781466675247) • US $225.00 (our price)

Evidence Discovery and Assessment in Social Work Practice
Margaret Pack (Australian Catholic University, Australia) and Justin Cargill (Victoria University of Wellington, New Zealand)
Information Science Reference • copyright 2015 • 402pp • H/C (ISBN: 9781466665637) • US $185.00 (our price)

Handbook of Research on Adult and Community Health Education Tools, Trends, and Methodologies
Victor C. X. Wang (Florida Atlantic University, USA)
Medical Information Science Reference • copyright 2014 • 485pp • H/C (ISBN: 9781466662605) • US $325.00 (our price)

List of Reviewers

Carlos Amabile-Cuevas, *Fundación Lusara, Mexico City, Mexico*
Marjorie Bardiau, *University of Brighton, UK*
Diana Bell, *University of East Anglia, UK*
Christopher Buse, *University of Toronto, Canada*
Ulisses Confalonieri, *Fundação Oswaldo Cruz (FIOCRUZ), Brazil*
Sally Cutler, *University of East London, UK*
Marie Ducrotoy, *The University of Edinburgh, UK*
James Ebdon, *University of Brighton, UK*
Millicent Eidson, *University at Albany, State University of New York, Albany, NY, USA*
Maria Esteve-Gassent, *Texas A&M University, TX, USA*
Maya Gislason, *Simon Fraser University, Canada*
Edmundo Grisard, *Universidade Federal de Santa Catarina, Brazil*
Claire Heffernan, *University of Bristol, UK*
Nynke Hofstra, *Wageningen University, The Netherlands*
Johanna Lindahl, *International Livestock Research Institute, Nairobi, Kenya*
Maura Manganelli, *Istituto Superiore di Sanità, Rome, Italy*
Michelle Mclean, *Bond University, Australia*
John McNeill, *Georgetown University, Washington D.C., USA*
Elizabeth Omukunda, *Kenya Medical Research Institute, Kisumu, Kenya*
Nicholas Ogden, *Public Health Agency of Canada, Saint-Hyacinthe, Quebec*
Jaclyn Paterson, *Health* Canada, Ottawa, Canada
Michel Pepin, *Université de Lyon, France*
Myriam Riyad, Université Hassan II de Casablanca, *Morocco*
Andrew Solow, *Woods Hole Oceanographic Institution, Woods Hole, MA, USA*
Yves Tourre, *Lamont-Doherty Earth Observatory of Columbia University, Palisades, New York, USA*
Dario Vezzani, *Universidad Nacional del Centro de la Provincia de Buenos Aires,* Argentina

Table of Contents

Detailed Table of Contents

This chapter provides a review of infectious disease to date and the challenges they may present in the future. The main pandemics that have driven the history of humanity are described, from the first to be recorded in 3180 BC to more recent ones such as AIDIS, SARS and others associated with emerging pathogens. The essential role of emerging scientific specialisms (particularly microbiology, public health and sanitary engineering) to our understanding of the causes of these diseases (and how they may be better monitored, controlled and prevented) is presented. Globalization and climate change, determining factors for the ecology of infectious diseases and their emergence and re-emergence, are discussed and point to the urgent need for research to deal with these threats that continue to have a significant impact on human development and wellbeing.

The prediction of emerging infectious diseases (EIDs) and the avoidance of their tremendous social and economic costs is contingent on the identification of their most likely drivers. It is argued that the drivers of global environmental change (and

In South America in the past decades several infectious diseases have emerged or re-emerged either as part of larger pandemics or as local processes involving autochthonous pathogens. These included arthropod-borne viral diseases, such as Dengue Fever, Chikungunya and Zika as well as viral hemorrhagic fevers, such as Hantavirus Pulmonary Syndrome, Junin, Machupo and Guanarito viruses. Parasitic disease was also important such as Malaria, endemic in the northern part of the continent, Leishmaniasis and Chagas Disease. Carrion disease, a bacterial infection originally from the Andes region, also seems to be expanding geographically. Several social and environmental processes have contributed to the emergence of these pathogens, including human migration, deforestation, road and dam building and climate shifts. Due to its high biological diversity of wildlife, arthropods and virus species in still untouched natural ecosystems in the Amazon has the greatest regional potential for the emergence of new human infections.

Climate change is expected to affect the distribution, prevalence and life cycle of several infectious diseases. This scenario is relevant to Morocco since the country is considered by many IPCC assessments reports as a climate change hotspot with a high vulnerability to many expected impacts. Given this existing vulnerability, this chapter aims to highlight relevant vector-borne diseases, the risks of their reemergence in many vulnerable regions and the pressing need to understand their dynamics within a context marked by knowledge gaps and limited scientific evidence; underline the problematic aspects of health adaptation to climate change and the

Preface

Infectious diseases have plagued humankind since the beginning of their existence, shaped their history and evolution, cause significant burden and claimed numerous lives and continue to do so today. Our understanding of infectious diseases is continuously evolving and several key steps helped this progress including the discovery of the microscope, antibiotics, vaccines, genetic manipulation, gene sequencing and bioinformatics (to mention just a few). At various stages of human history, it was thought that humans won the battle and were able to control and even eradicate some very serious and devastating infectious diseases. Unfortunately, while some pathogens have indeed been efficiently controlled, new threats to public health are constantly appearing, commonly known as emerging pathogens.

An emerging or re-emerging infectious disease is defined as a disease that just appeared or whose incidence has increased considerably in the last two decades or likely to increase in the near future according to the Centers for Disease Control and Prevention (CDC). While according to the World Health Organisation (WHO), an emerging disease is defined as appearing for the first time in a population or if it is already present, is rapidly increasing in incidence or geographical distribution. Therefore, the main feature of emerging pathogens is sudden appearance, which tends to take the world in general and public health officials in particular by surprise. This happened far too many times for health agencies and governments to feel infallible and resort to complacency. Throughout history, numerous infectious diseases emerged and some caused pandemics (epidemic infectious diseases that spread through the human population across a large region), typically associated with devastating and unforgettable losses such as plague, poliomyelitis, HIV/ AIDS, Spanish influenza, SARS, MERS and Ebola. However, this arms race between humans and pathogens is everlasting and both are evolving to ensure survival of the fittest as supported by the Red Queen Hypothesis "to survive in an evolving system, one must co-evolve with the best traits to survive". Therefore, these interactions are highly complex and under the influence of a multitude of intrinsic and extrinsic factors, making the infectious diseases field a fascinating domain. A lot of drivers intervene in the relationship between infectious diseases and human hosts and though we made significant headway in understanding the drivers and outcomes of these interactions, particularly in terms of disease occurrence, we are still poorly equipped to control and especially prevent new disease emergence. Whilst, the world started to forget about the devastating Ebola outbreak in West Africa, international health agencies praised for the efficient control of the outbreak, survivors attempting to go back to normal life and countries celebrating Ebola free status, a new virus steals the spotlight and media headlines. The new "kid on the block" is the Zika Virus, so far a mild arbovirus limited to endemic locations in Africa, Asia and South Pacific but currently responsible for a fast spreading outbreak in South and Central America (22 countries so far) and because of possible links to microcephaly in newborn babies, it has been declared a "Public Health Emergency of International Concern" by the World Health Organisation on 1 February 2016. The reasons for this wildfire like spread and the surge of neurological disease are still unknown but further resource allocation and research would allow to decipher the drivers of this new epidemic. It took time and resources to solve the Ebola emergence conundrum and several factors have contributed to the outbreak in West Africa including bushmeat consumption, burial practices, civil unrest, human mobility and behavioural and cultural practices, though urban settings and socioeconomic conditions are believed to be the main factors. In time, the drivers and mechanisms of the Zika virus emergence are likely to be unravelled. In the meantime, it is a valuable reminder to health officials, NGOs and governing bodies that, despite major advances, we cannot drop our guard and we need to keep investing in new vaccines and therapies and more efficient and novel mosquito control strategies.

The link between environment and health was noticed by humans earlier on. It has been reported that Roman aristocrats spent their summers in hill resorts to avoid malaria and that in south Asia, strongly curried food was likely to be protective of diarrhoea especially in hot summers. Another example of the effect of the environment on infectious diseases is the earlier belief that bad air (miasma) caused diseases such as the Black Death, cholera and malaria. Since then, advances in microbiology and medicine established that the causes of these diseases are infectious agents (germ theory of disease), particularly the work of Louis Pasteur and Robert Koch in the mid- late 1800s. Nevertheless, the influence of the environment on human health is paramount as everything around us (air, water, soil) can contribute to our good health and conversely potentially transmit disease causing particles (whether infectious or not). Therefore, changes in the living environment are bound to affect human health. The climate change field, though some opponents still exist, is firmly established and the main cause is related to anthropogenic activity. Climate change is projected to increase temperature, droughts, floods and extreme weather events and is associated with significant adverse impacts on human health through various mechanisms. While the burden of climate change could be mediated through non-infectious causes such as heat and cold related morbidity and mortality, risk of injury, death and mental illness following extreme weather events as well as impacts of air pollution; infectious diseases contribute to significant increased burden. The environment is a key player in disease emergence phenomenon. This book aims to examine the impact of environmental change on emerging infectious diseases and pandemics, in particular, lessons learnt from previous epidemics and pandemics, comprehensive investigation of significant drivers of disease emergence, climate sensitive diseases (vector-borne and waterborne), climate vulnerability in cities and the health sector, governance and adaptation strategies, and perspective of public health officials.

The target audience for this book is wide, encompassing people with prior biological, medical and environmental knowledge as well as the general public. This book would be of interest to environmental scientists, academics, students, researchers, microbiologists, stakeholders, policy makers, public and environmental health officials, advocates, practitioners, NGOs, health charity workers and interested laypersons.

This comprehensive and timely publication aims to be an essential reference, a valuable addition to the field and a useful resource for the interested reader. It will serve as a succinct yet comprehensive overview of the topic allowing the readers to capture the main facts and confer them the knowledge needed for meaningful discussion of the topic.

Maha Bouzid
University of East Anglia, UK
February 2016

Acknowledgment

I would like to thank all the contributors who committed to this project and dedicated their time to write fascinating chapters and share their expertise and insightful knowledge.

I express my deep appreciation to the reviewers who generously agreed to review the chapters and provided valuable feedback that contributed to the improvement of the publication.

Finally, I would like to express my gratitude to IGI staff who offered me this opportunity, assisted with the development and production of this publication and believed in my editorial capabilities. I thoroughly enjoyed the experience.

Chapter 1
History of Infectious Diseases

Maria Ines Zanoli Sato
CETESB, São Paulo, Brazil

ABSTRACT

This chapter provides a review of infectious disease to date and the challenges they may present in the future. The main pandemics that have driven the history of humanity are described, from the first to be recorded in 3180 BC to more recent ones such as AIDIS, SARS and others associated with emerging pathogens. The essential role of emerging scientific specialisms (particularly microbiology, public health and sanitary engineering) to our understanding of the causes of these diseases (and how they may be better monitored, controlled and prevented) is presented. Globalization and climate change, determining factors for the ecology of infectious diseases and their emergence and re-emergence, are discussed and point to the urgent need for research to deal with these threats that continue to have a significant impact on human development and wellbeing.

INTRODUCTION

In recent decades, concern about the quality and sustainability of the environment has been increasing and this is now a critical global issue, involving both industrialised and low-income countries.

DOI: 10.4018/978-1-5225-0553-2.ch001

The influence of the environment on human health dates back to antiquity, when human life expectancy was shorter than today as a result of the hostile environment in which they lived. Infectious diseases have been a health hazard faced by humanity since human beings first appeared. These diseases spread from person to person or from animals to humans, often through water, food or insect vectors and so transmission may be directly affected by environmental changes (Yassi et al., 2001). Hippocrates, in his treatise "Airs, Water and Places", described the importance of these environments to the spread of diseases and made an association between human infection and the characteristics of climate, soil, water, diet, and living conditions (Nelson & Williams, 2014). The impact of infectious diseases on human populations has influenced the history of the world, not just because of their devastating physical, economic, cultural, political and social consequences, but also because they have driven the development of several scientific and technological areas, such as microbiology, epidemiology, immunology, medicine, engineering, biotechnology, food safety and occupational infection control.

In the second half of the last century, with the advent of antibiotics and vaccines, some people believed that infectious diseases had been controlled, but this confidence declined with the appearance of emerging and re-emerging pathogens in 1980s. The emergence of these infectious diseases caused alarm because of their potential global spread and resulting health impact and coincided with economic globalization and climate change, both of which represented determining factors in their spread. Consequently, understanding the ecology of these pathogens is increasingly considered an essential component of management and control strategies for these diseases.

The aim of this chapter is to describe the history of infectious diseases and major human epidemics, as well as the significant features and discoveries that supported the control and prevention of these diseases. The historical report outlines ancient history, the post-classical era (Middle Ages) and modern history (early, middle and contemporary periods). The establishment of regulations/environmental quality criteria and health education, as important measures to deal with and reduce infectious diseases, are also highlighted. The final part covers the role of the environment in the transmission of these infectious diseases.

HISTORICAL PERSPECTIVES

Ancient History (3600 BC-500 AD)

The aggregations of humans into sedentary agricultural settlements along important rivers, such as the Tigris, Euphrates, Indus and Niles gave rise to the first cities and the now common diseases of humanity began to appear at that time. Since these prehistoric times, the continuous expansion of the human population has led to successive invasions by increasing numbers of different pathogens (Dobson & Carper, 1996).

The first recorded epidemic in human history occurred in Egypt in 3180 BC, during the reign of Pharaoh Mempses in the First Dynasty and it was referred to as the "great pestilence". Manetho, a contemporary priest and historian, in his listing of the pharaohs, mentioned "Mempses, for eighteen years. In his reign many portents and a great pestilence occurred" (Choi & Pak, 2001). Infectious diseases, which were often referred to as "pestilance" or "plagues", in early writings, have undoubtedly played an important role in human history.

The Ebers papyrus, a medical document dating from around 1550 BC., which was discovered in a tomb in Thebes, Egypt, describes "epidemic fevers" and identifies a disease that "has produced a bubo, and the pus has petrified, the disease has hit". Today this disease is known as bubonic plague and is recognized to be caused by the bacterium *Yersinia pestis* (Panagiotakopulu, 2004). In the Old Testament, when the Philistines captured the Ark of the Covenant, the text states that "the Lord was angry and rats appeared in their land.... He afflicted them with tumours in the groin ..." (1 Samuel 5:9). Also, the Iliad, Homer's epic poem, tells of a mysterious epidemic that decimated the Greek army outside Troy during the Trojan War (1194 BC-1184 BC), which is now believed to have been bubonic plague.

One of the greatest wartime epidemics in classical times was the plague of Athens, as described by the Greek historian Thucydides (c. 460 B.C. – c. 395 B.C.) in his book "A History of the Peloponnesian War" (431 BC – 404 BC). "The Plague of Thucydides", as it was also known, began in Ethiopia and then spread to Egypt, the Persian Empire and Athens. The Athenian mortality rate was estimated to be 25% of the whole population (Smallman-Raynor & Cliff, 2004). This was probably a mixed epidemic caused by both typhus and measles (Choi, 2012).

During the initial Christian era, Rome faced two of the major pandemics of early European history. The plague of Antonines (AD 165-80), associated with the Roman Eastern War, was spread by soldiers returning to Rome from the Parthian War and made tens of thousands of victims. Descriptions of disease symptoms suggest smallpox, but typhus fever, measles, or a mixture of both were also suspected. During the Roman-Gothic Wars (249 AD – 268 AD), the Mediterranean Basin was severely affected by other plagues, as described by Saint Cyprian, Bishop of Carthage, which became known as the "Plague of Cyprian" (AD 251-66). The pestilence devastated the Roman legions and is reported to have claimed five thousand victims a day in Rome. According to the symptoms, the disease may have been measles or smallpox (Smallman-Raynor & Cliff, 2004).

Hippocrates (460-377 BC) was one the first to record his theories on the occurrence of disease and it was he that defined the terms "endemic" and "epidemic" diseases. In his treatise on "Airs, Water and Places", Hippocrates dismissed supernatural explanations of disease and instead attributed illness to characteristics of the climate, soil, water, mode of life, and nutrition surrounding the patient (Nelson & Williams, 2014).

The Middle Ages (500-1500)

During this period, two of the three most devastating pandemics to hit the human race were recorded, namely the "Plague of Justinian" and the "Black Death". Both of these have been confirmed to be bubonic plague (Wagner et al., 2014). Genetic studies suggested the Orientalis biovar (China) of *Yersinia pestis* to have been the primary source of contagion (Haensch et al., 2010).

The Justinian plague began in 541 AD and spread from Egypt to the Eastern Roman Empire (Byzantine Empire), especially its capital Constantinople, during the reign of Emperor Justinian in the sixth century AD. He contracted the disease himself, yet survived. The plague returned periodically until the 8th century (767 AD) and is estimated to have killed between 25 (initial period) to 50 million people (two centuries later), probably about half the global population at that time (CDC, 2014; Rosen, 2007; Wagner et al, 2014).

The "Black Death" or the "Great Plague" appeared 600 years later, originating in Central Asia in 1347, possibly because of climate changes (William, 1997). The disease spread from the Caspian Sea to almost all European countries, causing the death of 60% of the European population (CDC, 2014; Cohn, 2005). However, despite the vast devastation caused by this pandemic, the labour scarcity that resulted from high mortality rates is thought to have accelerated the development of many economic, social, and technical innovations (CDC, 2014). It was even considered to have been a factor in the emergence of the Renaissance in the late 14th century.

During the period 664-689 AD, a widespread and fatal epidemic, known as the "Yellow Plague" because jaundice was a characteristic and predominant symptom, struck Medieval Britain and Ireland. This is now thought to have been an example of what was later called 'relapsing fever', a vector-borne disease caused by bacteria of the genus *Borrelia* (Choi & Park, 2001).

In addition to the diseases already mentioned, there are early accounts of rabies, anthrax, dysentery, smallpox, ergotism, botulism, measles, typhoid fever, typhus fever, diphtheria, and syphilis (Engelkirk & Duben-Engelkirk, 2011).

Although, since antiquity there have been examples of societal rules for isolating lepers and bubonic patients, the first formal system of quarantine (the separation of the diseased from the healthy) was established in the 14th century AD in Venice during the Black Death, requiring ships to lay at anchor for forty days before landing (the word "quarantine" means forty in Latin).

Modern History (1500-Present)

1. Early Modern Period (1500-1750)

From this period, it is important to highlight the "Great Plague of London". Plague had previously existed in England for centuries but in the spring and summer of 1665 an outbreak of Bubonic Plague spread from parish to parish until one fifth of the population was dead (around 60,000 people). The poor sanitary conditions, dense population and overcrowded housing, along with the very hot summer experienced that year, were a perfect breeding ground for rats and hence for dissemination of the disease. In 1666, London suffered another catastrophic event, the "Great Fire of London", which destroyed much of the centre of the city, including those areas most affected by plague. Despite this disaster, the fire helped to kill the rodents (black rats or gerbils) and fleas that carried the *Yersinia pestis* bacteria and finally brought the disease under control. The 1665 epidemic resulted in the enactment of important changes to public health policy by the English Government to confront epidemic diseases, including a new set of regulations. It also contributed to evolution in concepts of nature and causes of infectious diseases (Choi, 2012; Harvard University Library, n.d.; Hays, 2005; Schmid et al., 2015).

Girolamo Frascatoro, a Veronese physician, made the first known allusion to the "germ theory", at this period. In his essay on contagion, "De contagione et contagiosis morbis et eorum curatione libri tres" (The three books on contagion, contagious diseases and their cure), published in 1546, he wrote about seminaria morbis – tiny seeds of disease that can pass from one individual to another, and cause infection and epidemics (Echeverría, 2010). The name for syphilis is derived from Fracastoro's 1530 epic poem in three books, *Syphilis sive morbus gallicus*.

The invention of the microscope during the 1590s by Zacharias Janssen (c.1580 - c.1638), a Dutch spectacle-maker, led to the first observations of the invisible world of microorganisms. He was credited with the invention of a single-lens optical and compound microscopes.

In 1676, Anton van Leeuwenhoek (1632 - 1723), a Dutch tradesman and scientist, who worked on the development and improvement of the microscope, was the first person to observe and describe single-celled organisms. Using handcrafted microscopes, he examined water, blood, sperm, diarrhoeal stools and scrapings from teeth and reported tiny living creatures, which he called "animalcules". As a result, he has been referred to as the "Father of Bacteriology" and "Father of Protozoology" (Brock, 1961, Engelkirk & Duben-Engelkirk, 2011).

Some years earlier, in 1665, Robert Hooke, an English scientist, and like Van Leeuwenhoek a Fellow of the Royal Society, reported in *Micrographia* the first description of a microorganism, the microfungus *Mucor*. He observed the cellular composition of cork and introduced the word 'cell' to science. Around 150 years later, microscopy became the backbone of scientific understanding of the roles of microbes as aetiologic agents of infectious diseases and in the recycling of chemical elements in the biosphere (biogeochemical cycles) (Gest, 2004).

2. Mid-Modern Period (1750-1914)

At the beginning of this period, Muller, in his *Vermium terrestrium et fluviatilum* (1773), made the first description of organisms, naming life forms such as *Vibrio* and *Monas*, now known to be bacteria. Muller's work represented the beginning of the descriptive phase of bacteriology.

Some important developments in science that would prove to be very helpful for water and wastewater treatment processes occurred at this time. In 1774, Carl Wilhem Scheele, a Swedish chemist, discovered the element chlorine and in 1783, Martinus Van Marum, a Dutch scientist, identified ozone. Later Guyton developed the use of chlorine as a disinfectant and wrote his *Treatise on Disinfecting the Air* in 1801. In the same year, the physicist Johann Wilhelm Ritter discovered UV radiation in Jena and became the creator of modern electrochemistry. Ozonation, chlorination and UV began to be applied to drinking water disinfection in 1893, 1896 and 1901, respectively. Nowadays, these are the most common processes employed in the disinfection of water and sewage.

This period was also marked by the impact of cholera pandemics in Europe. Although the first known reference to an outbreak of cholera was recorded in an inscription in a temple near Gujarat, western India, dating from the fourth century BC, it was in the 19th century that this disease generated profound social effects, and contributed significantly to the fields of epidemiology and bacteriology. During the period 1817 to 1875, the world was swept by four cholera pandemics - 1817–1823, 1826–1837, 1846–1863, 1863–1875 – exacerbated by international steamship traffic and mass migration during the Industrial Revolution. India (1817, 1821), China (1820), Japan (1822), Russia (1823), England (1831), Canada and the USA (1832), Africa (1837), Central America (1863), and South America (1875) were all affected. In 1849, John Snow, who latterly became known as the "father of modern epidemiology", published an essay "On the Mode of Communication of Cholera", which posited a link between cholera cases and contaminated water. Five years later (1854), Snow proved his theory, when mapping cholera cases in the Soho district in London. During a local outbreak of cholera, he was able to identify a public water pump in Broad Street as the source of the disease and his evidence led to the formulation of the Public Health Act of 1875 in England (Kumar & Raul, 2014).

In 1856, William Budd demonstrated that typhoid fever had an almost identical transmission pattern to cholera; this contagious disease was transmitted via water sources that were faecally contaminated (NRC, 2004, Parry, 2005). By this time, large-scale chlorination of wastewater was applied in London to de-odourize the city's sewers, and later to disinfect the faeces of typhoid patients before disposal to the sewer. In 1893, in Hamburg, Germany, chlorination of wastewater was undertaken at treatment plant-scale to control a waterborne epidemic of typhoid fever.

At the end of 19th century (1894) the third known plague pandemic began in the Yunnan region of China and spread from Hong Kong to almost the entire world via shipping. During the next five decades, the pandemic persisted and caused over 15 million deaths, most of them in India. During this last pandemic, *Yersinia pestis* was identified as the aetiological agent of the disease; the transmission vector and the first effective treatments were discovered (Barry, 2008, Frith, 2012).

The period 1830 to 1900 was a turning point for microbiology, medicine and epidemiology. Although cells had already been discovered, it was during these years that cell theory was consolidated by three German scientists. In 1938, Matthias Schleiden, a botanist, reported that plants were composed of cells; a year later Theodor Schwan discovered that animals were also composed of cells. In 1855 Rudolph Virchow, a physician, proposed the theory of biogenesis and expressed its third dictum as "omnis cellula e cellula" (every cell originates from another existing cell). He also originated the concept of cellular pathology when he established that all diseases involve changes to normal cells (Myron, 2008).

It was, however, Louis Pasteur, a French chemist and microbiologist, who definitively proved the theory of biogenesis, i.e., that "life comes only from life". Details of the elegant experiments conducted by Pasteur using flasks with long s-curved necks, which disproved spontaneous generation by demonstrating that contamination of beef extract in the flasks came from air, can be found in several basic microbiological books (Pelczar et al., 2010; Tortora et al., 2010; Engelkirk and Duben-Engelkirk, 2011). Pasteur is also known for other important contributions that are considered fundamental to microbiology and modern medicine, such as the development of vaccines, fermentation and the pasteurisation.

Another scientist who made numerous contributions to the science of microbiology during this period was Robert Koch (1843–1910), a German physician. One major contribution was the consolidation of the germ theory of disease through an experimental procedure that demonstrated that a specific microbe was the cause of a specific infectious disease. This procedure was published by Koch and his colleagues in 1884, and has become known as Koch's Postulates. Also attributed to Robert Koch: the discovery of spores in *B. anthracis* (resistant forms), the development of solid media to culture bacteria, and the discovery of the bacteria that cause tuberculosis (*M. tuberculosis*) and cholera (*Vibrio cholera*).

In 1906, the first case of an 'asymptomatic carrier' was recorded. Mary Mallon, later known as 'Typhoid Mary', was the first person in the United States to be identified as an asymptomatic or healthy carrier of the pathogen *Salmonella typhi*. She worked as a cook and it is presumed that she directly infected 53 people, three of whom died. By 1907, about 3,000 New Yorkers had been infected by *Salmonella typhi*, and Mary was probably the main reason for the outbreak (Marineli et al., 2013).

In 1914, one of the most extensive studies ever undertaken on the characterisation of bacterial pollutants was conducted in the Great Lakes region. More than 18,000 water samples from 1,500 sampling sites were analysed for indicator bacteria by seventeen collaborating laboratories. The study was instigated by an interest in water quality and its impacts on human waterborne disease (Edge et al., 2013).

At that time, the germ theory of disease having been elucidated, it was clear that water contaminated with human sewage could harbour microorganisms that threaten public health. The concern was focused on water quality and the fate of pathogens in the environment, with the aim of protecting public health (Maier et al, 2009). A new field of 'water microbiology' was thereby established, which will be used as an example to illustrate disease control and prevention options as well as the effect of climate change.

3. Contemporary Period (1914-Present)

At the beginning of the contemporary period (1918-1919) the last of the three most devastating pandemics yet to strike humanity occurred, namely "Spanish Influenza". The disease, caused by influenza A (H1N1) virus, killed an estimated 40 million people, far more than the 10 million deaths that had been caused by the First World War (1914-1918). The human influenza virus was not isolated until 1933 and vaccine production began in 1957. Facemasks and hand washing were also adopted as preventive measures (Choi & Pack, 2001; Taubenberg, 2006; Choi, 2012). Pandemic influenza viruses have emerged twice more during the 20th century (1957, Asian influenza H2N2; 1968, Hong Kong influenza, H3N2) and in 2009 a new pandemic of influenza virus H1N1 emerged in the USA and subsequently affected more than 70 countries (CDC, 2010; Taubenberg, 2006).

During the first half of the 20th century, poliomyelitis was one of the most feared diseases in industrialised countries, reaching pandemic proportions in Europe, North America, Australia and New Zealand and paralysing hundreds of thousands of children every year. The first vaccine against polio, developed in 1955 by Dr Jonas Salk, was an injectable vaccine from an inactivated virus. Six years later Dr. Albert Sabin developed an oral vaccine, using an attenuated poliovirus, which is now extensively used in immunisation programmes around the world. In the second half of the century, poliomyelitis was also recognised to be a major problem in low-income countries and in 1988 the 'Global Polio Eradication Initiative' was launched. Consequently, global incidence of polio has decreased by 99% - in 2013, 416 cases were reported for the entire year as opposed to over 350,000 in 1988 (Polio Global Eradication Initiative, http://www.polioeradication.org/). The major problem faced today in relation to poliomyelitis is the outbreaks associated with vaccine-derived poliovirus.

The discoveries and advances in the fields of microbiology, physics, chemistry and epidemiology that occurred during the 1800s directly led to actions to prevent and control infectious diseases. In the 1920s, vaccines for diphtheria, tetanus, whooping cough and tuberculosis became available for immunisation of the world population and in the same decade Sir Alexander Fleming discovered penicillin, thus initiating the antibiotic era. All these measures, along with improvements in sanitary practices, personal hygiene and public health education decreased the prevalence and incidence of many infectious diseases and reduced mortality rates (Brachman, 2003; Satcher, 1995).

The 20th century also witnessed the rapid development of new microbial detection technologies that would subsequently support the rapid and accurate diagnosis of infectious diseases, which led to the award of Nobel Prizes to several scientists.

The invention of the electronic microscope by the German physicists Max Knoll and Ernst Ruska in 1931 brought higher visual resolution to the microbial world. In 1952, Renato Dulbecco, an Italian physician and researcher in the phage group of Caltech, USA, adapted the plaque assay, previously used to enumerate bacteriophages, to the enumeration of infectious virus particles in cultured host cells, opening up new horizons for virology studies (Baltimore, 2014). The following year James Watson, an American biologist, and Francis Crick an English physicist, described the double helix molecular structure of DNA, a milestone in the history of science that gave rise to modern molecular biology.

In 1983, Kary Mullis, a scientist from the Cetus Corporation (Berkeley, CA) invented the polymerase chain reaction (PCR), now recognised to be one of the most important scientific advances of the 20th century. This was a fast and simple methodology to amplify sections of DNA *in vitro* that is now widely employed in clinical, environmental and food laboratory settings, as well as in basic and applied molecular biology research. Publications in this field consequently grew rapidly and by 2001 around 3.5% of all publications listed on the website Pubmed included the phrase "polymerase chain reaction" (Bartlett & Stirling, 2003). Nowadays several variation of conventional PCR are available, such as nested PCR, multiplex PCR, RT-PCR, PCR integrated with culture cell (ICC/PCR), real-time PCR, microarrays and digital PCR (Cupples et al, 2010).

The early part of the 20th century is widely looked upon as a golden age in public health, though mainly for the industrialised world. Technology development in the field of microbial diagnostics led to the identification of new aetiologic agents associated with these diseases. Although large epidemics were controlled, many pathogens have persisted endemically and re-emerging diseases have become new threats to the world population. According to the report of the Expert Committee of US National Research Council, published in 1992, six categories of factors might explain the emergence or re-emergence of infectious diseases: human demographics and behaviour; technology and industry, economic development and land use, international travel and commerce, microbial adaptation and change and breakdown of public health measures (Brachman, 2003).

From the 1970s, a number of emerging and re-emerging diseases affected specific populations and geographical areas of the world. Human immunodeficiency virus (HIV) responsible for the acquired immunodeficiency syndrome (AIDS), identified in 1981, is the fourth leading cause of death in the world and has consequently caused considerable alarm during recent decades. HIV continues to be a major global public health issue, having claimed more than 40 million lives to date.

In the 1990s, the seventh cholera pandemic arrived in Peru, and spread to the countries of Central and South America. Only Uruguay was spared. From 1991 to 1997, the Latin American epidemic resulted in 1.2 million cases and around 12,000 deaths. Some have suggested bilge or ballast water from a ship from Asia or Africa (released in inshore waters) as the likely source of cholera infections. Haiti has faced a serious cholera outbreak following its devastating earthquake in 2010 (magnitude 7.0 and greater) with 738,528 cases reported, of which 421,013 have been hospitalised (57% cumulative hospitalisation rate), and 8,925 have died (PAHO, 2015). Today there are an estimated 3–5 million global cases and 100 000–120 000 deaths resulting from cholera every year.

In 1993, the United States witnessed the largest waterborne disease outbreak ever documented in the country, caused by *Cryptosporidium.* Around 400,000 cases and 100 deaths were reported. Ten years later, the severe acute respiratory syndrome (SARS), caused by a coronavirus, alarmed the world and demonstrated how quickly global travel can lead to the spread of an infectious disease. SARS started in Asia and over the following few months, the illness spread to twenty six countries in North America, South America, Europe and Asia (NIH, 2013).

Viral haemorrhagic fevers, including Ebola, although to date more geographically localized, are of great concern, considering the high mortality rate and the fact that no effective preventative vaccine or treatment is available yet.

Tuberculosis (TB) re-emerged over the past two decades in industrialised countries in association with emigration, and in Africa as a consequence of the HIV epidemic (Borgdorff & van Soolingen, 2013). Drug-resistant TB is a major threat worldwide, as is the increasing prevalence of antibiotic resistance in medically relevant bacteria including strains of *Streptococcus, Enterococcus, Klebsiella pneumoniae, Staphylococcus, Pseudomonas aeruginosa,* etc.

Emerging and re-emerging infectious diseases continue to take public authorities and general population by surprise, despite the technological developments that have become available in the world. As reported by Satcher (1995): ... *the complete history of infectious diseases remains to be written.*

CASE OF WATER MICROBIOLOGY AND INFECTIOUS DISEASES: AN EXAMPLE TO ILLUSTRATE CONTROL AND PREVENTION MEASURES AND THE EFFECT OF CLIMATE CHANGE

Control and Prevention Measures: History and Significant Developments

The first engineering concepts in the field of water treatment also date back to early days. Ancient Sanskrit and Greek writings dating from around 2000 BC recommended water treatment methods, such as charcoal filtration, exposure to sunlight, boiling and straining. The focus, however, at this time was to improve the aesthetic quality, i.e., the taste and odour of drinking water because people could not yet distinguish between microbially contaminated and uncontaminated water. Turbidity was the main driving force behind the earliest water treatments and little or nothing was known on microbial or chemical contaminants until much later.

At around 1500 BC, the Egyptians first discovered the principle of coagulation and applied chemical alum to settle suspended particles. Representations of this purification technique have been discovered on the wall of the tomb of Amenophis II and Ramses II. After 500 BC, Hippocrates is reported to have made use of the first filter bag, which has been referred to as the 'Hippocratic sleeve'. The main purpose of the bag was to trap sediments that caused bad tastes or odours. In 300-200 BC, Rome built its first aqueducts and Archimedes invented his water screw.

The first structure used to transport water was probably built by the Assyrians in the 7th century BC, but it was the Romans that applied and implemented this knowledge to build aqueducts to carry water over long distances and supply water to the cities of the Roman Empire. In the city of Rome, over 400 km of aqueducts were built over a period of 500 years. Most aqueducts were constructed underground, to protect them in times of war and to prevent pollution. Today, aqueducts can still be found in some parts of France, Germany, Spain and Turkey (Enzler, 2015).

During the Middle Ages, little progress was made in the field of water supply and sanitation and this period has been referred to as the "Dark Ages" because of the relative lack of scientific innovations.

In the 1700s, filtration was established as an effective means to remove particles from water, and this process was applied for domestic use. By the early 1800s, slow sand filtration was applied at the first municipal water treatment plant in Scotland. Initially, water distribution was by horse and cart, but soon this practice was replaced by the use of water pipes. The proposal was that every home should have access to safe drinking water but it would take more than a century to achieve this in many countries and even today, millions of people still lack access to safe drinking water (Enzler, 2015; USEPA, 2000).

During the mid to late nineteenth century, with the discovery of the microbial world and the link between outbreaks of enteric infectious diseases and drinking water contaminated by sewage demonstrated, it was clear that microscopic organisms could transmit disease through environmental media, such as water, and concerns about drinking water quality continued to focus mostly on disease-causing microbes (pathogens) in public water supplies. Scientists discovered that particulate material suspended in source water, particularly faecal matter, could harbour microbial contaminants, so by reducing turbidity by filtration, pathogens could be removed. Since then, the design of most drinking water treatment systems has included a filtration stage (slow and/or rapid sand filtration) to reduce turbidity. The filtration process was improved when the technique of coagulation and sedimentation was introduced to precede it. However, it was chlorination that probably made the greatest contribution to reductions in waterborne disease outbreaks during the early twentieth century (Enzler, 2015; USDHEW, 1963; USEPA, 2000). Figure 1 demonstrates the relationship between the introduction of filtration and chlorination of drinking water in the early 20th century and the reduction of typhoid fever death rate in the United State (EHC, 2010; USDHEW, 1963).

This discovery led governments to start installing municipal water treatments with filters and disinfection processes. Consequently, there was a need for government regulation of public water, as well as the establishment of microbial indicators of water contamination.

The use of bacteria as an indicator of the sanitary quality of water probably dates back to the 1880s, when *Klebsiella* was described as a microorganism characteristically present in human faeces (Ashbolt et al., 2001). Table 1 provides a summary of the major landmarks in the development of microbial faecal indicators for drinking water quality and the first regulations (Ashbolt et al, 2001, Edberg et al., 2000; Leclerc et al, 2001).

Table 1. Landmarks in the development of microbial faecal indicators and drinking water criteria

Year	Statement
1880	Von Fritsch described *K. pneumoniae* and *K. rhinoscleromatis* as microorganisms characteristically found in human faeces.
1885	• Perci and Grace Frankland started the first routine bacteriological examination of water in London, using Robert Koch's solid gelatin media to count bacteria. • Theodor Escherich described *Bacillus coli* (later renamed *Escherichia coli*) from the faeces of breast-fed infants.
1891	• Theobold Smith of the State of New York Department of Health developed a presumptive test for *E.coli* using a lactose-based fermentation test tube. • The Franklands originated the concept that organisms typically found in sewage must be identified to provide evidence of the potential presence of pathogens
1893	• Sanitary bacteriologists began to use the Wurtz method for *B. coli* enumeration by direct plating of water samples on litmus lactose agar, using the concept of acid production from lactose as a diagnostic test.
1897	U.S. American Public Health Association adopted standard procedures for coliform test.
1901	Horrocks in the U.K., suggested the term "coliform bacteria", for those bacteria resembling *Bacillus coli*
1904	Eijkman developed what become known as the 'faecal coliform test.' This method was modified in the early 1900s by the elimination of elevated temperature incubation, to comprise all members of the *Enterobacteriaceae* (lactose fermentation and acid gas production)
1909	Coliform procedures were published in the first edition of Standard Methods for the Examination of Water and Wastewater.
1914	Drinking water standards were implemented eventually for drinking water supplies in public transit vehicles (buses, trains, ships), by the U.S. Public Health Service, based on coliform growth.
1934	Formal guidance on the bacteriological examination of water supplies and the interpretation of results was first published by the United Kingdom Ministry of Health in 1934, which was to become universally known as "Report 71".
1940's	U.S. Drinking water standards widely applied to municipal drinking water systems.
1948	In the U.K., Mackenzie and collaborators distinguished "faecal coliforms" from "coliforms" through thermotolerance and indole production.
1958	WHO published the "International Standards for Drinking-Water" concerned with the guarantee of water quality because of the increase of air travel around the world (globalisation). These stardards came to be adopted mainly by developing countries.
1972	Clean Water Act was passed in the USA.
1974	The U.S. Safe Drinking Water Act (SDWA) was formulated. The general principle in the industrialised world was that every person has the right to have 'safe' drinking water.
1980s	The defined substrate technology (DST) for direct detection and identification of *E. coli* in drinking water was established.
1995	DST was included in Standard Methods for the Examination of Water and Wastewater, 19th edition.
1990s	Quantitative Microbial Risk Assessment was established for Drinking Water Criteria
2004	Water Safe Planning – WHO Guidelines for Drinking Water Quality

Figure 1. Typhoid fever death rate in United States (1900-1960) after the introduction of filtration system and chlorination in public water supplies
EHC, 2010.

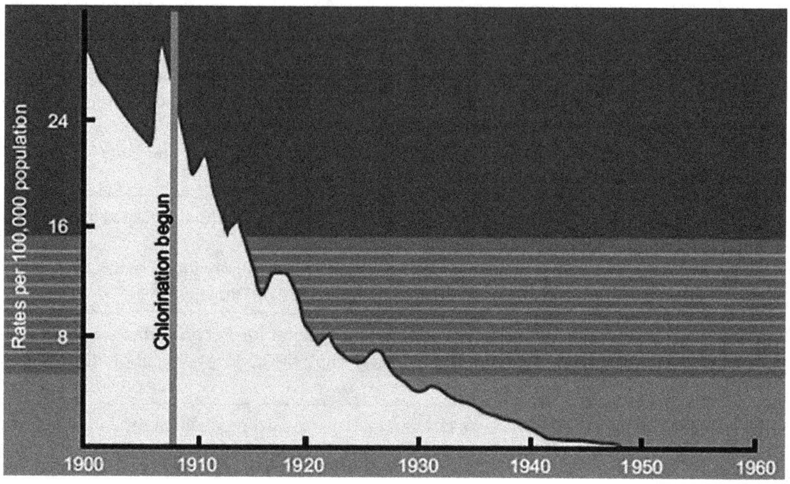

Although Theobald Smith proposed *E. coli* as a primary indicator of drinking water quality in the 1890s, it only became possible to detect this indicator directly by routine examination of water in the 1970's with the development of culture media containing defined substrates. During this period total coliforms and faecal coliforms were used as faecal indicators. Today, it is known that these groups include genera that are widely found in environments that are not contaminated by faeces (i.e., source waters, vegetation and soils), and so they can no longer legitimately be used as surrogates of *E. coli* or indeed indicators of faecal contamination (Edberg et al., 2001).

The third edition of the WHO Guidelines for Drinking Water Quality, published in 2004, established a new concept of ensuring the microbial safety of drinking water through a multiple-barrier approach, from catchment to consumer, highlighting the importance of source water protection. The preferred strategy is a management approach that places the primary emphasis on preventing or reducing the entry of pathogens into water sources and reducing reliance on treatment processes for removal of pathogens (WHO, 2004). According to the WHO Guidelines, the basic and essential requirement to ensure the safety of drinking-water is the implementation of a "framework for safe drinking-water" as a preventive management approach comprising three key components:

1. Health-based targets, based on an evaluation of health risks, with the application of quantitative microbial risk assessment

2. Water safety plans (WSPs), and
3. A system of independent surveillance that verifies that (1) and (2) are operating properly (WHO, 2004).

This framework is very important because the safety of drinking water for public health protection no longer focuses only on the monitoring of indicators for faecal pollution and standards compliance, but also on source protection, the efficacy of treatment processes to remove pathogens and source-to-consumer risk assessment. However, in practice, this remains a reality only for industrialised countries, and a few low-income countries have established regulations based on the new WHO Guidelines.

Climate Change and Water-Related Infectious Diseases

Climate change is today one of the major issues of concern to microbiologists and health and environmental authorities. Although this is a complex issue with many uncertainties, it is clear that climate change is likely to induce changes to fate and transport of pathogens within the environment, though the extent of these effects is not as yet clear. The potential direct impacts of climate change include changing patterns of rainfall, flooding, sea level rise, an increase in water temperature, drought and elevated levels of UV radiation among others (Braks & Husman, 2013). Considering the importance of the environment to the ecology of human pathogens, these events may lead to indirect effects, including changes in the distribution and incidence of infectious diseases. Therefore, climate change may have an impact on both the quality and the quantity of available water and will require planning and management to minimise adverse impacts on drinking-water supplies (WHO, 2011).

Water-related pathogens can infect humans through several routes and can be classified into four groups: water-borne, water-washed, water-based and water-related insect vector, which are each affected in different way by climate change (Table 2) (Bradley et al., 1977). The increased frequency of heavy rainfall with associated flooding, and increased temperature with large drought periods are considered the determining factors influencing the impact on society of these diseases. Table 2 presents the principal climatic determinants that can affect water-related infectious diseases.

Within the category of water-borne diseases, although higher temperatures and UV radiation can inactivate enteric viruses such as noroviruses, heavy rainfall events can produce peak concentrations of these viruses in surface water as a result of runoff and sewer overflows. Cholera transmission has been shown to be directly affected by environmental factors; being associated with monsoon seasons, warm temperatures, heavy rainfall, and the increase of plankton populations that result

Table 2. Water-related infectious diseases and climatic determinants according to their transmission route

Category	Transmission	Climatic Determinants	Disease Examples
Water-borne	Ingestion of water contaminated by human or animal faeces or urine containing pathogenic bacteria, viruses or parasites	Water temperature, Rainfall / Drought, UV radiation.	Gastroenteritis, cholera, enteric hepatitis, amoebic and bacillary dysentery, leptospirosis, poliomyelitis, typhoid/ paratyphoid fever
Water-washed	Skin, ear or eye contact with contaminated water and poor personal hygiene	Flooding, Drought, Water temperature, Rainfall	Conjunctivitis, trachoma, intestinal helminthic infections, leprosy, scabies.
Water-based	Organisms that originate in the water or spend part of their life cycle in aquatic animals and come in direct contact with humans in water or by inhalation	Water temperature, Relative humidity, Rainfall/Drought, UV radiation.	Darcunculiasis, schistosomiasis, (tricho)bilharziasis, legionellosis, algal diseases
Water-related insect vector	Organisms with life cycles associated with insects that live or breed in water	Water/air temperature, Flooding, Relative humidity, Rainfall/ Drought, UV radiation	Dengue, yellow fever, malaria, West Nile Fever, Rift Valley fever

Source: Adapted from Braks & Husman, 2013.

from elevated nutrient levels in aquatic environment (Braks & Husman, 2013). Colwell and Patz (1988) demonstrated the relationship between increased sea-surface temperatures and cholera incidence in Bangladesh.

The survival rates of water-washed pathogens are enhanced in warm moist environments and flooding facilitates the environmental transport of these microorganisms, raising the risk of outbreaks. However, drought may also result in water-washed diseases because of poor personal hygiene resulting from water scarcity (Braks & Husman, 2013). Incidence of water-based diseases is influenced by several climatic determinants. Rising temperatures favour the growth of cyanobacteria and the risk of toxic algal blooms, as well as accelerate the growth of freshwater snails, increasing the shedding of cercariae. An increase in atmospheric temperature leads to warmer water in plumbing systems, creating an environment for *Legionella* proliferation and for the establishment of biofilms (Braks & Husman, 2013).

Vector-borne diseases generally are sensitive to climate change, mainly because the arthropod vector is directly affected by climate-related abiotic factors, such as ambient temperature and humidity. These diseases are also influenced by rainfall patterns since the vector depends on water bodies for reproduction (Braks & Husman, 2013).

Over longer periods, climate change may cause a greater number of extreme weather events, including more frequent incidents of extreme high temperatures, droughts, heavy precipitation and violent storms. These events will affect more significantly areas with growing populations, where the provision of water supplies is already under considerable pressure, and where there is little, if any, water supply margin available to them in these situations (WHO, 2011). This entire scenario is highly conducive to the establishment and proliferation of human water-related infectious diseases.

FUTURE RESEARCH DIRECTIONS

Although infectious diseases have had a major effect on the course of human history since ancient times, scientific understanding of the issues has only advanced significantly during the past one hundred and fifty years, as a result of rapid advances in the microbiological, molecular and health sciences. Much progress has been achieved in the control and prevention of human infectious diseases, but scientists and public authorities have many remaining challenges ahead.

One of the principal challenges is to understand the ecology of these diseases, the dynamics of human population interactions with the pathogen and the interaction of the pathogen with different habitats. There is an urgent need for further studies into the origin of new and old infectious diseases to ascertain which factors have favoured the persistence of some diseases since antiquity, how new pathogens have emerged in animal and human populations and how new pathogens may cross the barriers that separate their natural reservoirs from human populations and ignite the epidemic spread of novel infectious diseases (Antia et al., 2003, Wolfe et al, 2007).

Emergency situations such as those encountered during infectious disease outbreaks need an immediate response, therefore, rapid and inexpensive methods are needed for pathogen detection. New methodologies, such as metagenomics and proteomics, have facilitated functional ecology studies of pathogens in the environment and in their host. Using these methodologies, prospects for new drugs to treat infectious diseases through the elucidation and exploitation of metabolic pathways and gene clusters involved in the biosynthesis of antibiotics are likely.

Both global and local efforts need to be conducted to enhance infectious diseases surveillance in low- and middle-income countries. Often, the epidemiological data are scarce in these countries, even for those diseases for which notification is compulsory. This information is essential to drive health programmes and to detect and control outbreaks at the local level, before they have a chance to spread globally. For many industrialised countries, global health is seen as a pillar of their foreign policy, and these countries are therefore expanding their investment in this field to support global security (Choi, 2012). Furthermore, bioterrorism is of great concern to many governments considering the potentially overwhelming impact of some waterborne and airborne pathogens on public health.

Additionally, microbial risk assessment can be seen as a potentially highly useful tool for decision-makers during the management of many infectious diseases, including water-borne, food-borne and air-borne diseases, helping to define the risks and costs and establishing benefits and appropriate interventions.

CONCLUSION

Infectious diseases were responsible for about 15% of the 56 million deaths that were estimated to have occurred globally in 2012 (WHO, 2015). Although this rate has dropped compared with the data from the previous two decades, it is still high, particularly in low and low to middle income countries, where infectious diseases continue to result from consumption of unsafe drinking water, inadequate sanitation and poor domestic and community hygiene. The statistics are worse for children below five years of age, for whom the mortality rate for the principal infectious diseases reaches 35%. These countries also suffer from other diseases associated with high rates of morbidity and mortality such as vector-borne diseases, TB, HIV and respiratory diseases (WHO, 2015).

The problem of emerging and re-emerging infectious diseases, that have plagued the world population at local, regional and global level during recent decades and worsened by the effects of climate change, demonstrates that humanity remains vulnerable to the effects of infectious diseases. It is essential to maintain our vigilance and to coalesce scientific and governmental efforts to confront threats that have had a significant impact on human development throughout its history. A starting point is to take into account the lessons learned from our ancestors and to integrate this understanding with the extensive and growing scientific knowledge available to us today.

The following chapters of this book present emerging knowledge to support a better understanding of how environmental changes can influence emerging infectious diseases and pandemics.

ACKNOWLEDGMENT

The author thanks Professor Huw Taylor from Brighton University, UK, for the final revision of the text.

REFERENCES

Antia, R., Regoes, R. R., Koella, J. C., & Bergstrom, C. T. (2003). The role of evolution in the emergence of infectious diseases[letters to nature]. *Nature, 426*(6967), 658–661. doi:10.1038/nature02104 PMID:14668863

Ashbolt, N. J., Grabow, W. O. K., & Snozzi, M. (2001). Indicators of microbial water quality. In L. Fewtrell & J. Bartram (Eds.), *Water Quality: Guidelines, Standards and Health* (pp. 289–316). London, UK: IWA Publishing.

Baltimore, D. (2014). *Renato Dulbecco 1914 – 2012. Biographical Memoirs.* National Academy of Sciences. Retrieved from http://www.nasonline.org/publications/biographical-memoirs/memoir-pdfs/dulbecco-renato.pdf

Barry, S. (2008). The plague of the third pandemic and its current reemergence. *Vesalius, 14*(2), 69–73. PMID:19579344

Bartlett, J. M. S., & Stirling, D. (2003). A short history of the polymerase chain reaction. PCR protocols. *Methods in Molecular Biology (Clifton, N.J.), 226*, 3–6. PMID:12958470

Borgdorff, M. W., & van Soolingen, D. (2013). The re-emergence of tuberculosis: What have we learnt from molecular epidemiology? *Clinical Microbiology and Infection, 19*(10), 889–901. doi:10.1111/1469-0691.12253 PMID:23731470

Brachman, P. (2003). Infectious diseases – past, present, and future[Editorial]. *International Journal of Epidemiology, 32*(5), 684–686. doi:10.1093/ije/dyg282 PMID:14559728

Bradley, D., Evison, L. M., James, A., Pescod, M. B., & White, A. U. et al. (1977). Health aspects of water supplies in tropical countries. In *Water, wastes and health in hot climates*. London: John Wiley.

Braks, M. A. H., & Husman, A. M. R. (2013). Dimensions of effects of climate change on water-transmitted infectious diseases. *Air & Water Borne Diseases*, 2, 109. doi:10.4172/2167-7719.10001092: 1000109.

Brock, T. (Ed.). (1961). *Milestones in Microbiology: 1546 - 1940.* Washington, DC: American Society for Microbiology.

CDC - Center for Disease Control and Prevention. (2010). *The 2009 H1N1 Pandemic: Summary Highlights, April 2009 - April 2010.* Retrieved from http://www.cdc.gov/h1n1flu/cdcresponse.htm

CDC - Center for Disease Control and Prevention. (2014, November18). *Plague*. Retrieved from www.cdc.gov/plague/histoty

Choi, B. C. K. (2012). The past, present, and future of public health surveillance. *Scientific, 2012*, 26 p.,10.6064/2012/875253

Choi, B. C. K., & Pak, A. W. P. (2001). Lessons for surveillance in the 21st century: A historical perspective from the past five millennia. *Sozial- und Praventivmedizin*, *46*(6), 361–368. doi:10.1007/BF01321662 PMID:11851070

Cohn, S. K. (2005). The Black Death 1346-1353: The Complete History (Book Reviews). *England Journal of Medicine*, *352*(10), 1054–1055. doi:10.1056/NEJM200503103521026

Colwell, R. R., & Patz, J. A. (1998). *Climate, infectious disease and human health: an interdisciplinary perspective*. Washington, USA: American Academy of Microbiology.

Cupples, A. M., Rose, J. B., & Xagoraraki, I. (2010). New Molecular Methods for Detection of waterborne Pathogens. In R. Mitchell & J.-D. Gu (Eds.), Environmental Microbiology (pp. 57-94). Hoboken, New Jersey: Wiley-Blackwell.

Dobson, A. P., & Carper, E. R. (1996). Infectious diseases and human population history. *Bioscience*, *46*(2), 115–126. doi:10.2307/1312814

Edberg, S. C., Rice, E. W., Karlin, R. J., & Allen, M. J. (2000). *Escherichia coli*: The best biological drinking water indicator for public health protection. *Journal of Applied Microbiology*, *88*(Suppl. 1), 1068–1168. doi:10.1111/j.1365-2672.2000.tb05338.x PMID:10880185

Edge, T. A., Khan, I. U. H., Bouchard, R., Guo, J., Hill, S., Locas, A., & Watson, S. et al. (2013). Occurrence of waterborne pathogens and *Escherichia coli* at offshore drinking water intakes in Lake Ontario. *Applied and Environmental Microbiology*, *79*(19), 5799–5813. doi:10.1128/AEM.00870-13 PMID:23835181

EHC - Research Triangle Environmental Health Collaborative. (2010). *Environmental Health Summit: Recommendations from the Research Triangle Environmental Health Collaborative*. Retrieved from http://www.pewtrusts.org/~/media/Assets/0001/EnvironmentalHealthResearchTriangleCollaborative2010EnvironmentalHealthSummit2010Recommendations.pdf

Engelkirk, P. G., & Duben-Engelkirk, J. (2011). *Burton's microbiology for the health sciences* (9th ed.). Philadelphia, PA: Wolters Kluwer / Lippincott Williams & Wilkins.

Enzler, S. M. (n. d.). *History of water treatment*. Lennteck. Retrieved from http://www.lenntech.com/history-water-treatment.htm

Frith, J. (2012). The history of plague – Part 1. The three great pandemics. *Journal of Military and Veterans'Health*, *20*(2), 11–16.

Gest, H. (2004). The discovery of microorganisms by Robert Hooke and Antoni van Leeuwenhoek, Fellows of The Royal Society.[PubMed]. *Notes and Records of the Royal Society of London*, *58*(2), 187–201. doi:10.1098/rsnr.2004.0055

Echeverría, V. I. (2010). Girolamo Fracastoro and the invention of syphilis. [PubMed]. *Historia, Ciencias, Saude--Manguinhos*, *17*(4), 877–884. doi:10.1590/S0104-59702010000400002

Haensch, S., Bianucci, R., Signoli, M., Rajerison, M., Schultz, M., & Kacki, S. … Bramanti, B. (2010). Distinct Clones of Yersinia pestis Caused the Black Death. Plos Pathogens, 6(10). Retrieved from http://www.plospathogens.org

Harvard University Library. (n. d.). *The Great Plague of London,1665. Contagium Historical Views of Disease and Epidemics*. Harvard University Library Open Collection Program. Retrieved from http://ocp.hul.harvard.edu/contagion/plague.html

Hays, J. N. (2005). *Epidemics and Pandemics: Their Impact in Human History*. Santa Barbara, California: ABC-CLIO.

Kumar, V., & Raul, D. (2014). History and evolution of surveillance in public health. *Global Journal of Medicine and Public Health*, 3(1). Retrieved from http://www.gjmedph.org/uploads/O6-Vo3No1.pdf

Leclerc, H., Mossel, D. A. A., Edberg, S. C., & Struijk, C. B. (2001). Advances in the bacteriology of the Coliform group: Their suitability as Markers of microbial water safety. *Annual Review of Microbiology*, *55*(1), 201–234. doi:10.1146/annurev.micro.55.1.201 PMID:11544354

Maier, R. M., Pepper, I. L., & Gerba, C. P. (2009). Introduction to Environmental Microbiology. In R. M. Maier, I. L. Pepper, & C. P. Gerba (Eds.), *Environmental Microbiology* (2nd ed., pp. 3–7). Burlington, MA: Elsevier/Academic Press. [etc.] doi:10.1016/B978-0-12-370519-8.00001-8

Marineli, F., Tsoucalas, G., Karamanou, M., & Androutsos, G. (2013). Mary Mallon (1869-1938) and the history of typhoid fever. *Annals of Gastroenterology*, *26*(2), 132–134. PMID:24714738

Myron, S. (2008). Rudolph Virchow. *Emerging Infectious Diseases*, *14*(9), 1480–1481. doi:10.3201/eid1409.086672

Nelson, K. E., & Williams, C. F. (2014). Early History of Infectious Disease. In K. E. Nelson & C. F. Williams (Eds.), *Infectious Diseases and Epidemiology* (pp. 3–21). Burlington, MA: Jones & Barlett Learning.

NIH - National Institute of Health. (2013). *Severe acute respiratory syndrome (SARS)*. National Institute of Health, US National Library of Medicine, MedlinePlus. Retrieved from http://www.nlm.nih.gov/medlineplus/ency/article/007192.htm

NRC - National Research Council. (2004). *Indicators for Waterborne Pathogens. Committee on Indicators for Waterborne Pathogens. Washington (DC)*. US: National Academies Press.

PAHO. (2015, 24 June). Epidemiological Update – *Cholera*. Pan American Health Organization (PAHO)/World Health Organization (WHO). Retrieved from http://www.paho.org/hq/index.php?option=com_docman&task=doc_view&gid=30752+&Itemid=999999&lang=en

Panagiotakopulu, E. (2004). Pharaonic Egypt and the origins of plague. *Journal of Biogeography, 31*(2), 269–275. doi:10.1046/j.0305-0270.2003.01009.x

Parry, C. M. (2005). Epidemiological and clinical aspects of human typhoid fever. In P. Mastroeni & D. Maskell (Eds.), *Salmonella Infections: Clinical, Immunological and Molecular Aspects* (pp. 1–24). Cambridge, New York: Cambridge University Press.

Pelczar, M. J., Chan, E. C. S., & Krieg, N. R. (2010). *Microbiology: An Application Based Approach. New Dehli, India*. Tata: Mac Graw Hill.

Rosen, W. (2007). *Justinian's Flea: The first great plague and the end of Roman Empire*. London, England: Penguin Books.

Satcher, D. (1995). Emerging infections: Getting ahead of the curve. *Perpectives, 1*(1), 1–6. PMID:8903147

Schmid, B. V., Büntgen, U., Easterday, W. R., Ginzler, C., Walløe, L., Bramanti, B., & Stenseth, N. C. (2015). Climate-drive introduction of the Black Death and successive plague reintroductions into Europe. *Proceedings of the National Academy of Sciences of the United States of America, 112*(10), 3020–3025. doi:10.1073/pnas.1412887112 PMID:25713390

Smallman-Raynor, M. R., & Cliff, A. D. (2004). *War Epidemics: An Historical Geography of Infectious Diseases in Military Conflict and Civil Strife, 1850-2000*. New York, New York: Oxford University Press.

Taubenberger, J. K. (2006). The Origin and Virulence of the 1918 "Spanish" Influenza Virus. *Proceedings of the American Philosophical Society*, *150*(1), 86–112. PMID:17526158

Tortora, G. J., Funke, B. R., & Case, C. L. (Eds.). (2010). *Microbiology: An Introduction*. San Francisco, CA: Benjamin Cummings.

USDHEW - U.S. Department of Health, Education, and Welfare. (1963). *Vital Statistics of United States* (Vol. 2). US Department of Health, Education and Welfare. Public Health Service, Washington, DC.

USEPA – U.S. Environmental Protection Agency. (2000). The history of drinking water treatment. United States Environmental Protection Agency, Office of Water, EPA-816-F-00-006.

Wagner, D. M., Klunk, J., Harbeck, M., Devault, A., Waglechner, N., Sahl, J. W., & Poinar, H. et al. (2014). *Yersinia pestis* and the Plague of Justinian 541–543 AD: A genomic analysis. *The Lancet Infectious Diseases*, *14*(4), 319–324. doi:10.1016/S1473-3099(13)70323-2 PMID:24480148

WHO - World Health Organization. (2004). *WHO Guidelines for drinking Water* (3rd ed.). Geneva, Switzerland: WHO press.

WHO - World Health Organization. (2011). *WHO Guidelines for drinking Water* (4th ed.). Geneva, Switzerland: WHO press.

WHO - World Health Organization. (2015). *World Health Statistics 2015*. Geneva, Switzerland: WHO press.

William, B. (1997). *Infectious Diseases in History: a guide to causes and effects*. Prepared by Brian Williams in 1997. For distribution within the Department of Economic and Social History at the University of Hull, England. http://urbanrim.org.uk/diseases.htm

Wolfe, N. D., Dunavan, C. P., & Diamond, J. (2007). Origins of major human infectious diseases. *Nature*, *447*(7142), 279–283. doi:10.1038/nature05775 PMID:17507975

Yassi, A., Kjellstrom, T., de Kok, T., & Guidotti, T. L. (2001). *Basic Environmental Health*. New York, New York: Oxford University Press. doi:10.1093/acprof:oso/9780195135589.001.0001

Chapter 2
Global Environmental Change and Emerging Infectious Diseases:
Macrolevel Drivers and Policy Responses

Catherine Machalaba
EcoHealth Alliance and Future Earth oneHEALTH, USA

Cristina Romanelli
Secretariat of the Convention on Biological Diversity, Canada & United Nations University - International Institute for Global Health, Malaysia

Peter Stoett
Concordia University, Canada

ABSTRACT

The prediction of emerging infectious diseases (EIDs) and the avoidance of their tremendous social and economic costs is contingent on the identification of their most likely drivers. It is argued that the drivers of global environmental change (and climate change as both a driver and an impact) are often the drivers of EIDs; and that the two overlap to such a strong degree that targeting these drivers is sound epidemiological policy. Several drivers overlap with the leading causes of biodiversity loss, providing opportunities for health and biodiversity sectors to generate synergies at local and global levels. This chapter provides a primer on EID ecology, reviews underlying drivers and mechanisms that facilitate pathogen spillover and spread, provides suggested policy and practice-based actions toward the prevention of EIDs in the context of environmental change, and identifies knowledge gaps for the purpose of further research.

DOI: 10.4018/978-1-5225-0553-2.ch002

INTRODUCTION

Recent outbreaks of emerging infectious diseases (EIDs) often seem to appear without warning, severely exacerbating public health burdens and spreading across regions (Heymann & Dar, 2014). Since 1940 the rate of infectious disease emergence events has risen, even after controlling for reporting bias, spiking in the 1980's with the human immunodeficiency virus (HIV) pandemic leading to acquired immunodeficiency syndrome (AIDS), which resulted in between 30 and 42 million deaths and untold costs in treatment (Jones et al., 2008; Nakagawa et al., 2015; Schwartlander et al., 2011; UNAIDS, 2013). More recently, as of February 2016, over 11,300 deaths had been recorded due to the West Africa Ebola outbreak that began in December 2013 (Baize et al., 2014; World Health Organization, 2016).

In this chapter we elaborate on the thesis that human modification of the environment serves as an underlying driver in EID risk: in particular, the processes associated with global environmental change cannot be neglected in outbreak investigations to identify disease origin and contribute to the development of effective policy responses. Most environmental degradation, such as deforestation or the contamination of water supplies, is most immediately apparent at the local level; yet broader-scale (macro) impacts also exist and influence local dynamics which both contribute to and are affected by global environmental change (World Health Organization and Convention on Biological Diversity, 2015). Impacts include changes in species abundance and pathogen prevalence, novel exposure to pathogens through increased or new types of exposure, selection pressures for rapid pathogen evolution, enablers for the spread of disease, and ultimately, the potential for increased associated human health burden. To fully grasp these dynamics and connections, a solid understanding of links between human, animal, and ecosystem health in a changing global environmental landscape is required- a concept described as One Health.

This chapter will begin by elaborating on this theme by introducing the concept of EIDs and pathogens, explicating some of their immense human and environmental costs, and then focusing on the drivers and impacts of global environmental change that are considered most pertinent to emerging disease risks. The chapter also proposes solutions to more effectively combat the direct and indirect causes of environmental change, thus lowering EID risk on a global scale. Finally, it puts forth suggestions for future research to enable more effective policy development.

BACKGROUND: A PRIMER ON PATHOGENS

We can examine infectious disease emergence on both micro- and macro-levels. On a microscopic (or single pathogen) level, emergence events may have complex mechanisms, prompting various theories about different factors that may lead to disease emergence and spread and their relation to changes in biodiversity or species composition in an ecosystem (Keesing, Holt, & Ostfeld, 2006; Randolph & Dobson, 2012). While microlevel analyses and their interactions with macrolevel processes are useful, the thematic focus of this chapter is on the macrolevel, where emergence is driven by the many anthropogenic impacts that are altering ecological dynamics on a large scale, leading to global environmental change. For example, land conversion for agricultural uses and livestock production and other anthropogenic activities have increased or led to new types of human or domestic animal contact with wildlife, thereby facilitating pathogen "spillover" to humans (Jones et al., 2008; Karesh et al., 2012). At the same time, the rapid expansion in trade and travel in recent decades also establishes new pathways for the spread of dangerous pathogens, including through the introduction of invasive species and the climatic and other habitat conditions that allow them to persist (Mazza, Tricarico, Genovesi, & Gherardi, 2014). To examine the broad drivers leading to disease emergence, a basic comprehension of infectious disease dynamics is needed.

Not all microbes are pathogenic to a given species – in fact, the majority are not, and many are even beneficial (as seen with humans' commensal bacteria, which may be protective against illness [for example see work out of the Human Microbiome Project (Human Microbiome Project Consortium, 2012)]) - but those that are can result in immense public health burden. Infectious health threats include certain bacteria, viruses, fungal infections, protozoa, helminths, and prions. They are typically classified by either their infectious agent (i.e. pathogenic microbes such as a rhinovirus) or their clinical manifestation(s) (i.e. symptoms or illnesses such as the common cold). Additionally, pathogens or diseases can be categorized by their relative presence in a given region, whether endemic (established) or newly seen (emerging). The so-called "emerging" pathogens or diseases (EIDs) have further layers of distinction, such as the first detection of a novel agent (for instance, in 2012 we saw the first detection of the coronavirus resulting in Middle East Respiratory Syndrome, or MERS-CoV) (Zaki, van Boheemen, Bestebroer, Osterhaus, & Fouchier, 2012), re-emerging (reappearing), or emerging in a new region or new host. Additionally, some pathogens may evolve to the point that they become classified as an entirely new pathogen (for example, an influenza virus that experiences reassortment in gene segments, or bacteria such as Methicillin-resistant Staphylococcus aureus, or MRSA, that develop resistance to antibiotics) (Enright et al., 2002; Steel & Lowen, 2014). Lastly, EIDs can be considered by their direct source of infection to humans, such as vector-borne (e.g. malaria), environmental (e.g. anthrax in soil spores), human (e.g. smallpox and poliovirus), or zoonotic (animal-borne, e.g. rabies virus). These distinctions may overlap (such as with Rift Valley fever, which has a mosquito source but is most commonly transmitted to humans via exposure to infected animals) (Bird, Ksiazek, Nichol, & MacLachlan, 2009). Of the known infectious agents pathogenic to humans, the majority (61%) are zoonotic, having an original or recurring animal source (Taylor, Latham, & Woolhouse, 2001). Among the zoonotic diseases emerging between 1940 and 2004, nearly three-fourths have originated from wildlife, including the appearance of novel infectious diseases such as HIV/AIDS, Severe Acute Respiratory Syndrome (SARS), and Ebola and Marburg virus diseases, all of which involved animal-to-human contact during at least the initial stage of transmission (Jones et al., 2008).

Essentially, the source of an EID that transmits the pathogen to a human is either a natural reservoir (typically, one that harbors the pathogen with low or no fitness burden), or an incidental host (one that does not naturally carry it, but is susceptible to infection with or without necessarily exhibiting clinical manifestations). This distinction between host type is important: intervention strategies may prove ineffective or inefficient if they mistakenly target incidental hosts rather than the ongoing source(s). For example, human cases of Ebola virus disease have been linked to great apes, yet these apes likely serve as incidental hosts with no evidence to date linking them to the maintenance cycle of this strain of the virus. Rather, certain species of fruit bats are now thought to be the actual reservoir (Olival & Hayman, 2014). Hosts can provide onward transmission (e.g. humans for some avian influenza strains), or may be a dead-end for a direct transmission cycle (e.g. human infection with Rift Valley fever virus). Throughout this chapter, however, the term 'host' is primarily used as an indication of any source – original (reservoir role) or incidental, reflecting the fact that a definitive reservoir for an EID is not always readily identified.

At its core, the basic pathogen-host(s) interaction is a natural process, and can be beneficial in terms of regulation of plant, animal and wildlife population(s) within an ecosystem, just as resource availability may serve to keep population numbers balanced in plants and wildlife. A pathogen will adapt to promote survival, potentially exploiting new niches such as a new host species. What has changed over recent decades, however, is the macro-level context: more specifically, the increase in widespread *opportunities*, created by anthropogenic activity, for both human and animal contact and for other factors enabling pathogen movement and adaptation to flourish (Karesh et al., 2012). These are sometimes apparent; other times, the associated causal factors are not immediately recognized. For example, in early 2016, scientists are faced with the challenge of elucidating the factors leading to the rapid spread of Zika virus now affecting parts of south and central America and extending upward toward the continental US (Hennessey, Fischer, & Staples, 2016).

It is important to note that the pathway to transmission is not unidirectional; humans may also transmit infections to animals in a process popularly known as reverse zoonosis, as seen with a metapneumonia outbreak introduced to endangered mountain gorilla populations from humans via ecotourism and the infection of captive elephants with tuberculosis from their human handlers (Mikota & Maslow, 2011; Palacios et al., 2011). In addition to direct contact transmission pathways, more subtle environmental movement of microbes can occur, as seen with antimicrobial resistance. Although the development of resistance is a natural process, with resistant strains found even in undisturbed habitats without human influence, human practices (i.e. in human clinical use, veterinary medicine, agriculture and aquaculture) are drastically speeding up bacterial evolution and dispersion (Allen et al., 2010). A study in Botswana comparing strains of *E. Coli* resistant to human antibiotics found high genetic similarity between strains found in banded mongoose within protected areas and those in human fecal waste, suggesting contamination of the mongoose via human feces (Pesapane, Ponder, & Alexander, 2013).

MAIN FOCUS OF THE CHAPTER

Human and Ecological Costs

Given the considerable human costs associated with EIDs, understanding the specific drivers of EIDs and developing concrete measures to prevent and curtail their spread is urgently needed in sustainable development discussions; endemic and enzootic (zoonoses maintained in animal populations) diseases alone account for approximately one billion human cases and millions of deaths each year (Karesh et al., 2012). While the incidence of some communicable diseases has shown significant declines related to so-called 'epidemiological transitions' due in large part to public health interventions (e.g. improved water sanitation systems, vaccination, therapeutics, and public education) and while non-communicable diseases continue to increase, infectious diseases continue to claim millions of lives worldwide each year, disproportionately affecting the most vulnerable populations (Dye, 2014). For example, recent data shows that in Sub-Saharan Africa, which has the highest concentration of poverty in the world, infectious diseases together with maternal causes of mortality and childhood illness present close to 70% of the disease burden; the same region experiences nearly half of the global mortality from cholera, as well as high incidence of infectious 'neglected' tropical diseases (so-called because of the relatively limited attention they gain in proportion to their burden) (GBD 2013 Mortality and Causes of Death Collaborators, 2015; Hotez & Kamath, 2009). However, an 'out of sight, out of mind' sentiment around infectious diseases provides false security; although infectious diseases may originate in a certain region, they could spread rapidly to other regions, rich or poor. Recently, the world witnessed this when passengers infected with Ebola virus traveled from the three heavily affected West African nations to countries such as Mali, Nigeria, Spain, and the United States in 2014, facilitating the progress of local transmission chains (Centers for Disease Control and Prevention, 2015). As this chapter went to press, cases of Zika virus in travelers coming from the Americas were being monitored in Europe with the European Union noting the potential for localized mosquito transmission during the summer months (European Centre for Disease Prevention and Control, 2016).

While some EID events have fortunately been short-term, others have the potential to become pandemics, as seen with HIV/AIDS, which has led to over 60 million cases globally, including 33 million at present, and has become a firmly entrenched health burden in heavily impacted areas, such as certain parts of Sub-Saharan Africa (Vitoria et al., 2009). Similarly, the outbreak in West Africa beginning in December 2013 of Ebola Zaire, which had never been reported previously in the region, caused exceedingly high morbidity, mortality, and financial impacts over two years into the outbreak, resulting in over 28,000 human cases. In addition to direct health burdens, associated costs and loss of productivity, there are wide-ranging indirect costs borne from societal disruption that affect commercial sectors such as hospitality and tourism, transportation, education, and the provisioning of public health services for other diseases (e.g. interruptions of vaccination campaigns), and the workers and users that rely on them (World Bank, 2014). Even emerging diseases that produce relatively low case incidence can be catastrophic; for example, the 2003 SARS outbreak, which led to fewer than 1,000 deaths, cost the global economy an estimated US$30+ billion (World Bank, 2012), disrupting both international tourism and commerce as well as local markets in China.

The burden of EIDs is not exclusive to human health; there is ample evidence that disease spillover can pose serious threats to food security and the conservation of non-human species (Daszak, Cunningham, & Hyatt, 2000; Nabarro & Wannous, 2014). For example, past Ebola virus outbreaks have led to major declines in endangered chimpanzee populations (Rouquet et al., 2005); their susceptibility suggests that rather than having a role in maintaining the virus, they are infected by another source, and may then incidentally infect humans that come into contact with infected non-human primates. Large-scale outbreaks of highly pathogenic avian influenza (HPAI) in poultry have been seen over the past fifteen years, related to production systems shifting from smallholder flocks to large-scale farms with high stocking rates without sufficient biosecurity advancements (Karesh et al., 2012). While minimally publicized, in the fall of 2014 some farms in Europe reported outbreaks of HPAI in domestic birds; shortly after, Canada and the United States followed suit. As of June 2015, the United States had reported deaths of over 48 million birds from infection or culling as an infection control measure. Cases have been primarily found in areas with major North American wild bird flyways, though specific sources of introduction between poultry farms have not been known. While not conclusive, a U.S. Department of Agriculture report suggests that, in some cases, insufficient biosecurity practices (e.g. not disinfecting vehicles moving between farms) and airborne transmission (e.g. wind-mediated) possibly enabled spread of the virus (USDA APHIS Veterinary Services, 2015). No associated human cases have been reported, but the financial impacts are seen across the value chain, including compensation to affected farms and high response costs from government intervention (estimated at $950 million) (USDA APHIS Veterinary Services, 2016).

Finally, while an in-depth analysis exceeds the scope of this chapter, even emerging diseases that are not transmissible to humans (e.g. white nose syndrome in bats or chytridiomycosis in amphibians) may pose threats to human health through the loss of health-supporting ecosystem services, such as pest control and pollination services for agriculture, and may pose species endangerment or extinction risks (Boyles, Cryan, McCracken, & Kunz, 2011; Daszak et al., 2000). Infectious disease impacts may add to pressures caused by other effects of global environmental change (e.g. extreme weather events) on food security and species survival. Immune system functions of wildlife may be reduced by other environmental change pressures, such as exposure to pollutants, potentially increasing susceptibility to infections (Acevedo-Whitehouse & Duffus, 2009). These forces acting together often disproportionately affect already vulnerable populations most directly reliant on ecosystems, potentially exacerbating the cycle of poverty and social inequity.

It is an increasingly apparent ecological reality that healthy ecosystems provide the essential resources upon which all life, including human life, depends, ranging from the provision of freshwater to food security to pest and disease regulation (Millennium Ecosystem Assessment, 2005). It is equally well established that biodiversity continues to be lost at alarming rates (Convention on Biological Diversity, 2014), and the rising pressures of biodiversity loss, climate change and human-induced land system changes can lead to the disruption of essential ecosystem functions, making ecosystems less resilient and more vulnerable to shocks and disturbances and increasingly compromising their ability to supply essential life-sustaining services (Chivian & Bernstein, 2008; Haines-Young & Potschin, 2010; McMichael & Beagehole, 2000; Naeem & Wright, 2003; World Health Organization and Convention on Biological Diversity, 2015), driving changes from a local to planetary scale (Whitmee et al., 2015).

EIDs and Global Environmental Change: Interrelated Drivers

We can utilize the lessons afforded from the Ebola outbreak in West Africa – including what we did *not* learn about its ecological underpinnings- to target future research and outbreak investigations. Despite a global response to mobilize medical services, attention to the underlying cause of the outbreak was underappreciated, with limited efforts to trace back to an original environmental source (and inconclusive, generating hypotheses but no concrete evidence) and any environmental pressures that may have driven risk (Mari Saez et al., 2015). Likewise, insufficient socio-economic dimensions, including cultural beliefs and behavioral practices, further curtailed the effectiveness of the response (Phua, 2015). While perhaps not an immediate concern during the height of the outbreak, future reintroduction of the virus from its environmental source remains a threat that will last even after international authorities declare West Africa as “Ebola-free” based on the end of the epidemic in humans. The West Africa Ebola outbreak raises a parallel with the emergence of HIV, though in the case of Ebola, the unprecedented scale of the outbreak is occurring decades later. Opportunities have been missed to explore more upstream dimensions that could have informed outbreak prevention methods. In this section we examine the macro-level drivers of EIDs related to environmental change to show how they facilitate disease emergence. While such pressures are presented in terms of their broad scales, we later present ways in which they can also be examined more holistically with more local level considerations for targeted risk assessment and response.

Although climate change is often used interchangeably as a term for global environmental change, the latter involves a much broader realm of processes and factors that are altering ecosystem dynamics. Human-induced environmental change has been broadly associated with globalization (Saker, Lee, Cannito, Gilmore, & Campbell-Lendrum, 2004). In their review of the ecology of infectious diseases, Karesh and colleagues highlight how human activities are contributing to a new phase in zoonotic diseases – both endemic and emerging – by changing the natural histories of pathogens (Karesh et al., 2012). The hundreds of outbreaks of emerging zoonotic diseases detected in the past half century have been linked to specific practices that enabled new pathways for transmission, created more frequent transmission opportunities, or increased the prevalence in host species (thus increasing opportunities for infection of humans). We limit our discussion to several leading drivers of EIDs that are part of global patterns of environmental change: land use or habitat change, agricultural developments, international trade and travel, climate and weather, and the hunting or consumption of wildlife (Kilpatrick & Randolph, 2012; Loh et al., 2015). This inventory is far from exhaustive, and many of these drivers are so intimately interrelated it is often difficult to separate them for analysis, a theme we specifically address below.

Land Use Change

Driven largely by economic development and population pressures, anthropogenic land use change is introducing fundamental and rapid changes in ecological dynamics and threatening biodiversity. Land conversion may be commonly associated with expansion of or establishment of new agricultural plots or plantations (witnessed in the explosion in palm oil plantations in Asia and elsewhere), natural resource extraction such as the timber, mining or oil industries, road building, and human settlements, among other reasons. On a global scale, the current discussion is often focused on conversion of native or otherwise undisturbed (sometimes referred to as 'pristine') natural habitat to landscapes serving human uses. The mechanism and timescale of conversion may differ by situation, ranging from scenarios with managed forests in which timber extraction is carefully planned to reduce environmental impact to slash-and-burn practices. Changes in habitat can affect ecosystems in many ways that may have disease implications, such as altering the species present and their relative abundance, movement, interactions with other species, access to resources, and immune responses (Gottdenker, Streicker, Faust, & Carroll, 2014).

Certain land conversions may result in landscape fragmentation, which not only threatens biodiversity but also provides potential for so-called "edge effects" (changes at the border of two ecosystems). This in turn has considerable repercussions on forest microclimate, tree mortality, carbon storage, fauna, and many other relevant dimensions of fragment dynamics (Laurance et al., 2011). Changes in community structure, function and composition can lead to an alteration of the type, abundance and distribution of species present. For example, the findings of a recent meta-analysis of studies on small rodent abundance and species richness response to land fragmentation in eight countries in the Americas suggest that fragmentation may serve as a risk factor for transmission of rodent-associated Hantaviruses to humans (Rubio, Avila-Flores, & Suzan, 2014). In addition to disease ecology changes, fragmentation may simply also increase the forest perimeter along which human or domestic animal contact with wildlife may occur.

Land conversion projections raise alarms for EID risks on the horizon. With rising agricultural demands, a total of 120 million hectares of expansion is expected in developing countries, accompanied by the conversion of a portion of arable land in developed nations for other uses (Food and Agriculture Organization of the UN, 2009). It has been estimated that some 25 million kilometers of roads will be added by 2050, with 90% occurring in developing nations (Laurance et al., 2014). These roads are not limited to urban areas; many are targeted for biodiversity-rich areas, providing passage through hundreds of miles of forest area stretches. With roads come people to provide travelers with services, further increasing the human footprint and placing humans on the edge of forests in which there is little existing sanitation, healthcare, or food provisions. These are but a few examples of the scale and magnitude of land conversions projected for the near future – projected demand for timber, metals, petroleum, palm oil, biofuels, and other industries are also likely to present major pressures on natural landscapes (Lee, Preston, Kooroshy, Bailey, & G., 2012).

Land use change has been the primary driver associated with recent emerging zoonotic disease events (Loh et al., 2015). The effects of disease transmission from land use change are not widely understood or acknowledged, in part because before and after incidence or prevalence studies are not necessarily routinely conducted across landscapes undergoing significant alteration. Yet, a review of studies published on land use and infectious disease reported that an increase in disease transmission was observed in more than half of the studies, though the authors note that outcomes are dependent on many factors, including the degree and type of land disturbance (Gottdenker et al., 2014).

Agricultural Developments

Smallholder farming is still common in some areas of the world, while livestock production has vastly intensified in others. Animal agriculture is growing to meet increased demand, and is linked to global environmental change. It has been estimated that agricultural activity- including both subsistence and industrial practices and the trade chain associated with it- is responsible for up to one-sixth to one third of global greenhouse gas emissions, and livestock production in particular makes up a large share of it (Foley et al., 2011; Gerber et al., 2013; McMichael, Powles, Butler, & Uauy, 2007), thereby further exacerbating pressures on climate and human health. In some cases, agricultural intensification has not been paired with adequate biosecurity – including in moderately intensive backyard poultry operations where wild and domesticated birds may come into contact and enable pathogen transmission (as observed with some avian influenzas); in comparison, in cases where veterinary services are strong, large-scale animal agriculture may operate with more controlled and contained conditions, reducing potential for disease introduction (Hosseini et al., 2013). At the same time, the scale of intensified production may result in high stocking density, often with limited genetic variability and reduced immune function from stress, and the value chain for livestock (as well as other species) may also present opportunities for so-called pathogen ‘pollution’, including in local and international market settings where their biosecurity may be insufficient (Perry, Grace, & Sones, 2013). Additionally, the increasing and often injudicious use of antimicrobials in agriculture can exert selection pressures in animal stocks for resistant strains, which may be transmissible to humans. For example, in the United States alone, nearly three-quarters of current antimicrobial use is for animal husbandry, including non-therapeutic uses such as growth promotion. Studies have suggested that a substantial amount (i.e. up to 90%) of antimicrobial agents, mostly unmetabolized, may be excreted by livestock, potentially entering the environment through soil and water dispersion (Marshall & Levy, 2011).

Pathogens can also find their way into livestock settings, and then potentially to people, through a range of routes. For example, the emergence of the first human Nipah virus outbreak in Malaysia between 1998 and 1999 spilled over to human populations via sick pigs, themselves infected by Pteropid fruit bats that served as natural hosts. The pigs were likely infected through feeding on fruit contaminated by bat saliva or urine from a tree on the swine farm. The dense stocking conditions, paired with the respiratory transmission potential of the Nipah virus, enabled efficient transmission among pigs and eventual spillover of the zoonosis to humans, causing many cases of encephalitis and over 100 human deaths (Chua, 2003; Luby, Gurley, & Hossain, 2009). Additionally, control measures included culling more than one million pigs (Lam & Chua, 2002).

International Trade and Travel

The rapidity and vast scale of trade and travel presents a serious risk of disease movement. Modern transport systems have enabled humans to move around the world in less than twenty-four hours, providing risk of importation and spread of so-called 'exotic' diseases across wide geographic areas. The 2003 SARS outbreak first detected in mainland China, which was later thought to have been transmitted by horseshoe bats (Ge et al., 2013), had spread to 29 countries in less than one year, resulting in interventions aimed at promoting "social distance" to limit the spread of the respiratory illness (Ahmad, Krumkamp, & Reintjes, 2009; Bell & World Health Organization Working Group on International and Community Transmission of SARS, 2004). More recently, air passenger data was rapidly investigated out of concern for potential international spread and establishment in response to the sudden Zika virus epidemic detected in Brazil in 2015 (Bogoch et al., 2016).

Risks associated with international trade are not limited to humans; animal trade also plays a role. For example, the accidental introduction of monkeypox into the United States in 2003, the first recorded occurrence in humans outside of its endemic range in Africa, is attributed to infection through prairie dogs that were housed with and infected by legally imported African rodents (Bernard & Anderson, 2006). In addition to legal wildlife trade, the illegal trafficking of wildlife - an underground industry estimated at upwards of US$20 billion (and upwards of US$100 billion per year with illegal logging and fishing included)- exacerbates the challenges of traceability and identification of unknown potential disease threats (United Nations Environment Programme, 2014; WWF / Dalberg, 2012). Other risks associated with wildlife transport include the purposeful or accidental import of invasive alien species or their introduction into the environment. Additionally, climate and ecological niche-related habitat suitability changes may assist in disease establishment for pathogen introduction enabled by trade or travel. For example, though it was discovered in Northern Uganda as early as 1937, West Nile virus was first detected in North America in 1999 and thought to have been introduced via international travel, likely from the Middle East, where a similar strain had been detected earlier (Sejvar, 2003).

Climate and Weather

Climate change from anthropogenic pressures is often mentioned as a likely contributor to some known and future outbreaks of zoonotic disease, but the relationships are complex (Machalaba et al., 2015a). Climate may have a significant role as changing climate conditions may cause shifts in host ranges, as seen with varying projections of contraction and expansion by region for the suitability of Nipah virus host species under different climate scenarios (Daszak et al., 2013). However, even accelerated climate change is likely to take decades to produce major ecological effects, and more immediate factors (e.g. global travel in <24 hours) may pose higher risks of an emergence or introduction event in a new region or species.

Temporal and spatial trends have shown increased reporting for certain diseases, including some vector-borne and waterborne diseases. For example, evidence suggests that rising temperatures and changing humidity and rainfall patterns have already altered the distribution of some waterborne illnesses and disease vectors (Intergovernmental Panel on Climate Change (IPCC), 2014). The suitable habitat for Lyme disease, the emergence of which was likely linked to reforestation that supported an increased population of its tick vector, is projected to grow over 200% in Canada (where it has been minimally established to date) by the 2080's under anticipated climate conditions (Brownstein, Holford, & Fish, 2005). These trends have not been universally consistent for all diseases, suggesting that the broad attribution of expanded disease ranges as a result of climate warming does not capture the complexities in disease dynamics (Kilpatrick & Randolph, 2012). In some cases, external changes to ecological systems can be expected to decrease or eliminate the local burden of some diseases as the transmission cycles become unsupported. For example, as Myers and Patz point out, malaria vectors in theory would not be maintained if a wetland is completely drained to remove mosquito breeding sites (Myers & Patz, 2009). However, in practice, such drastic scenarios rarely happen (for both feasibility and the importance of ecosystem services provided by wetlands). In an example of ecological variance, a study in the Brazilian Amazon found that malaria incidence may be negatively correlated to areas with widespread wetlands compared to areas with low wetland coverage (Olson et al., 2009). Rather than a simple precipitation-incidence model, there are many factors involved, including soil, landscape and rainfall determinants, availability of breeding sites, and flooding impacts to mosquito larvae (Olson et al., 2009) – and also human exposure opportunities.

The impact of climate and weather conditions on factors conferring virulence traits (e.g. basic reproductive rate, maintenance in a reservoir or intermediate host species) also warrants further study. However, since climate shifts are widely expected to extend the ranges of invasive species, which can in turn serve as vectors for disease, the connections must be taken seriously and investigated (Mazza et al., 2014). A recent study in Europe links drivers of global change to the distribution and abundance of invasive alien species with corresponding impacts on human health (Bayliss, Schindler, Essl, Rabitsch, & Pullin, 2015).

Climate change is likely to play a distinct role in emerging disease outbreaks, particularly in the case of extreme weather events. The latter are more variable in frequency and particularly difficult to predict in modeling scenarios. While climate change has not generally been linked to increased direct contact between humans and pathogen hosts, it may indirectly affect susceptibility to infection, for example as a result of fresh water scarcity or by potentially hindering food safety and availability (Tirado, Clarke, Jaykus, McQuatters-Gollop, & Frank, 2010). An estimated 80% of the global human population faces water security shortages (Vorosmarty et al., 2010) which may lead some populations to travel greater distances, potentially increasing exposure to pathogens or compounding ecological disturbances (World Health Organization and Convention on Biological Diversity, 2015). Water scarcity is also projected to compromise freshwater biodiversity (Vorosmarty et al. 2010), which may lead to loss of pathogen-regulating services and contribute to waterborne and water-related disease outbreaks such as cholera (Ali et al., 2012) or diarrheal diseases (Ebi, Woodruff, von Hildebrand, & Corvalan, 2007).

Future climate trends may yield a more pronounced role of climate on outbreak trajectory as new pathogens become endemic to a region. Perhaps the impacts of climate change on pathogens may best be considered on both micro- and macro-levels. On a micro level, temperature and precipitation-dependent pathogen development cycles may affect the basic reproductive rate (R_0), thus determining pathogen survival and spread in a population, and may similarly affect the suitable habitat range of its competent host(s). On a macro level, EIDs are dependent on a wide range of factors, and climate cannot be viewed in isolation when assessing risks. Rapid changes to ecosystems and resulting alterations in composition and species abundance, as well as anthropogenic practices that are facilitating increased and novel human-animal contact, pose complex biotic dynamics that interface with abiotic climate variability.

Wildlife Hunting, Butchering, or Consumption

Other emergence factors (or drivers) have also played significant roles in zoonotic disease outbreaks. For example, the hunting, butchering or consumption of certain wild animals in the tropics ("bushmeat") has been linked to significant public health threats. For instance, the detection in humans of Human Immunodeficiency Virus (HIV), whose precursor SIV (Simian Immunodeficiency Virus) is found in non-human primates, was linked to the hunting of chimpanzees in Sub-Saharan Africa (Faria et al., 2014); the hunting or butchering of non-human primates has also been associated with Ebola virus outbreaks (Leroy et al., 2004). Some studies have estimated the volume of annual consumption of bushmeat at approximately one billion kilos in central Africa alone (Karesh, Cook, Bennett, & Newcomb, 2005; Wilkie & Carpenter, 1999).

Bushmeat hunting is a means of subsistence for some populations in many developing countries, and can make an important contribution to the micro and macronutrient requirements. For example, a study in Madagascar reported that bushmeat consumption in northeastern Madagascar had a significant impact (by approximately 30%) in lowering the incidence of child anaemia particularly in poorer households (Golden, Fernald, Brashares, Rasolofoniaina, & Kremen, 2011). It is also worth noting, however, that a significant amount of wildlife may also be harvested for urban demand where it is in some cases associated with wealth (e.g. as a specialty food) or for non-food trade (e.g. exotic pet trade), with some of it ending up in countries or continents far away and posing pressure on hunted species, as well as potential pathogen introduction (Smith et al., 2012; Wilkie & Carpenter, 1999). Certain taxonomic groups (for example, bat and non-human primate species) may in general present elevated risk for disease transmission. Risks are not limited to species from the tropics, however; for example, foodborne outbreaks from the *Trichinella* parasite have been associated with consumption of undercooked pigs, wild boar and some wild game (Karesh et al., 2012).

Interrelatedness

While outbreaks may be classified by primary drivers, it is important to recognize that emergence events almost invariably involve several interlinked factors. For example, land use change may occur concurrently with agricultural industry changes in a given location. Human food demands at extractive industry sites (e.g. timber logging, mining, oil extraction, which fall under land use change) may not only potentially hinder availability and supply of food derived from fisheries, but also increase the potential for wildlife hunting and trade (Suarez et al., 2009). International trade or travel may enable pathogen introduction, and climate change-enabled suitable habitat changes may allow it to persist. These and other drivers may intersect with human susceptibility to infection, another leading driver of infectious disease emergence in humans from wildlife (e.g. opportunistic infections linked to compromised immunity from co-infection, genetic pre-disposition, antimicrobial resistance development from human clinical use or agriculture/food exposures). Additionally, while beyond the scope of this chapter, population-based or individual socio-economic and behavioral factors such as human demography, migration or poor sanitation, may promote spread of an outbreak in a population. The overlapping causes of EID outbreaks present a case for integrated risk analysis and prevention strategies that consider risks under different scenarios. We turn now to an introductory discussion of such policy directions and practical solutions.

SOLUTIONS AND RECOMMENDATIONS

Toward Integrated Policy Development

Thankfully, the multifarious relationships between global environmental change and disease emergence present numerous opportunities for the development of harmonized, coherent and sustainable policies and strategies that jointly promote co-benefits of biodiversity and ecosystem conservation (and restoration) and public health (World Health Organization and Convention on Biological Diversity, 2015). They also offer opportunities to address public health challenges associated with climate change (Watts & al., June 23, 2015), and to jointly improve public and planetary health outcomes (Whitmee et al., 2015).

Several of the leading causes of disease emergence overlap with five of the six leading direct drivers of biodiversity loss. These include habitat loss, degradation and fragmentation; overexploitation of biological resources; unsustainable production and consumption; invasive species introductions; and compounding pressures such as climate change and ocean acidification (Convention on Biological Diversity, 2014). Despite the breadth of these challenges, these shared drivers also provide new opportunities for the health and biodiversity sectors to work together synergistically across a wide range of areas. Several of these are explored in depth in the recent State of Knowledge review on the links between biodiversity and human health jointly led by the Convention on Biological Diversity and the World Health Organization (World Health Organization and Convention on Biological Diversity, 2015). The Rockefeller Foundation-Lancet Commission report on Planetary Health (Whitmee et al., 2015) also examines opportunities to jointly address disturbances to the structure and function of the Earth's natural systems on human health in ways that ensure intergenerational equity. Opportunities to yield compound benefits across sectors in the context of sustainable development are not always apparent but they remain critical and solutions achievable. For example, policy measures aimed at addressing issues such as deforestation and climate change can yield co-benefits for disease prevention by reducing ecological pressures, increasing social and ecological resilience and avoiding new pathways for human exposure to infectious disease transmission (Machalaba et al., 2015a).

Much of the earlier discussion on global environmental change and health has focused on outcomes, such as projections on climate change and disease incidence, once ecological changes have *already* occurred. Mitigation strategies also tend to emphasize broad contributors to health outcomes (for example, CO_2 emissions or temperature). Less attention in policy arenas has been devoted to the underlying drivers of ecosystems change that are enabling disease emergence as well as causing other health impacts. In the case of climate change, global greenhouse gas emissions are broken down by broad source – e.g. energy, transport, building, industry, agriculture, forestry and other land use changes (Blanco et al., 2014). As indicated above, these drivers have strong links to the underlying drivers of disease emergence. Within the health sector, the lack of awareness, emphasis on discrete cause-effect relationships, and frequently limited input into the processes to address environmental causes of ill health, which traditionally characterized the health sector response, are insufficient to address new and emerging global public health challenges or the tools required to evaluate and respond to them (Campbell-Lendrum, 2005; Keune et al., 2013; Oosterbroek, de Kraker, Huynen, & Martens, 2016). Developing preventive strategies, practices and tools aimed at the sustainable use, management and trade of resources essential for food and nutrition security, as well as assessing the potential unintended consequences of unsustainable practices, will also be critical to addressing drivers of disease and biodiversity loss (World Health Organization and Convention on Biological Diversity, 2015). Examples of such opportunities are explored below.

Addressing Future Drivers of Emerging Infectious Diseases

Projections suggest that rapid population growth, rising incomes and urbanization will increase food demands, especially for cereals and meat, in such a way that food production may need to be increased by up to 70% by 2050 (Food and Agriculture Organization of the UN, 2009). If current trends continue, this production increase could be detrimental to natural ecosystems, especially in Latin America and Sub-Saharan Africa where the majority of currently uncultivated arable land is found (Bruinsma, 2009). Given past links from land use changes and agricultural transitions (without adequate biosecurity measures) to disease emergence and outbreaks, risk-mitigating strategies can be employed going forward. For example, instead of converting undisturbed landscapes, agricultural policies can prioritize already-converted land. In Brazil, a country with both the highest expected agricultural production over the coming decades as well as rich terrestrial biodiversity, Strassburg and colleagues suggest that pastureland production efficiency is approximately 1/3rd of its potential; increasing it to ~50% of its production potential at existing agricultural sites could meet rising demands through 2040 or later, helping to avoid development of additional land for production (Strassburg et al., 2014). The authors acknowledge that attaining such production levels could potentially face constraints from soil acidity or livestock disease, but nonetheless production efficiencies can be explored further. Other farming practices, such as more efficient irrigation systems, have also been developed over past decades and may be employed to heighten production at current farm sites. Similarly, more effective disease control in livestock may also boost outputs of existing production systems - an important factor in light of global environmental change, as livestock raising under 2050 demands is projected to result in pressures on fresh water supplies and contribute to carbon emissions (Herrero & Thornton, 2013). Humans may be better protected in the process if zoonotic diseases are addressed in livestock. As agriculture is also susceptible to environmental impacts, climate-smart investments may promote resiliency in areas with more pronounced drought or flooding risks.

Requiring health impact assessments (HIAs), which consider disease risks and any disproportionate impacts to vulnerable populations, prior to the approval of development projects can help better anticipate EID risks, thus enabling preventive or mitigation-related actions. HIAs should be conducted in concert with other assessment tools including environmental, strategic and risk impact assessments to help determine coordinated solutions that benefit multiple sectors (World Health Organization and Convention on Biological Diversity, 2015). They are also most effective when used in conjunction with supporting legislative frameworks. While HIA may be routine in many high-income countries, its application has not historically extended to low- and middle-income countries, or is incomplete (namely, covering community health and safety infrastructure but not considering infectious disease emergence risk) (Seifman et al., 2015; Winkler et al., 2013). Knowledge gained from ecological niche modeling, hazard identification (such as surveys of wildlife on the site premises and the human practices that could elevate transmission risk), assessment of exposure pathways, and potentially ongoing active or passive surveillance, as well as strategic environmental assessment could add significant information about the levels of (both ecological and public health) risks and guide risk mitigation opportunities. The added value of comprehensive or integrated assessment processes is not only to better inform preventive action but to create synergies and complementarity with broader policies, such as energy policy, which may help generate stakeholder support (Patz, Campbell-Lendrum, Gibbs, & Woodruff, 2008).

Given the financial impact of emerging diseases, tools for risk prediction and mitigation, as well as valuation of heath-protective benefits, can be advantageous to industry, governments, and local populations – as all three sectors may face costs associated with disease outbreaks – and should be applicable to many market mechanisms and existing economic valuation tools; an example is the United Nations (UN) Framework Convention on Climate Change Clean Development Mechanism, designed to mitigate climate change while promoting sustainable development. Under the Convention's Kyoto Protocol, the Clean Development Mechanism allows countries to offset emissions by investing in carbon-reducing projects in low- to middle-income countries. Similarly, the UN's Reducing Emissions from Deforestation and forest Degradation (REDD+) program provides financial incentives for developing countries to preserve forest areas. While these scenarios assign financial value to environmental protection, the financing levels are less lucrative than could be yielded from immediate gains from industry-run development, which may be especially important in regions in dire need of investments. While research on ecosystem services has begun to yield economic valuation figures, the protective value of *not* cutting down a forest – in terms of disease transmission risk avoidance – is not calculated when making land use planning decisions. A mechanism is needed to incorporate the value of health-promoting ecosystem services and, likewise, to calculate the cost of emerging disease risks from ecosystem modification. From an economic perspective, economic valuation models that have gained considerable prominence in the conservation arena, such as The Economics of Ecosystems and Biodiversity (TEEB) initiative, offer critical opportunities for advancing this objective which can be achieved through a more systematic integration of the monetary benefits associated with both public health and conservation (World Health Organization and Convention on Biological Diversity, 2015). Moreover, more systematic evaluation and integration of social as well as economic factors and their impact on health outcomes are also critical to the development of more robust data, models and tools aimed at evaluating ecosystem-health-disease risks (see for example Oosterbroek et al., 2016), their associated human and environmental costs, and policy responses.

Porous borders present a risk of increased international movement of pathogens; the introduction and spread of MERS-CoV in South Korea provides a recent example (Cowling et al., 2015), and the first detection of Marburg virus, a filovirus that can cause severe hemorrhagic fever, occurred when African green monkeys imported for laboratory use transmitted it to laboratory workers (Feldmann, Slenczka, & Klenk, 1996). While surveillance measures can be feasibly implemented for human travelers, trade of wildlife is not typically regulated for disease risks, nor are such risks assessed for many species, which is of concern given the potential for pathogen "pollution" across species and populations in trade situations (Rosen & Smith, 2010). Expanding existing multilateral frameworks, such as the Convention on International Trade in Endangered Species of Wild Fauna and Flora, to include regulations based on health risk, or placing the burden of proof of low-risk on the importer, could help mitigate risks. Given the lucrative illegal wildlife trade, penalties should also be increased significantly to deter wildlife trafficking, especially given traceability challenges (WWF / Dalberg, 2012). On a local level where wildlife is sought for subsistence, risks can be minimized while still maintaining an important source of nutrition by working with hunters and local communities to avoid certain high-risk species (e.g. non-human primates) or report die-offs observed in wildlife to enable early detection of disease threats.

Pandemic risks (i.e., where and how the next pandemic may emerge) have been largely unknown given the lack of surveillance effort for pathogen detection (especially novel pathogens) in environmental sources. The United States Agency for International Development (USAID) Emerging Pandemic Threats program has embraced a One Health approach to guide low and middle-income countries that are ‘hotspots’ for disease emergence in early prediction and detection, and ideally, ultimately prevention, of novel disease outbreaks and their negative health and economic effects. Its ‘PREDICT’ program, which began in 2009 in 20 countries, built capacity for conducting sampling of wildlife – primarily in bats, rodents and primates identified as priority taxa based on factors including genetic similarity to humans, past spillover events, and frequency of contact with humans – and viral screening, yielding discovery of 820 novel viruses (PREDICT Consortium, December 2014). Surveillance findings are shared among the ministries responsible for human, livestock and wildlife ministries to promote coordinated interpretation and action. The project, now in its second phase, is conducting coordinated human, wildlife and livestock sampling with partners to assess viral circulation between species. The project also examines disease spillover interfaces – essentially, the characteristics of a setting including the presence of drivers of EIDs– and human behavioral risk factors to understand what may enable inter-species pathogen movement (USAID, 2014). Through the sharing of its sampling and diagnostic protocols and surveillance findings, the project can provide a model for expansion elsewhere.

A More Integrated Approach

As a starting point for the public health community to address emerging diseases in the context of global environmental change, more integrated surveillance is needed to precisely determine the long-term impacts of factors such as climate change or biodiversity loss on pathogen evolution and transmissibility. As a long-term direction, a paradigm shift is likely needed to catalyze public health to go beyond the scope of human health to the forces driving health risks to more fully protect population and ecosystem health. A broader view is needed that recognizes the connections between humans, animal and the environment (i.e. the ecosystem as a whole) and the interfaces that may lead to disease risks from their interaction, a concept often referred to as the “One Health” approach (also known as EcoHealth) (Karesh & Cook, 2005; Machalaba & Karesh, 2015).

One Health and EcoHealth encourage ecological systems thinking to improve understanding of how the individual components contribute to the much larger sum of their parts. For example, once a forest is converted to farmland or urban space, it becomes much harder to restore the original ecology because of the loss of the interconnections between components. Soil quality may rely on certain biomass and plant species that may have particular shading or sunlight requirements. These, in turn, may be factors in determining habitat suitability for pathogens or their vectors; changes in ecosystems may result in changes in species abundance and composition. The use of slash-and-burn approaches versus timber logging may yield different ecological outcomes, and over different temporal scales.

One Health also shares many affinities with other dimensions of complexity science. EcoHealth, adaptive management and landscape epidemiology also provide critical insights useful to the development of public policies that are not constrained by the habitual confines of reductionist thinking (see for example (Berbés-Blázquez, Oestreicher, Mertens, & J., 2014; Meentemeyer, Haas, & Vaclavik, 2012)). By examining social and ecological processes as coupled systems - or social-ecological systems - drawing from these approaches can contribute to more robust analyses of feedback loops between ecological, epidemiological and social processes while identifying areas of potential intervention to strengthen social-ecological resilience. As transdisciplinary literature in this area continues to develop (see for example (De Vos et al., 2016) on the mutual influences between wildlife, domestic and human diseases in the context of social-ecological systems within protected areas) but further integrative research in this area is needed. Further scientific investigation in this area necessitates the concomitant development of tools and heuristics to bridge our understanding of complex socio-ecological systems and disease emergence in ways that can be usefully operationalized as public policies and interventions. Linking ecological and epidemiological models of disease exposure to social processes and economic processes such as poverty, inequity, violence and gender is an important yet underdeveloped dimension of disease risk (see for example (Janes, Corbett, Jones, & Trostle, 2012; Nading, 2013)). Integrative research that draws from the social sciences provides critical opportunities for the development of effective behavioral and structural disease control strategies and intervention.

The complexities in anticipating disease outcomes also requires the development of new tools to increase our understanding and improve monitoring of the underlying mechanisms that lead to their emergence. As we transition toward the Sustainable Development Goals these challenges can also be harnessed as opportunities for the development of robust indicators to monitor disease patterns. They are also an opportunity to evaluate changes in exposure to a variety of other health-related risks and to assess progress towards environmental sustainability more broadly (Dora et al., 2015; World Health Organization and Convention on Biological Diversity, 2015).

Existing tools can also be used to enhance our understanding and, ideally, prevention of EIDs. Integration of climate and weather data in outbreak risk analyses from historical events may help better inform future disease projections. For example, in comparing land surface temperature and vegetation indices from satellite imaging, Anyamba and colleagues show the correlation of weather extremes with agricultural outcomes and vector-borne disease outbreaks (Anyamba et al., 2014). Rift Valley Fever outbreaks have been found to occur in weather-dependent cycles in Kenya, allowing for prediction up to five months ahead and enabling proactive risk prioritization for targeted resource allocation (e.g. such as vaccination) (Linthicum et al., 1999). Given variations in weather conditions, ecosystems, and priority pathogens, efforts to anticipate epidemic cycles may require local or regional analysis, but can benefit from systematic data collection for comparability across sites as well as provide flexibility for upscaling for global trend analysis, and may be used to look at several different health outcomes (Patz, 2005).

Expansion in the scope of disciplines also may be a long-term strategy for the development of more integrated cross-sectoral approaches. The One Health approach can help us more fully understand the links and complex interactions between ecological dynamics and determinants of EIDs, but requires professionals coming together to work in new ways. Our national and international agencies can help move this forward through coordinated policy development that works toward co-benefits, or that at least considers implications for other sectors prior to policy decisions to minimize negative impacts. For example, country delegates or National Focal Points of the World Health Organization, Convention on Biological Diversity, and World Organisation for Animal Health (OIE) can look for areas of integration in their regulations, within their particular country's priorities; they can consult each other's expertise on proposed policies to gain a sense of the positive or negative outcomes to another sector (Machalaba et al., 2015a). Similarly, development donors can require coordinated baseline capacity assessments across multiple sectors – such as veterinary services, public health, and environment – to identify ways that investments can be made to benefit multiple disciplines and help bridge disciplinary silos (i.e. a single-discipline focus of scope, based on the literal term from agricultural grain silos, which are vertical rather than horizontally-integrated). Working with multiple disciplines within and outside of the health sector can help assess multiple or cumulative health impacts from observed or anticipated environmental change, such as dual impacts on infectious diseases emergence and food security from ecosystem alteration (Myers et al., 2013). Moreover, comprehensive strategies aimed at jointly addressing the root causes of ill health and anthropogenic ecosystem degradation must not only draw on formal scientific disciplines (including ecological, epidemiological, health and other sciences), but also on the social sciences, other forms of knowledge (such as traditional knowledge), engage in more inclusive participatory processes, supportive legal frameworks and include both systemic and behavioral interventions (World Health Organization and Convention on Biological Diversity, 2015).

In resource-limited countries, public and veterinary health capacity is often substantially lacking. The International Health Regulations (IHR) were developed by the World Health Organization (WHO), which is the public health entity for the United Nations, to ensure some level of core universal health capacity across all nations. At the end of 2013, more than half of WHO's 196 member countries had requested second extensions to meet the requirements (which took effect in 2007) (World Health Organization, 2014). In addition to capacity, stronger governance mechanisms are critically needed in other sectors to alleviate the actual causes that lead to public health risks- the drivers of disease identified earlier in the chapter. Examples include more prudent use of antimicrobials in animal production, such as phasing out their non-therapeutic use (e.g. in growth promoters), requiring veterinary oversight, and regulating antimicrobial classes acceptable for uses (strategies applied by the European Union and even more strictly by Denmark), and land use planning that optimizes- or at least finds a balance with- both public health and development (Patz et al., 2004; Wagenaar & Scott, 2015). Industry policies, too, can reduce risk; for example, the Collaborative Partnership on Sustainable Wildlife Management, a consortium of UN and other international organizations, recommends that extractive industry companies (e.g. mining, timber logging, which often operate in remote settings) provide their workers with a reliable protein source to reduce hunting demands on wildlife in order to protect health and safety of their employees (Collaborative Partnership on Sustainable Wildlife Management (CPW), March 18, 2015).

Financial Justification

The long-term impacts of global environmental change and EIDs are expected to present 'grand challenges' for our societies and especially our economies. Thus, in addition to the public health burden, the financial toll from emerging diseases also makes the burden relevant to finance ministries and other payers. In 2015 the World Economic Forum Global Risks Perceptions Survey ranked the threat of "rapid and massive spread of infectious diseases" as the second largest risk based on impact (World Economic Forum, 2015). Extreme weather events, natural catastrophes, environmental risk from insufficient adaptation to climate change and loss of biodiversity and ecosystem collapse were also identified as major risks in terms of likelihood, impact, or both, and we argue they should not be viewed in isolation, as they may play a role in each others' outcomes. Like earthquakes and tsunamis, infectious disease outbreaks may be relatively rare compared to other risks, but exceedingly catastrophic. The cost of emerging disease outbreaks in the past two decades has been calculated as hundreds of billions of dollars (Karesh et al., 2012). The World Bank estimates that the cost of six zoonotic diseases alone – some well publicized, such as Avian Influenza in Asia and Europe, some less known, such as Rift Valley Fever in three African nations – accounted for US$80 billion or more from 1997-2009 (World Bank, 2012).

The high costs of EIDs present a strong cost-benefit argument for prevention strategies. Implementing pathogen surveillance in wildlife can help identify and characterize novel pathogens, and over time, may help with risk prioritization of potential zoonotic threats. While some projects have been implemented (including the USAID PREDICT project, now operating in 32 countries, highlighted above), there is a still a major deficit of information on viral diversity. Known pathogens have been mostly identified from obvious die-offs or morbidity in species, or traced back from human outbreaks to other mammalian hosts or reservoirs. Minimal surveillance has been conducted to determine the full extent of viral diversity; this is the case even in humans, with undiagnosed or misdiagnosed febrile illnesses likely in some parts of the world (Susilawati & McBride, 2014). Anthony and colleagues' projections of viral richness based on viral screening in one wildlife host suggest there are at least 320,000 mammalian viruses from nine viral families- of which less than 2% has been detected to date (Anthony et al., 2013). The cost of the majority (85%) of mammalian viral discovery is estimated at US$1.4 billion, or four-and-a-half times that for complete discovery. While a significant investment, it constitutes a small portion of the costs seen from recent viral EIDs. Over the long term, tracking pathogen diversity may also allow for understanding and ongoing detection of pathogen evolution, which may be especially important for pathogens with reassortment potential (e.g. influenza viruses), or for development and refinement of clinical therapies, including to prevent or respond to resistant strains.

Outbreak response operations can be highly resource-intensive, with resource mobilization potentially challenging, as seen in the West Africa Ebola crisis (Gostin & Friedman, 2015). Rather than adapting to the threat of EIDs, risk reduction or prevention strategies have a clear benefit. Economic optimization projections suggest that implementing a One Health approach, which takes into account human, animal and environmental health links by working across disciplines, to help prevent pandemic threats would yield cost savings of more than US$300 billion over the next century compared to adaptation programs (Pike, Bogich, Elwood, Finnoff, & Daszak, 2014). A separate cost-benefit analysis projects global benefits valued at US$30 billion per year from an annual investment of US$3.4 billion in preventive One Health systems that emphasize attainment of veterinary services and human health national performance standards with at least some preventive capacity (based on the assumption that a once-a-century pandemic would total US$3 trillion) (World Bank, 2012). However, as a global public good, this preventive benefit is not "sellable," and there is currently no clear investor to bring it about.

FUTURE RESEARCH DIRECTIONS

The Role of Interdisciplinarity

Scientific research itself has an important role to play in addressing both global environmental change and emerging infectious diseases, including at their intersection. In the near term, research is needed to enhance understanding of the ecological dynamics and mechanisms that lead to emerging infectious disease events and their burden on human populations, in particular among vulnerable populations. To do this effectively, scientific investigation (for example ecological, epidemiological and conservation sciences) must also draw upon social science research (such as economics, anthropology, and behavioural sciences) to address the full breadth of risks of and develop effective responses to disease emergence. Research investigations must also step beyond the commonly used variables in epidemiological modeling of past and future events to incorporate additional parameters, such as those in ecological niche modeling (Daszak et al., 2013; Peterson, 2006). This approach uses estimates of species distributions and the geographic range of abiotic and biotic factors to assess the spatial 'niche' for a given species, and can incorporate possible shifts, such as from climate change or trade-mediated introduction. It can be a useful tool for improving understanding of pathogens that could be present at a location currently or in the future, based on the niche of their host(s) species. Ecological niche modeling has been employed in response to recent outbreaks, including Ebola virus in Africa and Zika virus in the Americas (Bogoch et al., 2016; Pigott et al., 2014).

The scale of macro-level drivers of EIDs does not translate seamlessly to targeted interventions. A growing body of research suggests we can obtain greater predictability of EID threats from past history of outbreaks (Morse et al., 2012). Based on an extensive analysis of over 300 EID events recorded between 1940-2004, Jones et al. generated spatial analyses of where emerging diseases are likely to originate – so-termed the "hotspots" of EIDs – which are typically mammalian biodiversity-rich areas in the tropics (Jones et al., 2008). These regions also happen to be located primarily within the legal territory of developing states already struggling with poor public health and sanitation infrastructure, including low reporting and testing capacity. Because of their exploitable natural resources, many of these hotspot areas are also targeted for major development projects. As such, EIDs must also be analyzed within the context of sustainable development policies and outcomes. Targeting hotspot areas to assess and identify risk factors (i.e. in the form of drivers and transmission risk interfaces) of EIDs may provide a starting point for surveillance on a country or regional level (indeed, as noted in the prior section, the PREDICT project has employed this approach in selecting surveillance sites). Specific assessment of local drivers can help in priority setting around local contexts of disease ecology, risk and burden. For example, Liu and colleagues highlight the factors identified as contributors to emerging and re-emerging diseases in China, including rapid increase in population, urbanization-associated migration, deforestation, and changes in livestock raising (Liu, Cao, & Zhu, 2014). Viewing drivers on this scale makes information more tangible, enabling government agencies and local partners to work together to address country-specific pressures that are anticipated to increase EID risks. A local approach to support global health can help identify and address unique cultural or behavioral practices of a population that drive risk, helping to work with communities to balance cultural preservation with risk reduction.

The nature of scientific research funding is typically short-term, limiting opportunities to design and conduct long-term field studies. Yet these studies are essential for elucidating temporal or spatial gradients of changing ecological dynamics and disease outcomes (Machalaba, Daszak, Karesh, & Shrivastava, 2015b). Integrated health and ecology studies implemented at sites with and without planned or natural changes to ecosystems can provide new and robust information, and can be expanded to tie into research on pressures to planetary boundaries. Perhaps most fundamentally, research must not stop at the data collection or publication step, but must be designed for full-spectrum use that clearly advises policy makers – many unfamiliar with the discipline – on its key messages and actionable next steps.

Promising new interdisciplinary models of research are emerging, including the recently established ten-year Future Earth scientific platform for global sustainability research, which brings together a range of disciplinary projects (ranging from ecosystem services, atmospheric science, health, urbanization, and others) all working toward generating knowledge and solutions to address impacts of global environmental change (Machalaba et al., 2015b). With a research agenda spanning different stages (observing and understanding change processes all the way to transformations for sustainability), Future Earth has heralded innovative trans-disciplinary research while supporting the 'co-design' and 'co-production' of knowledge, in which sustainable scientific pursuits are oriented toward continual refinement and balance of social responses, instead of purely technical measurements of planetary boundaries *per se* (Cornell, 2013; Future Earth, 2014). The initiative also seeks to engage stakeholders outside of traditional research parameters, including governments, intergovernmental agencies, and importantly, the private sector, youth as well as stakeholders with different cultural perspectives including traditional knowledge holders. Consequently, benefits could be conferred to those communities by making science relevant and usable. Ideally, research should be formulated to directly inform policy and practice.

Knowledge Gaps

In identifying the broad drivers of EIDs above, we recognize that disease dynamics operate in highly complex and variable social-ecological systems. The particular mechanisms (i.e. the exact macro- and micro-scopic variables) that lead to a given outbreak have yet to be deciphered. The field of disease ecology seeks to look at these specific factors among the host, pathogen and their interactions, including evolutionary and ecological determinants (Kilpatrick & Altizer, 2010). However, more integrated research is needed to jointly address the social and ecological determinants of health.

Research on human pathogens has primarily concentrated on terrestrial animal and vector-borne sources or transmission cycles, and our chapter reflects this scope of work. However, marine dynamics are also an important area for further and more integrated study, especially considering rising demands on marine resources and major oceanic changes underway such as warming, acidification, and accumulation of pollutants, all of which may change the composition and population ranges of species, including marine microbes (Burge et al., 2014; Sandifer & Sutton-Grier, 2014; Secretariat of the Convention on Biological Diversity, 2014). Plant pathogens are also highly relevant to both human health and to broader conservation considerations, including in the context of food safety and security, with plants potentially serving as vectors for pathogen transmission or pathogens affecting crop yields (Erickson & Doyle, 2012; Strange & Scott, 2005). Further study in this area can help elucidate the impact of anthropogenic environmental change on marine and plant pathogen dynamics (Pautasso, Döring, Garbelotto, Pellis, & Jeger, 2012; World Health Organization and Convention on Biological Diversity, 2015).

In addition to social and ecological knowledge gaps, limited infrastructure within the public health realm also presents challenges in the early identification of health risks from global environmental change. First, many public health systems do not routinely conduct pathogen surveillance in wildlife proactively, preventing knowledge of pathogens that could potentially emerge in humans or other species. Instead, pathogen screening in wildlife is often reactive, employing a wait-and-see approach until an outbreak occurs. Passive surveillance strategies (i.e. sampling and pathogen screening when disease is suspected, usually based on clinical signs or mortality) are commonly used for animal diseases compared to active surveillance (periodic sampling and screening), and the different approaches may also affect findings (Hadorn & Stark, 2008). The limitations of passive surveillance include difficulties in accessing and acquiring wildlife samples (for example, sick or deceased animals may not be reported to authorities, carcasses may decompose rapidly in some ecosystems, and cold chains to proper sample preservation may not be easily maintained in some remote location) (Institute of Medicine and National Research Council, 2008). Furthermore, animals that may appear healthy may still harbor and shed pathogens. Increased screening of healthy wildlife in combination with analysis of morbidity or mortality events would add to the knowledge base on viral diversity (Levinson et al., 2013). While active surveillance may not be feasible or cost-effective for all sites and species given resource limitations, targeted active surveillance can be determined through risk assessments (informed by outputs from ecological niche modeling, past pathogen detection and disease severity, and risk factors for transmission).

In many cases, even when pathogen surveillance in wildlife takes place, it is rarely systematically implemented; for example, a review of bat viral diversity publications found major gaps in bat species coverage and highlighted how different observed sample types (e.g. urine versus tissue specimens) and sample sizes may affect viral discovery yield (Young & Olival, 2016). Surveillance systems for Avian Influenza in wild birds could arguably be viewed as the most robust globally in terms of recent pathogen surveillance coverage in wildlife, with some large-scale programs in place, but taken together these programs typically lack the methodological (including extent of subtype screening), spatial, temporal, diagnostic and reporting consistency and continuity needed to monitor viral diversity, and with some exceptions most are more coordinated to research than to early detection (Bevins et al., 2014; Machalaba et al., 2015c).

Public health has conventionally focused on human transmission dynamics, affording little consideration of the interactions between species, ecological dynamics, and climate factors, thus preventing a holistic view of the root causes. Such narrow focus reduces the potential for understanding the full transmission cycle dynamics that could help in finding effective solutions to prevent future threats (Machalaba & Karesh, 2015). For example, factors such as seasonality (whether tied to weather, migration, reproductive cycles, etc.) may affect viral prevalence levels (Amman et al., 2012; Ip et al., 2008); similarly, the behaviors that may facilitate transmission may also have seasonal determinants (e.g., reliance on wild meat hunting may increase during periods of 'lean' agricultural yield) (Schulte-Herbruggen, Cowlishaw, Homewood, & Rowcliffe, 2013).

Global environmental change will manifest in both expected and uncertain ways, with likely resulting effects on disease transmission and spread risks. Thus, there is high importance of anticipating future challenges in near-term decision making to mitigate risks, and building in resiliency strategies to adapt to uncertainty. In order to enable systems to address EID threats, investments should be strategic to promote cross-disciplinary collaboration, including data sharing across ministries, resources to conduct predictive models for different land use, climate, and other scenarios, and develop long-term road maps for EID prevention. As knowledge of the mechanisms of disease emergence increases, more evidence-based action focused on prevention can be successfully implemented.

CONCLUSION

While global environmental changes pose risks for emerging infectious disease outbreaks, the two share common underlying drivers that can be targeted to mitigate detrimental outcomes. While still speculative, expected increases in EID events may even be considered tipping points for our ecological systems. The impacts of EIDs may be more immediately evident than more "subtle" changes in ecosystem dynamics. Yet the latter can have major impacts, even if these are only visible at later stages, as ecosystem services continue to erode. Building capacity to better understand the dynamics and synergies between global environmental change and disease outbreaks, in ways that enhance our understanding of their complex multi-scalar linkages, is critical to effectively addressing some of the major sustainability challenges of this century. Doing so can also strengthen our capacity for managing the social and environmental determinants of other disease risks including those related to pollution, endemic water-borne and water-related diseases, as well as those associated with the growing burden of non-communicable diseases.

Despite major advances in pathogen screening, our overall capacity for the prevention of emerging diseases remains insufficient. While solutions are within reach, they require phase changes in our public health systems and our development processes in ways that are more proactive, preventive, and synergistic. Addressing the underlying causes of both disease emergence and global environmental change may yield risk avoidance benefits in both realms, helping to find sustainable and healthy solutions to meet or reshape the demands of a growing global population. The courageous decisions made now can shape the trajectory of global environmental change and emerging infectious diseases in the future.

ACKNOWLEDGMENT

We thank Mo Usavage, Peter Wanfried, Daniel Wang and Linda Oliwa for their review of the chapter and constructive feedback. We are grateful for the generous support of the DIVERSITAS-Future Earth oneHEALTH project. This chapter also benefited from intellectual developments from the PREDICT project of the United States Agency for International Development (USAID) Emerging Pandemic Threats program.

REFERENCES

Acevedo-Whitehouse, K., & Duffus, A. L. (2009). Effects of environmental change on wildlife health. *Philosophical Transactions of the Royal Society of London. Series B, Biological Sciences, 364*(1534), 3429–3438. doi:10.1098/rstb.2009.0128 PMID:19833653

Ahmad, A., Krumkamp, R., & Reintjes, R. (2009). Controlling SARS: A review on China's response compared with other SARS-affected countries. *Tropical Medicine & International Health, 14*(Suppl. 1), 36–45. doi:10.1111/j.1365-3156.2008.02146.x PMID:19508440

Ali, M., & Lopez, A. L., AeYou, Y., EunKim, Y., Sah, B., Maskery, B., & Clemens, J. (2012). The global burden of cholera. *Bulletin of the World Health Organization, 90*(3), 209–218. doi:10.2471/BLT.11.093427 PMID:22461716

Allen, H. K., Donato, J., Wang, H. H., Cloud-Hansen, K. A., Davies, J., & Handelsman, J. (2010). Call of the wild: Antibiotic resistance genes in natural environments. *Nature Reviews. Microbiology, 8*(4), 251–259. doi:10.1038/nrmicro2312 PMID:20190823

Amman, B. R., Carroll, S. A., Reed, Z. D., Sealy, T. K., Balinandi, S., Swanepoel, R., & Towner, J. S. et al. (2012). Seasonal pulses of Marburg virus circulation in juvenile Rousettus aegyptiacus bats coincide with periods of increased risk of human infection. *PLoS Pathogens, 8*(10), e1002877. doi:10.1371/journal.ppat.1002877 PMID:23055920

Anthony, S. J., Epstein, J. H., Murray, K. A., Navarrete-Macias, I., Zambrana-Torrelio, C. M., Solovyov, A., & Lipkin, W. I. et al. (2013). A strategy to estimate unknown viral diversity in mammals. *mBio, 4*(5), e00598–e00513. doi:10.1128/mBio.00598-13 PMID:24003179

Anyamba, A., Small, J. L., Britch, S. C., Tucker, C. J., Pak, E. W., Reynolds, C. A., & Linthicum, K. J. et al. (2014). Recent weather extremes and impacts on agricultural production and vector-borne disease outbreak patterns. *PLoS ONE, 9*(3), e92538. doi:10.1371/journal.pone.0092538 PMID:24658301

Assessment, M. E. (2005). Ecosystems and Human Well-being. *Synthesis.*

Baize, S., Pannetier, D., Oestereich, L., Rieger, T., Koivogui, L., Magassouba, N., & Gunther, S. et al. (2014). Emergence of Zaire Ebola virus disease in Guinea. *The New England Journal of Medicine, 371*(15), 1418–1425. doi:10.1056/NEJMoa1404505 PMID:24738640

Bayliss, H. R., Schindler, S., Essl, F., Rabitsch, W., & Pullin, A. S. (2015). What evidence exists for changes in the occurrence, frequency or severity of human health impacts resulting from exposure to alien invasive species in Europe? A systematic map protocol. *Environmental Evidence*, *4*(1), 10. doi:10.1186/s13750-015-0037-4

Bell, D. M. (2004). Public health interventions and SARS spread, 2003. *Emerging Infectious Diseases*, *10*(11), 1900–1906. doi:10.3201/eid1011.040729 PMID:15550198

Berbés-Blázquez, M., Oestreicher, J. S., Mertens, F., & Saint-Charles, J. (2014). Ecohealth and resilience thinking: A dialog from experiences in research and practice. *Ecology and Society*, *19*(2), 24. doi:10.5751/ES-06264-190224

Bernard, S. M., & Anderson, S. A. (2006). Qualitative assessment of risk for monkeypox associated with domestic trade in certain animal species, United States. *Emerging Infectious Diseases*, *12*(12), 1827–1833. doi:10.3201/eid1212.060454 PMID:17326932

Bevins, S. N., Pedersen, K., Lutman, M. W., Baroch, J. A., Schmit, B. S., Kohler, D., & DeLiberto, T. J. et al. (2014). Large-scale avian influenza surveillance in wild birds throughout the United States. *PLoS ONE*, *9*(8), e104360. doi:10.1371/journal.pone.0104360 PMID:25116079

Bird, B. H., Ksiazek, T. G., Nichol, S. T., & MacLachlan, N. J. (2009). Rift Valley fever virus. *Journal of the American Veterinary Medical Association*, *234*(7), 883–893. doi:10.2460/javma.234.7.883 PMID:19335238

Blanco, G., Gerlagh, R., Suh, S., Barrett, J., de Coninck, H. C., Diaz Morejon, C. F., . . . Zhou, P. (2014). Drivers, Trends and Mitigation. In O. Edenhofer, R. Pichs-Madruga, Y. Sokona, E. Farahani, S. Kadner, ... J.C.M. (Eds.), Climate Change 2014: Mitigation of Climate Change. Contribution of Working Group III to the Fifth Assessment Report of the Intergovernmental Panel on Climate Change. New York, NY, USA: Cambridge University Press.

Bogoch, I. I., Brady, O., Kraemer, M. U. G., German, M., Creatore, M. I., Kulkarni, M. A., & Khan, K. et al. (2016). Anticipating the international spread of Zika virus from Brazil. *Lancet*, *387*(10016), 335–336. doi:10.1016/S0140-6736(16)00080-5 PMID:26777915

Boyles, J. G., Cryan, P. M., McCracken, G. F., & Kunz, T. H. (2011). Conservation. Economic importance of bats in agriculture. *Science*, *332*(6025), 41–42. doi:10.1126/science.1201366 PMID:21454775

Brownstein, J. S., Holford, T. R., & Fish, D. (2005). Effect of Climate Change on Lyme Disease Risk in North America. *EcoHealth*, *2*(1), 38–46. doi:10.1007/s10393-004-0139-x PMID:19008966

Bruinsma, J. (2009). The resource outlook to 2050: By how much do land, water use and crop yields need to increase by 2050? *Paper presented at the Expert Meeting on How to Feed the World in 2050*, Rome.

Burge, C. A., Mark Eakin, C., Friedman, C. S., Froelich, B., Hershberger, P. K., Hofmann, E. E., & Harvell, C. D. et al. (2014). Climate change influences on marine infectious diseases: Implications for management and society. *Annual Review of Marine Science*, *6*(1), 249–277. doi:10.1146/annurev-marine-010213-135029 PMID:23808894

Campbell-Lendrum, D. H. (2005). How much does the health community care about global environmental change? *Global Environmental Change*, *15*(4), 296–298. doi:10.1016/j.gloenvcha.2005.09.001

Centers for Disease Control and Prevention. (2015, June 26). 2014 Ebola Outbreak in West Africa - Outbreak Distribution Map. Retrieved from http://www.cdc.gov/vhf/ebola/outbreaks/2014-west-africa/distribution-map.html

Chivian, E., & Bernstein, A. (2008). *Sustaining life: How human health depends on biodiversity*. New York: Oxford University Press.

Chua, K. B. (2003). Nipah virus outbreak in Malaysia. *Journal of Clinical Virology*, *26*(3), 265–275. http://www.ncbi.nlm.nih.gov/pubmed/12637075 doi:10.1016/S1386-6532(02)00268-8 PMID:12637075

Collaborative Partnership on Sustainable Wildlife Management (CPW). (March 18, 2015). *Bushmeat Sourcebook*.

Convention on Biological Diversity. (2014). *Global Biodiversity Outlook 4*.

Cornell, S., Berkhout, F., Tuinstra, W., Tàbara, J. D., Jäger, J., Chabay, I., & van Kerkhoff, L. et al. (2013). Opening up knowledge systems for better responses to global environmental change. *Environmental Science & Policy*, *28*, 60–70. doi:10.1016/j.envsci.2012.11.008

Cowling, B. J., Park, M., Fang, V. J., Wu, P., Leung, G. M., & Wu, J. T. (2015). Preliminary Epidemiological Assessment Of Mers-Cov Outbreak In South Korea, May To June 2015. *Eurosurveillance*, *20*(25), 21163. doi:10.2807/1560-7917.ES2015.20.25.21163 PMID:26132767

Daszak, P., Cunningham, A. A., & Hyatt, A. D. (2000). Emerging infectious diseases of wildlife--threats to biodiversity and human health. *Science*, *287*(5452), 443–449. http://www.ncbi.nlm.nih.gov/pubmed/10642539 doi:10.1126/science.287.5452.443 PMID:10642539

Daszak, P., Zambrana-Torrelio, C., Bogich, T. L., Fernandez, M., Epstein, J. H., Murray, K. A., & Hamilton, H. (2013). Interdisciplinary approaches to understanding disease emergence: The past, present, and future drivers of Nipah virus emergence. *Proceedings of the National Academy of Sciences of the United States of America*, *110*(Suppl. 1), 3681–3688. doi:10.1073/pnas.1201243109 PMID:22936052

De Vos, A., Cumming, G., Cumming, D., Ament, J., Baum, J., Clements, H., & Moore, C. et al. (2016). Pathogens, disease, and the social-ecological resilience of protected areas. *Ecology and Society*, *21*(1), art20. doi:10.5751/ES-07984-210120

Dora, C., Haines, A., Balbus, J., Fletcher, E., Adair-Rohani, H., Alabaster, G., & Neira, M. et al. (2015). Indicators linking health and sustainability in the post-2015 development agenda. *Lancet*, *385*(9965), 380–391. doi:10.1016/S0140-6736(14)60605-X PMID:24923529

Dye, C. (2014). After 2015: Infectious diseases in a new era of health and development. *Philosophical Transactions of the Royal Society of London. Series B, Biological Sciences*, *369*(1645), 20130426. doi:10.1098/rstb.2013.0426 PMID:24821913

Earth, F. (2014). *Strategic Research Agenda 2014*. Retrieved from http://www.futureearth.org/media/strategic-research-agenda-2014

Ebi, K. L., Woodruff, R., von Hildebrand, A., & Corvalan, C. (2007). Climate change-related health impacts in the Hindu Kush–Himalayas. *EcoHealth*, *4*(3), 264–270. doi:10.1007/s10393-007-0119-z

Enright, M. C., Robinson, D. A., Randle, G., Feil, E. J., Grundmann, H., & Spratt, B. G. (2002). The evolutionary history of methicillin-resistant Staphylococcus aureus (MRSA). *Proceedings of the National Academy of Sciences of the United States of America*, *99*(11), 7687–7692. doi:10.1073/pnas.122108599 PMID:12032344

Erickson, M. C., & Doyle, M. P. (2012). *Plant Food Safety Issues: Linking Production Agriculture with One Health Improving food safety through a One Health approach*. Washington, DC: The National Academies Press.

Faria, N. R., Rambaut, A., Suchard, M. A., Baele, G., Bedford, T., Ward, M. J., & Lemey, P. et al. (2014). HIV epidemiology. The early spread and epidemic ignition of HIV-1 in human populations. *Science*, *346*(6205), 56–61. doi:10.1126/science.1256739 PMID:25278604

Feldmann, H., Slenczka, W., & Klenk, H. D. (1996). Emerging and reemerging of filoviruses. *Archives of Virology. Supplementum*, *11*, 77–100. Retrieved from http://www.ncbi.nlm.nih.gov/pubmed/8800808 PMID:8800808

Flint, H. S., Franson, P. L., Dusek, J. C., Derksen, R. J., Gill, D. V., Jr, R. E., & Rothe, T. C. et al. (2008). Prevalence of Influenza A viruses in wild migratory birds in Alaska: Patterns of variation in detection at a crossroads of intercontinental flyways. *Virology Journal*, *5*(1), 71. doi:10.1186/1743-422X-5-71 PMID:18533040

Foley, J. A., Ramankutty, N., Brauman, K. A., Cassidy, E. S., Gerber, J. S., Johnston, M., & Zaks, D. P. et al. (2011). Solutions for a cultivated planet. *Nature*, *478*(7369), 337–342. doi:10.1038/nature10452 PMID:21993620

Food and Agriculture Organization of the UN. (2009). *How to Feed the World in 2050*. Retrieved from http://www.fao.org/fileadmin/templates/wsfs/docs/expert_paper/How_to_Feed_the_World_in_2050.pdf

GBD 2013 Mortality and Causes of Death Collaborators. (2015). Global, regional, and national age-sex specific all-cause and cause-specific mortality for 240 causes of death, 1990-2013: a systematic analysis for the Global Burden of Disease Study 2013. *Lancet, 385*(9963), 117-171. doi:10.1016/S0140-6736(14)61682-2

Ge, X. Y., Li, J. L., Yang, X. L., Chmura, A. A., Zhu, G., Epstein, J. H., & Shi, Z. L. et al. (2013). Isolation and characterization of a bat SARS-like coronavirus that uses the ACE2 receptor. *Nature*, *503*(7477), 535–538. doi:10.1038/nature12711 PMID:24172901

Gerber, P. J., Steinfeld, H., Henderson, B., Mottet, A., Opio, C., Dijkman, J., . . . Tempio, G. (2013). *Tackling climate change through livestock – A global assessment of emissions and mitigation opportunities.*

Golden, C. D., Fernald, L. C., Brashares, J. S., Rasolofoniaina, B. J., & Kremen, C. (2011). Benefits of wildlife consumption to child nutrition in a biodiversity hotspot. *Proceedings of the National Academy of Sciences of the United States of America*, *108*(49), 19653–19656. doi:10.1073/pnas.1112586108 PMID:22106297

Gostin, L. O., & Friedman, E. A. (2015). A retrospective and prospective analysis of the west African Ebola virus disease epidemic: Robust national health systems at the foundation and an empowered WHO at the apex. *Lancet*, *385*(9980), 1902–1909. doi:10.1016/S0140-6736(15)60644-4 PMID:25987158

Gottdenker, N. L., Streicker, D. G., Faust, C. L., & Carroll, C. R. (2014). Anthropogenic land use change and infectious diseases: A review of the evidence. *EcoHealth*, *11*(4), 619–632. doi:10.1007/s10393-014-0941-z PMID:24854248

Hadorn, D. C., & Stark, K. D. (2008). Evaluation and optimization of surveillance systems for rare and emerging infectious diseases. *Veterinary Research*, *39*(6), 57. doi:10.1051/vetres:2008033 PMID:18651991

Haines-Young, R., & Potschin, M. (2010). The links between biodiversity, ecosystem services and human well-being. In D. C. F. Raffaelli (Ed.), *Ecosystem Ecology: a new synthesis* (pp. 110–139). Cambridge. doi:10.1017/CBO9780511750458.007

Hennessey, M., Fischer, M., & Staples, J. E. (2016). Zika Virus Spreads to New Areas - Region of the Americas, May 2015-January 2016. *MMWR. Morbidity and Mortality Weekly Report*, *65*(3), 55–58. doi:10.15585/mmwr.mm6503e1 PMID:26820163

Herrero, M., & Thornton, P. K. (2013). Livestock and global change: Emerging issues for sustainable food systems. *Proceedings of the National Academy of Sciences of the United States of America*, *110*(52), 20878–20881. doi:10.1073/pnas.1321844111 PMID:24344313

Heymann, D. L., & Dar, O. A. (2014). Prevention is better than cure for emerging infectious diseases. *BMJ (Clinical Research Ed.)*, *348*, g1499. PMID:24563451

Hosseini, P. R., Fuller, T., Harrigan, R., Zhao, D., Arriola, C. S., Gonzalez, A., & Daszak, P. et al. (2013). Metapopulation dynamics enable persistence of influenza A, including A/H5N1, in poultry. *PLoS ONE*, *8*(12), e80091. doi:10.1371/journal.pone.0080091 PMID:24312455

Hotez, P. J., & Kamath, A. (2009). Neglected tropical diseases in sub-saharan Africa: Review of their prevalence, distribution, and disease burden. *PLoS Neglected Tropical Diseases*, *3*(8), e412. doi:10.1371/journal.pntd.0000412 PMID:19707588

Huttenhower, C., Gevers, D., Knight, R., Abubucker, S., Badger, J. H., Chinwalla, A. T., & White, O. et al.Human Microbiome Project Consortium. (2012). Structure, function and diversity of the healthy human microbiome. *Nature*, *486*(7402), 207–214. doi:10.1038/nature11234 PMID:22699609

Institute of Medicine and National Research Council. (2008). *Current Surveillance Systems for Detecting Zoonoses in Animals*.

Intergovernmental Panel on Climate Change (IPCC). (2014). *Summary for policymakers*.

IOM (Institute of Medicine) European Centre for Disease Prevention and Control. (2016). *Rapid Risk Assessment. Zika virus disease epidemic: potential association with microcephaly and Guillain–Barré syndrome.*

Janes, C. R., Corbett, K. K., Jones, J. H., & Trostle, J. (2012). Emerging infectious diseases: The role of social sciences. *Lancet*, *380*(9857), 1884–1886. doi:10.1016/S0140-6736(12)61725-5 PMID:23200487

Jones, K. E., Patel, N. G., Levy, M. A., Storeygard, A., Balk, D., Gittleman, J. L., & Daszak, P. (2008). Global trends in emerging infectious diseases. *Nature*, *451*(7181), 990–993. doi:10.1038/nature06536 PMID:18288193

Karesh, W. B., & Cook, R. A. (2005). The Human-Animal Link. *Foreign Affairs*, July/August.

Karesh, W. B., Cook, R. A., Bennett, E. L., & Newcomb, J. (2005). Wildlife trade and global disease emergence. *Emerging Infectious Diseases*, *11*(7), 1000–1002. doi:10.3201/eid1107.050194 PMID:16022772

Karesh, W. B., Dobson, A., Lloyd-Smith, J. O., Lubroth, J., Dixon, M. A., Bennett, M., & Heymann, D. L. et al. (2012). Ecology of zoonoses: Natural and unnatural histories. *Lancet*, *380*(9857), 1936–1945. doi:10.1016/S0140-6736(12)61678-X PMID:23200502

Keesing, F., Holt, R. D., & Ostfeld, R. S. (2006). Effects of species diversity on disease risk. *Ecology Letters*, *9*(4), 485–498. doi:10.1111/j.1461-0248.2006.00885.x PMID:16623733

Keune, H., Kretsch, C., De Blust, G., Gilbert, M., Flandroy, L., Van Den Berge, K., & Bauler, T. et al. (2013). Science–policy challenges for biodiversity, public health and urbanization: Examples from Belgium. *Environmental Research Letters*, *8*(2), 025015. doi:10.1088/1748-9326/8/2/025015

Kilpatrick, A. M., & Altizer, S. (2010). Disease Ecology. *Nature Education Knowledge*, *3*(10), 55. PMID:19836101

Kilpatrick, A. M., & Randolph, S. E. (2012). Drivers, dynamics, and control of emerging vector-borne zoonotic diseases. *Lancet*, *380*(9857), 1946–1955. doi:10.1016/S0140-6736(12)61151-9 PMID:23200503

Lam, S. K., & Chua, K. B. (2002). Nipah Virus Encephalitis Outbreak in Malaysia. *Clinical Infectious Diseases*, *34*(Suppl. 2), S48–S51. doi:10.1086/338818 PMID:11938496

Laurance, W. F., Camargo, J. L., Luizão, R. C., Laurance, S. G., Pimm, S. L., Bruna, E. M., & Vasconcelos, H. L. et al. (2011). The fate of Amazonian forest fragments: A 32-year investigation. *Biological Conservation*, *144*(1), 56–67. doi:10.1016/j.biocon.2010.09.021

Laurance, W. F., Clements, G. R., Sloan, S., O'Connell, C. S., Mueller, N. D., Goosem, M., & Arrea, I. B. et al. (2014). A global strategy for road building. *Nature*, *513*(7517), 229–232. doi:10.1038/nature13717 PMID:25162528

Lee, B., Preston, F., Kooroshy, J., & Bailey, R. (2012). *G., L.* Resources Futures.

Leroy, E. M., Rouquet, P., Formenty, P., Souquiere, S., Kilbourne, A., Froment, J. M., & Rollin, P. E. et al. (2004). Multiple Ebola virus transmission events and rapid decline of central African wildlife. *Science*, *303*(5656), 387–390. doi:10.1126/science.1092528 PMID:14726594

Levinson, J., Bogich, T. L., Olival, K. J., Epstein, J. H., Johnson, C. K., Karesh, W., & Daszak, P. (2013). Targeting surveillance for zoonotic virus discovery. *Emerging Infectious Diseases*, *19*(5), 743–747. doi:10.3201/eid1905.121042 PMID:23647732

Linthicum, K. J., Anyamba, A., Tucker, C. J., Kelley, P. W., Myers, M. F., & Peters, C. J. (1999). Climate and satellite indicators to forecast Rift Valley fever epidemics in Kenya. *Science*, *285*(5426), 397–400. http://www.ncbi.nlm.nih.gov/pubmed/10411500 doi:10.1126/science.285.5426.397 PMID:10411500

Liu, Q., Cao, L., & Zhu, X. Q. (2014). Major emerging and re-emerging zoonoses in China: A matter of global health and socioeconomic development for 1.3 billion. *International Journal of Infectious Diseases*, *25*, 65–72. doi:10.1016/j.ijid.2014.04.003 PMID:24858904

Loh, E. H., Olival, K. J., Zambrana-Torellio, C., Bogich, T. L., Johnson, C. K., Mazet, J. A. K., & Daszak, P. et al. (2015). Targeting transmission pathways for emerging zoonotic disease surveillance and control. *Vector Borne and Zoonotic Diseases (Larchmont, N.Y.)*, *15*(7), 432–437. doi:10.1089/vbz.2013.1563 PMID:26186515

Luby, S. P., Gurley, E. S., & Hossain, M. J. (2009). Transmission of human infection with Nipah virus. *Clinical Infectious Diseases*, *49*(11), 1743–1748. doi:10.1086/647951 PMID:19886791

Machalaba, C., Romanelli, C., Stoett, P., Baum, S. E., Bouley, T. A., Daszak, P., & Karesh, W. B. (2015a). Climate Change and Health: Transcending Silos to Find Solutions. *Ann Glob Health*, *81*(3), 445–458. doi:10.1016/j.aogh.2015.08.002 PMID:26615080

Machalaba, C. C., Daszak, P., Karesh, W. B., & Shrivastava, P. (2015b). Future Earth and EcoHealth: A New Paradigm Toward Global Sustainability and Health. *EcoHealth*, *12*(4), 553–554. doi:10.1007/s10393-015-1076-6 PMID:26631383

Machalaba, C. C., Elwood, S. E., Forcella, S., Smith, K. M., Hamilton, K., Jebara, K. B., & Karesh, W. B. et al. (2015c). Global avian influenza surveillance in wild birds: A strategy to capture viral diversity. *Emerging Infectious Diseases*, *21*(4), e1–e7. doi:10.3201/eid2104.141415 PMID:25811221

Machalaba, C. C., & Karesh, W. B. (2015). Envisioning a World without Emerging Disease Outbreaks. *Solutions*, *6*(2), 63–71.

Mari Saez, A., Weiss, S., Nowak, K., Lapeyre, V., Zimmermann, F., Dux, A., & Leendertz, F. H. et al. (2015). Investigating the zoonotic origin of the West African Ebola epidemic. *EMBO Molecular Medicine*, *7*(1), 17–23. doi:10.15252/emmm.201404792 PMID:25550396

Marshall, B. M., & Levy, S. B. (2011). Food animals and antimicrobials: Impacts on human health. *Clinical Microbiology Reviews*, *24*(4), 718–733. doi:10.1128/CMR.00002-11 PMID:21976606

Mazza, G., Tricarico, E., Genovesi, P., & Gherardi, F. (2014). Biological Invaders are Threats to Human Health: An Overview. *Ethology Ecology and Evolution*, *26*(2-3), 112–129. doi:10.1080/03949370.2013.863225

McMichael, A. J., & Beaglehole, R. (2000). The changing global context of public health. *Lancet*, *356*(9228), 495–499. doi:10.1016/S0140-6736(00)02564-2 PMID:10981904

McMichael, A. J., Powles, J. W., Butler, C. D., & Uauy, R. (2007). Food, livestock production, energy, climate change, and health. *Lancet*, *370*(9594), 1253–1263. doi:10.1016/S0140-6736(07)61256-2 PMID:17868818

Meentemeyer, R. K., Haas, S. E., & Vaclavik, T. (2012). Landscape epidemiology of emerging infectious diseases in natural and human-altered ecosystems. *Annual Review of Phytopathology*, *50*(1), 379–402. doi:10.1146/annurev-phyto-081211-172938 PMID:22681449

Mikota, S. K., & Maslow, J. N. (2011). Tuberculosis at the human-animal interface: An emerging disease of elephants. *Tuberculosis (Edinburgh, Scotland)*, *91*(3), 208–211. doi:10.1016/j.tube.2011.02.007 PMID:21397564

Morse, S. S., Mazet, J. A., Woolhouse, M., Parrish, C. R., Carroll, D., Karesh, W. B., & Daszak, P. et al. (2012). Prediction and prevention of the next pandemic zoonosis. *Lancet*, *380*(9857), 1956–1965. doi:10.1016/S0140-6736(12)61684-5 PMID:23200504

Myers, S. S., Gaffikin, L., Golden, C. D., Ostfeld, R. S., Redford, K. H., Ricketts, T. H., & Osofsky, S. A. et al. (2013). Human health impacts of ecosystem alteration. *Proceedings of the National Academy of Sciences of the United States of America*, *110*(47), 18753–18760. doi:10.1073/pnas.1218656110 PMID:24218556

Myers, S. S., & Patz, J. A. (2009). Emerging Threats to Human Health from Global Environmental Change. *Annual Review of Environment and Resources*, *34*(1), 223–252. doi:10.1146/annurev.environ.033108.102650

Nabarro, D., & Wannous, C. (2014). The potential contribution of Iivestock to food and nutrition security: The application of the One Health approach in livestock policy and practice. *Revue Scientifique et Technique (International Office of Epizootics)*, *33*(2), 475–485. Retrieved from http://www.ncbi.nlm.nih.gov/pubmed/25707178 doi:10.20506/rst.33.2.2292 PMID:25707178

Nading, A. M. (2013). Humans, animals, and health: From ecology to entanglement. *Environment and Society: Advances in Research*, *4*(1), 60–78. doi:10.3167/ares.2013.040105

Naeem, S., & Wright, J. P. (2003). Disentangling biodiversity effects on ecosystem functioning: Deriving solutions to a seemingly insurmountable problem. *Ecology Letters*, *6*(6), 567–579. doi:10.1046/j.1461-0248.2003.00471.x

Nakagawa, F., Miners, A., Smith, C. J., Simmons, R., Lodwick, R. K., Cambiano, V., & Phillips, A. N. et al. (2015). Projected Lifetime Healthcare Costs Associated with HIV Infection. *PLoS ONE*, *10*(4), e0125018. doi:10.1371/journal.pone.0125018 PMID:25901355

Olival, K. J., & Hayman, D. T. (2014). Filoviruses in bats: Current knowledge and future directions. *Viruses*, *6*(4), 1759–1788. doi:10.3390/v6041759 PMID:24747773

Olson, S. H., Gangnon, R., Elguero, E., Durieux, L., Guegan, J. F., Foley, J. A., & Patz, J. A. (2009). Links between climate, malaria, and wetlands in the Amazon Basin. *Emerging Infectious Diseases*, *15*(4), 659–662. doi:10.3201/eid1504.080822 PMID:19331766

Oosterbroek, B., de Kraker, J., Huynen, M. M., & Martens, P. (2016). Assessing ecosystem impacts on health: A tool review. *Ecosystem Services*, *17*, 237–254. doi:10.1016/j.ecoser.2015.12.008

Palacios, G., Lowenstine, L. J., Cranfield, M. R., Gilardi, K. V., Spelman, L., Lukasik-Braum, M., & Lipkin, W. I. et al. (2011). Human metapneumovirus infection in wild mountain gorillas, Rwanda. *Emerging Infectious Diseases*, *17*(4), 711–713. doi:10.3201/eid1704.100883 PMID:21470468

Patz, J. (2005). Satellite remote sensing can improve chances of achieving sustainable health. *Environmental Health Perspectives*, *113*(2), A84–A85. Retrieved from http://www.ncbi.nlm.nih.gov/pubmed/15687039 doi:10.1289/ehp.113-a84 PMID:15687039

Patz, J., Campbell-Lendrum, D., Gibbs, H., & Woodruff, R. (2008). Health impact assessment of global climate change: Expanding on comparative risk assessment approaches for policy making. *Annual Review of Public Health*, *29*(1), 27–39. doi:10.1146/annurev.publhealth.29.020907.090750 PMID:18173382

Patz, J. A., Daszak, P., Tabor, G. M., Aguirre, A. A., Pearl, M., & Epstein, J. et al. (2004). Unhealthy landscapes: Policy recommendations on land use change and infectious disease emergence. *Environmental Health Perspectives*, *112*(10), 1092–1098. Retrieved from http://www.ncbi.nlm.nih.gov/pubmed/15238283 doi:10.1289/ehp.6877 PMID:15238283

Pautasso, M., Döring, T. F., Garbelotto, M., Pellis, L., & Jeger, M. J. (2012). Impacts of climate change on plant diseases—opinions and trends. *European Journal of Plant Pathology*, *133*(1), 295–313. doi:10.1007/s10658-012-9936-1

Perry, B. D., Grace, D., & Sones, K. (2013). Current drivers and future directions of global livestock disease dynamics. *Proceedings of the National Academy of Sciences of the United States of America*, *110*(52), 20871–20877. doi:10.1073/pnas.1012953108 PMID:21576468

Pesapane, R., Ponder, M., & Alexander, K. A. (2013). Tracking pathogen transmission at the human-wildlife interface: Banded mongoose and Escherichia coli. *EcoHealth*, *10*(2), 115–128. doi:10.1007/s10393-013-0838-2 PMID:23612855

Peterson, A. T. (2006). Ecologic niche modeling and spatial patterns of disease transmission. *Emerging Infectious Diseases*, *12*(12), 1822–1826. doi:10.3201/eid1212.060373 PMID:17326931

Phua, K. L. (2015). Meeting the Challenge of Ebola Virus Disease in a Holistic Manner by Taking into Account Socioeconomic and Cultural Factors. *Infectious Diseases*, *8*, 39–44. PMID:26604778

Pigott, D. M., Golding, N., Mylne, A., Huang, Z., Henry, A. J., Weiss, D. J., & Hay, S. I. et al. (2014). Mapping the zoonotic niche of Ebola virus disease in Africa. *eLife*, *3*, e04395. doi:10.7554/eLife.04395 PMID:25201877

Pike, J., Bogich, T., Elwood, S., Finnoff, D. C., & Daszak, P. (2014). Economic optimization of a global strategy to address the pandemic threat. *Proceedings of the National Academy of Sciences of the United States of America, 111*(52), 18519–18523. doi:10.1073/pnas.1412661112 PMID:25512538

PREDICT Consortium. (2014, December). *Reducing Pandemic Risk, Promoting Global Health.*

Randolph, S. E., & Dobson, A. D. (2012). Pangloss revisited: A critique of the dilution effect and the biodiversity-buffers-disease paradigm. *Parasitology, 139*(7), 847–863. doi:10.1017/S0031182012000200 PMID:22336330

Rosen, G. E., & Smith, K. F. (2010). Summarizing the evidence on the international trade in illegal wildlife. *EcoHealth, 7*(1), 24–32. doi:10.1007/s10393-010-0317-y PMID:20524140

Rouquet, P., Froment, J. M., Bermejo, M., Kilbourn, A., Karesh, W., Reed, P., & Leroy, E. M. et al. (2005). Wild animal mortality monitoring and human Ebola outbreaks, Gabon and Republic of Congo, 2001-2003. *Emerging Infectious Diseases, 11*(2), 283–290. doi:10.3201/eid1102.040533 PMID:15752448

Rubio, A. V., Avila-Flores, R., & Suzan, G. (2014). Responses of small mammals to habitat fragmentation: Epidemiological considerations for rodent-borne hantaviruses in the Americas. *EcoHealth, 11*(4), 526–533. doi:10.1007/s10393-014-0944-9 PMID:24845575

Saker, L., Lee, K., Cannito, B., Gilmore, A., & Campbell-Lendrum, D. H. (2004). *Globalization and infectious diseases: a review of the linkages* Retrieved from http://apps.who.int/iris/handle/10665/68726

Sandifer, P. A., & Sutton-Grier, A. E. (2014). Connecting stressors, ocean ecosystem services, and human health. *Natural Resources Forum, 38*(3), 157–167. doi:10.1111/1477-8947.12047

Schulte-Herbruggen, B., Cowlishaw, G., Homewood, K., & Rowcliffe, J. M. (2013). The importance of bushmeat in the livelihoods of West African cash-crop farmers living in a faunally-depleted landscape. *PLoS ONE, 8*(8), e72807. doi:10.1371/journal.pone.0072807 PMID:23977355

Schwartlander, B., Stover, J., Hallett, T., Atun, R., Avila, C., Gouws, E., & Padian, N. et al. (2011). Towards an improved investment approach for an effective response to HIV/AIDS. *Lancet, 377*(9782), 2031–2041. doi:10.1016/S0140-6736(11)60702-2 PMID:21641026

Secretariat of the Convention on Biological Diversity. (2014). *An Updated Synthesis of the Impacts of Ocean Acidification on Marine Biodiversity.*

Seifman, R., Kornblet, S., Standley, C., Sorrell, E., Fischer, J., & Katz, R. (2015). Think big, World Bank: Time for a public health safeguard. *Lancet Glob Health, 3*(4), e186–e187. doi:10.1016/S2214-109X(15)70012-4 PMID:25678307

Sejvar, J. J. (2003). West nile virus: an historical overview. *Ochsner J, 5*(3), 6-10. Retrieved from http://www.ncbi.nlm.nih.gov/pubmed/21765761

Smith, K. M., Anthony, S. J., Switzer, W. M., Epstein, J. H., Seimon, T., Jia, H., & Marano, N. et al. (2012). Zoonotic viruses associated with illegally imported wildlife products. *PLoS ONE, 7*(1), e29505. doi:10.1371/journal.pone.0029505 PMID:22253731

Steel, J., & Lowen, A. C. (2014). Influenza A virus reassortment. *Current Topics in Microbiology and Immunology, 385*, 377–401. doi:10.1007/82_2014_395 PMID:25007845

Strange, R. N., & Scott, P. R. (2005). Plant disease: A threat to global food security. *Annual Review of Phytopathology, 43*(1), 83–116. doi:10.1146/annurev.phyto.43.113004.133839 PMID:16078878

Strassburg, B. B. N., Latawiec, A. E., Barioni, L. G., Nobre, C. A., da Silva, V. P., Valentim, J. F., & Assad, E. D. et al. (2014). When enough should be enough: Improving the use of current agricultural lands could meet production demands and spare natural habitats in Brazil. *Global Environmental Change, 28*, 84–97. doi:10.1016/j.gloenvcha.2014.06.001

Suarez, E., Morales, M., Cueva, R., Utreras Bucheli, V., Zapata-Ríos, G., Toral, E., & Vargas Olalla, J. et al. (2009). Oil industry, wild meat trade and roads: Indirect effects of oil extraction activities in a protected area in north-eastern Ecuador. *Animal Conservation, 12*(4), 364–373. doi:10.1111/j.1469-1795.2009.00262.x

Susilawati, T. N., & McBride, W. J. (2014). Acute undifferentiated fever in Asia: A review of the literature. *The Southeast Asian Journal of Tropical Medicine and Public Health, 45*(3), 719–726. Retrieved from http://www.ncbi.nlm.nih.gov/pubmed/24974656 PMID:24974656

Taylor, L. H., Latham, S. M., & Woolhouse, M. E. (2001). Risk factors for human disease emergence. *Philosophical Transactions of the Royal Society of London. Series B, Biological Sciences, 356*(1411), 983–989. doi:10.1098/rstb.2001.0888 PMID:11516376

Tirado, M. C., Clarke, R., Jaykus, L. A., McQuatters-Gollop, A., & Frank, J. M. (2010). Climate change and food safety: A review. *Food Research International*, *43*(7), 1745–1765. doi:10.1016/j.foodres.2010.07.003

UNAIDS. (2013). 2013 Global Fact Sheet. Retrieved from http://www.unaids.org/en/resources/campaigns/globalreport2013/factsheet

United Nations Environment Programme. (2014). Illegal Trade in Willife *UNEP Year Book 2014 emerging issues update.*

USAID. (2014). *PREDICT 2*. Retrieved from https://www.usaid.gov/sites/default/files/documents/1864/Predict2-factsheet.pdf

USDA APHIS Veterinary Services. (2015). *Epidemiologic and Other Analyses of HPAI-Affected Poultry Flocks: June 15, 2015 Report.*

USDA APHIS Veterinary Services. (2016). *2016 HPAI Preparedness and Response Plan*. Retrieved from https://www.aphis.usda.gov/animal_health/downloads/animal_diseases/ai/hpai-preparedness-and-response-plan-2015.pdf

Vitoria, M., Granich, R., Gilks, C. F., Gunneberg, C., Hosseini, M., Were, W., & De Cock, K. M. et al. (2009). The Global Fight Against HIV/AIDS, Tuberculosis, and Malaria- Current Status and Future Perspectives. *American Journal of Clinical Pathology*, *131*(6), 1844–1848. doi:10.1309/AJCP5XHDB1PNAEYT PMID:19461091

Vorosmarty, C. J., McIntyre, P. B., Gessner, M. O., Dudgeon, D., Prusevich, A., Green, P., & Davies, P. M. et al. (2010). Global threats to human water security and river biodiversity. *Nature*, *467*(7315), 555–561. doi:10.1038/nature09440 PMID:20882010

Wagenaar, J. A., & Scott, H. M. (2015). Reducing antimicrobial usage in agriculture and aquaculture: Beyond regulatory policy. *Zoonoses and Public Health*, *62*(Suppl. 1), 1–2. doi:10.1111/zph.12197 PMID:25903490

Watts, N., & al., e. (June 23, 2015). Health and climate change: policy responses to protect human health. *Lancet Commission on Health and Climate Change.*

Whitmee, S., Haines, A., Beyrer, C., Boltz, F., Capon, A., de Souza Dias, B. F., & Yach, D. et al. (2015). Safeguarding Human Health in the Anthropocene Epoch: Report of The Rockefeller Foundation-Lancet Commission on Planetary Health. *Lancet*, *386*(10007), 1973–2028. doi:10.1016/S0140-6736(15)60901-1 PMID:26188744

Wilkie, D. S., & Carpenter, S. F. (1999). Bushmeat hunting in the Congo Basin: An assessment of impacts and options for mitigation. *Biodiversity and Conservation*, *8*(7), 927–955. doi:10.1023/A:1008877309871

Winkler, M. S., Krieger, G. R., Divall, M. J., Cisse, G., Wielga, M., Singer, B. H., & Utzinger, J. et al. (2013). Untapped potential of health impact assessment. *Bulletin of the World Health Organization, 91*(4), 298–305. doi:10.2471/BLT.12.112318 PMID:23599554

World Bank. (2012). *People, Pathogens and Our Planet: The Economics of One Health.* Retrieved from https://openknowledge.worldbank.org/handle/10986/11892

World Bank. (2014). *The economic impact of the 2014 Ebola epidemic: short and medium term estimates for West Africa.*

World Economic Forum. (2015). *The Global Risks report 2015.* Retrieved from http://reports.weforum.org/global-risks-2015/

World Health Organization. (2014). *Information to States Parties regarding determination of fulfilment of IHR national Core Capacity requirements and potential extensions.* Retrieved from http://apps.who.int/iris/bitstream/10665/112789/1/WHO_HSE_GCR_2014.9_eng.pdf?ua=1

World Health Organization. (2016). *Ebola Situation Report - 17 February 2016.* Retrieved from http://apps.who.int/ebola/current-situation/ebola-situation-report-17-february-2016

World Health Organization and Convention on Biological Diversity. (2015). *Connecting Global Priorities: Biodiversity and Human Health, a State of Knowledge Review.*

WWF / Dalberg. (2012). *Fighting illicit wildlife trafficking: A consultation with governments.*

Young, C. C., & Olival, K. J. (2016). Optimizing Viral Discovery in Bats. *PLoS ONE, 11*(2), e0149237. doi:10.1371/journal.pone.0149237 PMID:26867024

Zaki, A. M., van Boheemen, S., Bestebroer, T. M., Osterhaus, A. D., & Fouchier, R. A. (2012). Isolation of a novel coronavirus from a man with pneumonia in Saudi Arabia. *The New England Journal of Medicine, 367*(19), 1814–1820. doi:10.1056/NEJMoa1211721 PMID:23075143

KEY TERMS AND DEFINITIONS

Anthropogenic: Human-caused activity or outcome, differentiated from naturally occurring factors, processes or results.

Biological Diversity: Frequently used in its contracted form, biodiversity, as defined in Article 2 of the Convention on Biological Diversity, biodiversity is understood as "the variability among living organisms from all sources including, inter alia, terrestrial, marine and other aquatic ecosystems and the ecological complexes of which they are a part; this includes diversity within species, between species and of ecosystems".

Emerging Infectious Disease: Newly identified pathogen/disease, or expanded range or species.

Endemic Infectious Disease: Established pathogen/disease in a region.

Global Environmental Change: Large-scale abiotic and biotic changes; individual changes may happen locally, but in aggregate, they are widely occurring globally and leading to fundamental changes in ecological systems.

One Health: A framework for recognizing and addressing the connections between human, animal and environmental health.

Spillover: An event in which a pathogen is transmitted from one species to another.

Surveillance: The practice of conducting screening (i.e. sampling and diagnostic testing) to determine presence or absence of a pathogen or its prevalence in a given population. Surveillance may be active (periodic collection of samples) or passive (collection of samples from animals or humans with suspected disease).

Underlying Driver: The root cause of an outcome.

Vector-Borne: Commonly applied to diseases transmitted from infected arthropods (such as mosquitoes, midges, and ticks), usually via a bite during a blood meal. Examples are malaria, Lyme disease, Schistosomiasis, West Nile fever, and Zika.

Zoonosis: An infection resulting from a pathogen transmitted from a vertebrate animal to a human. Some zoonoses have then spread via human-human transmission.

Chapter 3
Climate Change-Associated Conflict and Infectious Disease

Devin C. Bowles
Australian National University, Australia

ABSTRACT

One of the least appreciated mechanisms by which climate change will affect infectious diseases is via increased violent conflict. Climate change will diminish agricultural and pastoral resources and increase food scarcity in many areas, including already impoverished equatorial regions. Many in the defence and public health fields anticipate that climate change will increase conflict by fuelling competition over scarce resources. Already, some commentators argue that the conflicts in Darfur and Syria were partially caused or exacerbated by climate change. Conflict facilitates a range of conditions conducive to the spread of many infectious diseases, including malnutrition, forced migration, unhygienic living conditions and widespread sexual assault. Flight or killing of health personnel inhibits vaccination, vector control and disease surveillance programs. Emergence of new diseases may go undetected and discovery of outbreaks could be suppressed for strategic reasons. These conditions combine to increase the risk of pandemics.

INTRODUCTION

This chapter examines the effect of climate change-associated conflict on infectious disease. It first reviews the nexus between climate, conflict and infectious disease historically. It then examines the literature linking resource scarcity and conflict. It

DOI: 10.4018/978-1-5225-0553-2.ch003

argues that climate change will increase resource scarcity in many areas and increase the risk of conflict and subsequent migration. Finally, it examines the consequences of elevated conflict risk on infectious disease. It concludes that the types of conflict made more likely by climate change could be particularly detrimental to efforts to fight infectious disease.

WAR, PEACE, AND INFECTIOUS DISEASE IN HISTORY

The evolutionary contest between humans and infectious diseases has been waged since the first *Homo sapiens*. Holding other factors constant, the increased population densities and high rates of international travel of recent decades are conducive to the transmission and spread infectious disease. Yet, even accounting for AIDS, the death rate from infectious disease has not risen as might be expected. In the US, the annual death rate from infectious disease was approximately 60 per 100,000 in the early 1990s, compared with approximately 500 per 100,000 in 1900 (Lederberg, 2000). Of course, not all countries fared so well, because the American gains were largely due to advances in nutrition, public health and sanitation infrastructure. In many conflicts, antagonists deliberately push back these advances, facilitating conditions that benefit the spread of infectious disease. Some scholars suggest that war and public health efforts are at odds (Toole, Galson, & Brady, 1993).

Historically, periods of conflict were often associated with infectious disease epidemics. The 1918 influenza pandemic caused 40 million deaths globally. The virus may have emerged on the Western Front in the winter of 1916 or 1917, where conditions would have fostered increasing virulence of the disease (Oxford, Sefton, Jackson, Innes, & Johnson, 2002). Many soldiers had lung irritation from the use of chemical weapons, and were exposed to the elements in conditions that were overcrowded and stressful. Under-nutrition was common in the civilian population, weakening immune response to disease (Erkoreka, 2009; Oxford, et al., 2002). Demobilisation at the end of the war saw massive population movement, which contributed to global disease spread (Erkoreka, 2009; Oxford, et al., 2002).

The conditions of encamped soldiers have long been a fertile ground for illness. Disease killed over 130,000 Allied and Austrian forces during the Crimean war between 1854 and 1856, far more than the approximately 25,000 killed in combat (Smallman-Raynor & Cliff, 2004). The rapid increase of US military personnel in preparation for and during the Spanish American war was concentrated at a number of assembly and training camps erected in the continental United States. While no fighting took place at these sites, massive personnel movements combined with hastily assembled camps and unhygienic conditions led to an epidemic of typhoid fever in 1898 (Smallman-Raynor & Cliff, 2001).

Epidemics also accompanied or followed many of the conflicts during the General Crisis of the seventeenth century (Parker, 2013). Conflict was, however, not the only factor contributing to an increase in disease incidence. The General Crisis occurred during part of the Little Ice Age, a period which experienced prolonged cooling in the Northern hemisphere, diminishing harvests and leading to numerous episodes of famine in Europe and Asia. Malnourished populations were more vulnerable to a range of diseases, especially during times of war.

CLIMATE CHANGE, RESOURCE SCARCITY, AND CONFLICT IN HISTORY

War itself was made more likely by the altered climate during the General Crisis (Parker, 2013). One way of conceptualising the risk of civil conflict is by engaging in a cost-benefit analysis of the status quo and likelihood of participation in civil conflict for members of population (Kahl, 2006). Personal risks that stem from participation in civil conflict tend to be high. Unless there is a high probability of success in the aims of the civil conflict, or the status quo is intolerable, this tends to discourage participation in civil conflict. Populations made hungry by cooler temperatures became less content with their governments, which themselves had fewer resources to pay troops to keep order. At the same time, worsening conditions made appeals to the divine right to rule more difficult to sustain. Together, these trends shifted people's decision making about participating in rebellion and helped increase civil conflict to one of its highest ever rates in Europe (Parker, 2013). Conversely, foreign conflicts allowed sovereigns the opportunity to distract their subjects from domestic problems, and were thus a tempting solution to discontent (Parker, 2013).

A series of quantitative analyses by Zhang and colleagues demonstrate that the association between climate change, which decreased food production and conflict extends beyond the General Crisis. There was a strong correlation of conflict in general and climate change in Eastern China throughout the last millennium. Dynastic change also tended to follow episodes of climate change, with the three longest cold phases ending the three longest dynasties (Song, Ming and Qing) (Zhang, et al., 2006). Reduced harvests and ensuing peasant revolts weakened the state, thus inviting foreign aggression (Zhang, Zhang, Lee, & He, 2007). Examining, Europe and the northern hemisphere over the last thousand years, a similar pattern emerges, with the climate predicting alternating "dark" and "golden" ages (Zhang, Brecke, Lee, He, & Zhang, 2007; Zhang, et al., 2011). Here, the sequential and causal relationships of elements of social crisis, observed in qualitative description, are confirmed with quantitative data, with climate change dampening bio-productivity, leading to decreased harvests and increased food prices (Zhang, Brecke, et al., 2007; Zhang, et al., 2011). This, in turn, led to social disturbance, war, famine, migration and population decline (Zhang, Brecke, et al., 2007; Zhang, et al., 2011). Notably, epidemics (operationalised by instances of the plague) also followed climate change, and in particular were linked with decreased nutritional status and migration (Zhang, et al., 2011). This suggests that political, social and economic consequences of climate change, also known as its tertiary effects (Butler, 2014b; Butler & Harley, 2010) could have important consequences of infectious disease apart from the effects of the climate on pathogens and vectors.

MODERN RESOURCE SCARCITY AND CONFLICT

The connection between climate change and conflict is not relegated to history. The idea that reduced availability of essential resources can increase the risk of conflict has a respected pedigree in the literature (Homer-Dixon, 1994; Rønnfeldt, 1997). Scarcity is often the result of demographic and ecological stress, though obviously mechanisms for resource distribution can create or contribute to conditions of scarcity among part of the population when resources are distributed inappropriately. Theory suggests than in the modern day, demographic and ecological stress increases the risk of civil conflict via two pathways. First, it weakens the state through a variety of mechanisms, including reduced revenue (Kahl, 2006). This increases the likelihood that rebellions and civil conflicts will be successful, and decreases the opportunity cost of participation. The two alternative scenarios for the potential rebel are rebellion and peace. If peace means a miserable status quo, the opportunity cost of rebellion is diminished. These factors contribute to enhance the appeal of rebellion and revolution to potential participants (Kahl, 2006). Second, political leaders often use scarcity as a means of justifying conflict that will further their own interests. Their incentive to do so is often increased by the need to distract the population from deteriorating social conditions, as occurred in the Kenyan ethnic clashes of the early 1990s (Kahl, 2006). The narrative of blaming scarcity on foreign countries and domestic minorities is a frequent response to demographic and ecological stress, and is common before war and genocide, as illustrated by Nazis' blame of Germany's economic decline on a Jewish conspiracy (Talmon, 1980).

Resource scarcity, however, always occurs in a political and social context, and this can either amplify or mitigate the risk of conflict from resource scarcity. Greater "institutional inclusivity", in which individuals face low levels of discrimination and possess relatively high political freedom, provides alternative pathways for people with grievances. This lowers the risk of conflict by increasing the opportunity cost of violence for the individual (Kahl, 2006). Additionally, societies that are not beset by strong social divisions, including along ethnic or religious lines, tend to be at lower risk of conflict (Kahl, 2006). Social and political efforts to promote peace are therefore important. Increased resource scarcity nevertheless enhances conflict risk, so that a world with greater resource scarcity will be more violent than a world with less resource scarcity, holding other factors constant.

ANTHROPOGENIC CLIMATE CHANGE AND RESOURCE SCARCITY

There is remarkable consensus that the climate is changing due to human interference. Globally, greenhouse gases trap heat in the atmosphere, causing it to warm. This is altering precipitation patterns, which in many places are becoming more intense but less regular, increasing the frequency of both flooding and drought (Intergovernmental Panel on Climate Change, 2013). The most recent Intergovernmental Panel on Climate Change (IPCC) report indicates that climate change is likely to increase variability and reduce global agricultural production; (Intergovernmental Panel on Climate Change, 2014). The situation is worse when viewed regionally. Agricultural productivity is expected to shift from lower to higher latitudes (Intergovernmental Panel on Climate Change, 2014). This is important because most small scale and subsistence agriculture and herding take place at lower latitudes, where a large proportion of the population is directly reliant on these activities. Many developing countries are located in low latitudes. Therefore, the decreased local agricultural production will exacerbate the poverty trap and further diminish the food security for millions who rely on subsistence food production and already have limited ability to engage with the global cash economy (Butler, 2010, 2014a).

Climate change will combine with other large scale environmental changes including overfishing and pollution to deplete many fish stocks (Bradbury, 2012; Hoegh-Guldberg, 2011; Hoegh-Guldberg, et al., 2007). This would have grave consequences for those who rely on fish and other marine resources as their primary source of protein, especially the world's poor who may be unable to compete in the global marketplace for fish or other protein sources. As increased carbon dioxide is emitted into the atmosphere, some of it is absorbed by the oceans, leading to ocean acidification. This has been shown to interfere with the development and growth of some commercially important fish species (Frommel, et al., 2012). More fundamentally, it disrupts the chemical processes used by a wide array of marine animals, to make their shells, and similarly disrupt calcification of corals (Hoegh-Guldberg, 2011; Hoegh-Guldberg, et al., 2007). Coral reef ecosystems are threatened from many sides, leading some to conclude that they will soon cease to be a globally functional ecosystem (Bradbury, 2012). Waters warmed by climate change will increase the incidence and severity of coral bleaching. Additionally, overfishing has already depleted a number of the world's fish stocks, and without substantial and rapid action, will further harm these important ecosystems (Cooley & Doney, 2009). Finally, agricultural and other nutrient-dense runoff is increasingly polluting the coastal waters in which many coral reefs are located (Bradbury, 2012). This increase in nutrient density alters the competitive balance in favour of algae at the expense of coral. Algal blooms can reduce oxygen availability and suffocate the animals that normally inhabit affected water. Due to the effects of all of these pressures, many coral reef ecosystems are likely to collapse, and be replaced by ecosystems dominated by algae and jelly fish. These will be ecologically poorer and much less economically useful to humans than the ecosystems they replace (Bradbury, 2012).

Climate change will reduce food security including both the calories and protein for many of the world's poorest. At the same time, it will undermine economic growth of regions and nations across much of the tropics, with potentially profound political and security implications.

ANTHROPOGENIC CLIMATE CHANGE AND CONFLICT

The historical link between climate change and conflict, combined with the knowledge that anthropogenic climate change is already depleting resource availability for millions has led a number of scholars to suggest that anthropogenic climate change is increasing conflict risk (Gleick, 2014; Hsiang, Meng, & Cane, 2011; Mazo, 2010). Among advocates of this idea, climate change is understood as a "threat multiplier" for conflict. It interacts with and elevates other, pre-existing risk factors, such as political instability. To be clear, climate change is not widely predicted to cause conflict on its own. Nevertheless, by acting as a threat multiplier across many areas with pre-existing risk factors for conflict, climate change will enhance the probability of conflict globally. If the dice are weighted in favour of conflict, overall incidence of conflict will increase with enough rolls of the dice. If climate change is a threat multiplier for conflict, it would have important consequences for public health.

Unfortunately, standard techniques for assessing this claim are all imperfectly suited to the task. The climate change, likely to occur in the next hundred years, has no direct analogue in human history. The speed of change, combined with its spatial and temporal extent and likely duration are unprecedented. Obviously, such a proposition cannot be tested in a laboratory, and even standard social science techniques for studying large populations are flawed, especially when used in isolation. One of the strengths of public health as a discipline is that, because of its practical focus on outcomes, its practitioners are unperturbed by adapting thinking and techniques from other disciplines. A number of disciplines have sought to shed light on the question of a link between climate change and conflict, each bringing with it a particular set of strengths and weaknesses. Within disciplines, various researchers have used quantitative and qualitative methods to investigate the issue.

Many scholars within the field of political science are investigating the causes of modern conflict. The discipline investigates which actors have participated in recent conflicts as well as their motivations, though there is rarely unanimous agreement between political scientists as to what drives conflict. There is a venerable, if contested, history of scholarship that argues that resource scarcity contributes to climate change (Homer-Dixon, 1994; Kahl, 2006), summarised above in the section on modern resource scarcity and conflict section. More recently, some scholars have extended this argument to climate change, which will cause resource scarcity (Mazo, 2010; Parenti, 2011; Welzer, 2012). Despite its expertise in modern conflict, political science is limited by the fact that the climate change to date is only a fraction of what will occur in the next century. This means there is no natural experiment that simulates the breadth and intensity of the effects of future climate change. Political scientists are left with proxy measures that they can substitute for future climate change. Typically, these include selected temporary perturbations in the weather, rather than broader, longer-lasting changes to the climate. By their nature, these underestimate the full effects of climate change. Nevertheless, a number of studies have found a link between weather events that serve as partial proxies of climate change, and increased conflict. Other studies, often using less comprehensive proxy measures, did not find a link, and for this reason some scholars are yet to be convinced of the connection between climate change and conflict or await further data (Buhaug, et al., 2014; Gleditsch, 2012; Salehyan, 2008). Yet this pattern of results is what might be expected if climate change increases conflict risk, given that most studies use a proxy measure of climate change that does not fully replicate the severity and scope of climate change.

The historical and archaeological records are useful because they document periods of natural climate change which reduced agricultural production in ways similar to that expected from future climate change. The affected societies obviously differ from contemporary ones, but their responses are nonetheless a valuable guide to the dynamics that could occur in future. A number of societies seem to have become substantially more prone to conflict during periods of climate change, including the Maya (Haug, et al., 2003; Kennett, et al., 2012), the civilisation of Chaco Canyon in the Southwest United States in the 12th century (Haas, 2009; Marlar, Leonard, Billman, Lambert, & Marlar, 2000), Bronze Age Ireland (Turney, Baillie, Palmer, & Brown, 2006), and in both Pacific Islands and East Timor over the past millennium (Field & Lape, 2010; Lape & Chin-Yung, 2008). The General Crisis in the seventeenth century affected much of Europe and Asia. It was a period of unparalleled social upheaval, in which Europe saw only three years of peace, and the Chinese,

Ottoman and Mughal empires fought wars almost continuously (Parker, 2013). There is compelling evidence that it was driven by the confluence of a particularly severe part of the Little Ice Age and political mismanagement (Parker, 2013). As noted above, qualitative studies of the historical record show a strong association between climate change and conflict (Zhang, Brecke, et al., 2007; Zhang, et al., 2011; Zhang, et al., 2006; Zhang, Zhang, et al., 2007).

On balance, the evidence from political science, history and archaeology is strongly suggestive that future climate change will act as a threat multiplier for conflict. However, to be absolutely certain of a causal link, humanity would need to endure several conflicts that were obviously attributable to climate change. Given the humanitarian consequences of such proof, action based on incomplete evidence rather than absolute proof is warranted. Those in the military are accustomed to situations necessitating action based on imperfect situational awareness. Perhaps for this reason, there is already remarkable consensus that anthropogenic climate change will act as a threat multiplier within the world's militaries and defense establishments (Department of Defense, 2014; Press, Bergin, & Gansey, 2013; Sullivan, et al., 2007; UK Ministry of Defence, 2014). Conversely, the public health literature on the topic is remarkable for its uniform conclusion that anthropogenic climate change will increase the risk of conflict (Anonymous, 1989; Bowles, Braidwood, & Butler, 2014; Bowles & Butler, 2014; Bowles, Butler, & Friel, 2014; Holdstock, 1995; Jarvis, Montgomery, Morisetti, & Gilmore, 2011; A. J. McMichael, 2013; Morisetti, 2012). It has also been suggested that climate change might also make some conflicts harder to conclude, acting as a "peace inhibitor" as well as a threat multiplier (Bowles, Braidwood, et al., 2014; Bowles, Butler, & Morisetti, 2015). Most conflicts end with a truce rather than a complete victory. Areas in which climate change has diminished the total resource pool could make a negotiated settlement more difficult, if there are not enough local resources to accommodate the minimum aspirations of all sides. The pathways through which the likelihood of conflict is elevated by climate change are similar to those set out above for resource scarcity generally.

CLIMATE CHANGE AND CONTEMPORARY CONFLICT: THE CASE OF SYRIA

While the effects of climate change will continue to worsen, further increasing the risk of conflict, there is strong evidence that climate change has already contributed to contemporary conflicts. Climate change has been only one among many contributing factors, and the nature and importance of its role can be disputed, but evidence suggests it contributed to or catalysed the conflict in Syria.

Anthropogenic climate change is the cause of a long-term drying trend in the Fertile Crescent, including Syria. Between 2007 and 2010, Syria experienced the worst drought on instrumental record (Kelley, Mohtadi, Cane, Seager, & Kushnir, 2015). A drought of this severity and duration was unlikely in the absence of the regional drying trend likely caused by climate change (Kelley, et al., 2015). The drought occurred in a particular political context which heightened its impact. Prior to the drought, the Assad regime encouraged a number of ecologically unsustainable policies to increase food production. Among the consequences were the depletion of groundwater supplies and a reliance of the entire economy on agricultural and pastoral production; agricultural production accounted for 25% of GDP prior to the drought in 2003 (Kelley, et al., 2015). Recent government efforts to liberalise the economy led to the end of fuel and food subsidies, a policy which persisted during the drought, when many small and medium scale farmers were virtually unable to produce food. The price of staple foods such as wheat and rice more than doubled in a year, while the cost of animal feed also increased substantially (Kelley, et al., 2015). Many herds were decimated. Urban areas were already stressed from the influx of Iraqi refugees, but a substantial rural to urban migration occurred as farmers and pastoralists sought relief. The swelling urban fringes were characterised by unemployment, illegal settlements, overcrowding, corruption, inequality and crime (Kelley, et al., 2015). Perhaps unsurprisingly, these areas became a hotbed of the unrest that has overtaken the country. A growing number of scholars conclude that climate change contributed to or catalysed the conflict (Ahmed, 2013; Femia & Werrell, 2013; Gleick, 2014; Kelley, et al., 2015).

CLIMATE CHANGE, CONFLICT, AND MIGRATION

Climate change increases conflict risk in part because it will lead to substantial migration as livelihood options decrease in many parts of the world. Agricultural and pastoral jobs are most obviously at risk, but large-scale emigration of people in these roles can have a cascade effect, with the supporting businesses also becoming untenable. Migration will be an adaptive response to climate change undertaken by individuals (Black, Bennett, Thomas, & Beddington, 2011), but it should be recognised that many will have little option but to migrate. For many future migrants, emigration will be the least unpalatable of a number of bad options.

Migration associated with climate change carries with it a number of negative consequences, including for health (Bowles, Reuveny, & Butler, 2014; C. McMichael, Barnett, & McMichael, 2012). Migration will often be to a nearby urban centre without the infrastructure to cope with the new arrivals. The substandard social conditions can foment discontent and conflict, especially in scenarios in which migration brings rival ethnic or religious groups into close proximity, forcing them to compete for a limited pool of jobs and resources. A number of studies have shown that environmental immigration is associated with conflict (Martin, 2005; Reuveny, 2007, 2008; Reuveny & Allen, 2007). Conversely, conflict can spur large scale migration, as seen out of Syria, with over 4 million refugees having fled the country (UNHCR, 2015a). Migration away from conflict can further heighten future conflict risk when migrants bring violent ideologies and weapons on their journeys. Cycles of migration and conflict are possible. The health consequences of such spirals are especially dire (Bowles, Reuveny, et al., 2014).

MODERN CONFLICT AND DISEASE

Many recent conflicts are different from the archetypal war fought between clearly defined armies of two or more countries. A number of ongoing and recent conflicts, including in Syria, Iraq, Democratic Republic of Congo, Afghanistan, and Sudan, are characterised by combatants from at least one side not representing a national government. Often, much of the violence is intentionally directed at civilians, often from other ethnic or religious groups. Goals of combatants can include mass migration or extermination of "other" groups (Kaldor, 2012). Therefore, it is unsurprising that the number of refugees and internally displaced persons is higher than it has ever been, and is now almost 60 million (UNHCR, 2015b). To finance their fighting, rebel or militant groups may "tax" international humanitarian aid, stealing or extorting supplies either from the donor agencies or aid recipients. This means that leaders have an economic incentive not to seek peace, but instead to create complex emergencies (Kaldor, 2012). In such situations, the effects of humanitarian crisis and weak institutions are multiplied the presence of conflict. The consequences of this trend have received little attention in the public health literature, however, complex emergencies foster conditions conducive to the spread of communicable disease.

Communicable disease, often paired with under-nutrition, is the cause of most deaths in complex emergencies (Connolly, et al., 2004). Mortality rates among refugees can be elevated 60 times above the baseline (Connolly, et al., 2004). War and ensuing forced migration reduce baseline health through a number of pathways, including overcrowded conditions in temporary camps and reduced shelter, nutrition, water, sanitation and economic status (Connolly, et al., 2004).

HEALTH INFRASTRUCTURE

Modern conflict destroys the health infrastructure that has allowed humanity to gain an upper hand in its perpetual contest with infectious disease. Health infrastructure is often destroyed or looted during conflict. This can include buildings, ambulances, drugs and other medical equipment (International Committee of the Red Cross, 2011; Medact, 2008). Occupation of buildings used for healthcare by combatant forces undermines their neutrality and can disrupt the fair provision of health care. Medical facilities can be used by military and paramilitary forces to locate and kill or kidnap their opponents, with recent cases occurring in Croatia, Lebanon, Sierra Leone, the Democratic Republic of Congo, Columbia and the occupied Palestinian territory (International Committee of the Red Cross, 2011). Such actions discourage others from seeking aid. When hospitals are occupied by combatants, it can result in threats to health personnel, impeding their work and discouraging returning to work (International Committee of the Red Cross, 2011).

Health Personnel

Many wars deplete health expertise. Widespread conflict can disrupt health training, interrupting the supply of new personnel and hindering the professional development of those already qualified (International Committee of the Red Cross, 2011). Compared with most other people caught in conflict zones, health professionals often have more highly valued skills and financial assets, which facilitate migration. Their incentive to escape may be heightened in conflicts in which medical neutrality is not respected and when health professionals are specifically targeted for kidnap and killing, as in the most recent war in Iraq (Medact, 2008). Together, these 'push' and 'pull' factors increase the likelihood of a shortage of qualified personnel. In Iraq, 75% of doctors, nurses and pharmacists left their jobs between 2003 and 2008, with half leaving the country entirely (Medact, 2008).

Disruption to Public Health Services

With harm to health infrastructure and personnel, it is no surprise that violence disrupts public health services, including those that help reduce chronic disease. Vaccination programs are often neglected or impossible to undertake. Global eradication campaigns have been hindered by fighting and complex emergency situations (Connolly, et al., 2004). Disease surveillance systems and vector control projects can be disrupted or jettisoned in the face of conflict, with predictable results.

Consequences for Communicable Diseases

The erosion of baseline health status and the diminution of health services from conflict lead to increased infectious disease morbidity and mortality. War facilitates the spread of sexually transmitted infections (STIs), including HIV. Exposure to conflict can alter people's tolerance for risk, leading them to engage in more dangerous behaviour, including sexual behaviour and intravenous drug use (Hankins, Friedman, Zafar, & Strathdee, 2002). Access to condoms may be diminished in conflict situations, due to supply chain issues and reduced purchasing power. War increases the rate of partner change, as couples are often separated for prolonged periods, while tumultuous social situations lead to a decrease in the normal social restraints on sexual activity (Hankins, et al., 2002). In complex emergencies, women may resort to transactional sex to feed themselves and their children. Circumstances with great disparities in wealth between soldiers and others particularly facilitate such an occurrence. Widespread prostitution can form core groups which facilitate HIV and other STI epidemics (Hankins, et al., 2002). Rape and sexual violence become more common, and are sometimes used as a deliberate tactic to terrorise and subjugate civilian populations and to spread disease (Hankins, et al., 2002; Stearns, 2011). Treatment of HIV to diminish the compromise to the immune system becomes more difficult in conflict situations. The spread of AIDS reduces baseline health, increasing a population's vulnerability to other infectious diseases.

Diarrheal diseases and acute respiratory infections both contribute to a substantial fraction of morbidity and mortality from infectious disease in complex emergencies. Investigations following diarrheal disease outbreaks have revealed water and food contaminated by faeces, lack of soap and sharing of drinking and food storage containers (Connolly, et al., 2004). Transmission of respiratory infections is enhanced by overcrowded conditions and indoor cooking, poor shelter and insufficient blankets. Respiratory infections increase transmission of meningococcal disease, with large outbreaks occurring in some recent complex emergencies (Connolly, et al., 2004). Increased emphasis on immunisation programs for measles in camp situations has reduced the relative impact of this disease (Connolly, et al., 2004), but continued vigilance is required.

Vector borne diseases are often amplified in complex emergencies, due to overcrowding, temporary shelters, diminished vector control programs and reduced personal protective measures such as mosquito netting. Particularly large malaria outbreaks have been observed in complex emergencies. Afghanistan had virtually eliminated malaria until chronic fighting undermined vector control programs, allowing the disease to regain a foothold in the country (Connolly, et al., 2004). Conflicts in Afghanistan and Africa led to migration and contributed to large outbreaks of Kala-azar (visceral leishmaniasis) (Despommier, Gwadz, Hotez, & Knirsch, 2006).

Forced migration from conflict led to a breakdown of control measures of African trypanosomiasis and also dramatically increased the infection rate (Despommier, et al., 2006).

EFFECTS OVER TIME AND IN OTHER COUNTRIES

The health consequences of conflict are not limited temporally or spatially to the geographical zone of fighting. Evidence suggests that the health effects of civil conflict persist for many years, including from diminished economic growth, lack of health investment, and continued morbidity and mortality from infectious disease, with HIV and malaria being especially important (Ghobarah, Huth, & Russett, 2004).

Even in countries not involved in conflict, a global increase in war and violence has negative public health impacts. Ghobarah and colleagues (2004) found that a civil war in a bordering country decreased health substantially, especially through infectious diseases including HIV and malaria. Conflict often drives refugees into other countries (UNHCR, 2015a). Regardless of the level of goodwill, destination countries do not have the infrastructure to accommodate their displaced neighbours to the standard that they enjoyed in pre-war times. In addition, the diversion of health resources may diminish the health of the original inhabitants of neighbouring countries. This is true even if such diversion is utilitarian, and promotes the average health of the whole population, including refugees (Bowles, Reuveny, et al., 2014).

Perhaps more insidiously, an increase in the level of conflict globally would erode feelings of security, including in countries not affected by the fighting. This would likely increase xenophobia, leading to isolationist policies, with a follow-on effect of reducing foreign humanitarian aid, leaving the world's most vulnerable even more isolated (Bowles, Braidwood, et al., 2014). Apart from aid, countries may become less willing to export food if there is a perception that its availability is limited. A number of countries restricted food exports in response to recent decreases in domestic production (Cribb, 2010). This would reduce the flow of food across borders, and inadvertently decrease food security globally (Bowles, Braidwood, et al., 2014; Bowles, Butler, et al., 2014).

Domestically, mounting fear due to escalating global violence would lead to diversion of governments' expenditure toward defence. In a time of limited economic growth and constrained government budgets, this would likely limit spending on health or its precursors such as education and economic safety nets (Bowles, Braidwood, et al., 2014).

Diseases and vectors do not respect national borders. International cooperation is essential to optimise global disease surveillance, as well as vector and other disease control programs. If increased conflict from climate change limits international cooperation, on-the-ground programs may suffer.

Emerging Infectious Diseases and Pandemics

A number of war's consequences can elevate the risk of novel zoonotic diseases passing to humans, including incursion into new environments, ecological disruption and contact with novel species. As outlined in other chapters of this book, climate change drives many of these risk factors, and their effects could be synergistic. Conflict depletes a population's baseline health, not least through the spread of infectious diseases discussed above. This could assist an emerging infectious disease to become established in a human population. The spread of AIDS in conflict situations is of particular concern (Connolly, et al., 2004).

The disruption to health systems caused by conflict would likely delay recognition of any emerging infectious diseases, allowing them to become better established in the human population. Neighbouring countries are unlikely to cooperate in outbreak control when at war. The initial outbreak could be kept secret if it were seen as providing the enemy with information about the combat strength (or weakness) of troops. Alternatively, there could be deliberate attempts to infect opposing forces with any new diseases.

Infectious diseases may also emerge outside combat zones, even when they are appropriately managed by the majority of the global community and could reach areas beset by conflict. This risk is heighted by the re-emerging trend for wars to be fought by mercenaries and volunteers drawn from all over the globe, as seen in Syria and Iraq. If an emerging infectious disease becomes established in a combat zone, the area could act as a reservoir for the disease, therefore undermining the efforts to control the disease elsewhere.

Climate change-associated conflict would enhance the probability of a disease becoming established in a human population and impede health measures to control the outbreak, locally and globally. Each of these factors would elevate the risk of a pandemic, especially if an emerging virulent pathogen is involved.

CONCLUSION

Many of the conditions of modernity such as high population densities and frequent international travel, would facilitate the emergence and spread of infectious diseases. In times of peace and prosperity, the effects of these conditions are counterbalanced by the many public health benefits of modernity, including access to clean water and safe food. Conflict robs humanity of many of the advantages enjoyed in the modern age, including good baseline health and effective health systems. In doing so, it provides pathogens with a decisive advantage in their age-old contest with humanity. A situation that is likely to increase the incidence of infectious diseases, facilitates the establishment of emerging pathogens and enhances the probability of pandemics is clearly within the purview of the public health community. Additionally, climate change-associated conflict will severely limit the measures available to preserve health. Therefore, now is the time to act. The public health community can take positive steps to demonstrate to the public and policy-makers that climate change is a public health issue of substantial importance. The risks of inaction are profound, while the co-benefits of action to reduce climate change would increase health substantially (Watts, et al., 2015).

REFERENCES

Ahmed, N. (2013). Peak oil, climate change and pipeline geopolitics driving Syria conflict. *The Guardian*. Retrieved from http://www.guardian.co.uk/environment/earth-insight/2013/may/13/1

Anonymous, . (1989). Health in the Greenhouse. *Lancet*, *333*(8642), 819–820. doi:10.1016/S0140-6736(89)92275-7 PMID:2564905

Black, R., Bennett, S. R. G., Thomas, S. M., & Beddington, J. R. (2011). Climate change: Migration as adaptation. *Nature*, *478*(7370), 447-449. Doi:10.1038/478477a

Bowles, D. C., Braidwood, M., & Butler, C. D. (2014). Unholy Trinity: Climate change, conflict and ill health. In C. D. Butler (Ed.), *Climate change and global health*. Wallingford: Cabi. doi:10.1079/9781780642659.0144

Bowles, D. C., & Butler, C. D. (2014). Socially, politically and economically mediated health effects of climate change: Possible consequences for Africa. *South African Medical Journal*, *104*(98). PMID:26307807

Bowles, D. C., Butler, C. D., & Friel, S. (2014). Climate change and health in Earth's future. *Earth's Future*(2), 60-67.

Bowles, D. C., Butler, C. D., & Morisetti, N. (2015). Climate change, conflict and health. *Journal of the Royal Society of Medicine*, *108*(10), 390–395. doi:10.1177/0141076815603234 PMID:26432813

Bowles, D. C., Reuveny, R., & Butler, C. D. (2014). Moving to a better life? Climate, migration and population health. In C. D. Butler (Ed.), *Climate change and global health*. Wallingford: CABI. doi:10.1079/9781780642659.0135

Bradbury, R. (2012, July 13). A world without coral reefs. *New York Times*.

Buhaug, H., Nordkvelle, J., Bernauer, T., Böhmelt, T., Brzoska, M., & … J.S.A. Link, (2014). One effect to rule them all? A comment on climate and conflict. *Climatic Change*, *127*, 391–397. doi:10.1007/s10584-014-1266-1

Butler, C. D. (2010). Climate change, crop yields, and the future. *SCN News*, *38*, 18–25.

Butler, C. D. (2014a). Famine, hunger, society and climate change. In C. D. Butler (Ed.), *Climate Change and Global Health*. Wallingford: Cabi. doi:10.1079/9781780642659.0124

Butler, C. D. (Ed.). (2014b). *Climate Change and Global Health*. Wallingford: CABI. doi:10.1079/9781780642659.0000

Butler, C. D., & Harley, D. (2010). Primary, secondary and tertiary effects of eco-climatic change: The medical response. *Postgraduate Medical Journal*, *86*(1014), 230–234. doi:10.1136/pgmj.2009.082727 PMID:20354046

Connolly, M. A., Gayer, M., Ryan, M. J., Salama, P., Spiegel, P., & Heymann, D. L. (2004). Communicable diseases in complex emergencies: Impact and challenges. *Lancet*, *364*(9449), 1974–1983. doi:10.1016/S0140-6736(04)17481-3 PMID:15567014

Cooley, S. R., & Doney, S. C. (2009). Anticipating ocean acidification's economic consequences for commercial fisheries. *Environmental Research Letters*, *4*(2), 024007. doi:10.1088/1748-9326/4/2/024007

Cribb, J. (2010). *The coming famine: The global food crisis and what we can do to avoid it*. Berkeley: University of California Press.

Department of Defense. (2014). *Quadrennial Defense Review 2014*. Washington, D.C.: Department of Defense.

Despommier, D. D., Gwadz, R. W., Hotez, P. J., & Knirsch, C. A. (2006). Parasitic Diseases (5th ed.). New York: Apple Tree Productions, LLC.

Erkoreka, A. (2009). Origins of the Spanish Influenza pandemic (1918-1920) and its relation to the First World War. *Journal of Molecular and Genetic Medicine*, *3*(2), 190–194. PMID:20076789

Femia, F., & Werrell, C. (2013). Climate change before and after the Arab Awakening: The cases of Syria and Libya. In C. E. Werrell & F. Femia (Eds.), *The Arab Spring and Climate Change*. Washington, DC: Center for American Progress.

Field, J. S., & Lape, P. V. (2010). Paleoclimates and the emergence of fortifications in the tropical Pacific islands. *Journal of Anthropological Archaeology, 29*(1), 113-124. doi:10.1016/j.jaa.2009.11.001

Frommel, A. Y., Maneja, R., Lowe, D., Malzahn, A. M., Geffen, A. J., Folkvord, A., et al. (2012). Severe tissue damage in Atlantic cod larvae under increasing ocean acidification. *Nature Clim. Change, 2*(1), 42-46. Doi:10.1038/nclimate1324

Ghobarah, H. A., Huth, P., & Russett, B. (2004). The post-war public health effects of civil conflict. *Social Science & Medicine*, *59*(4), 869–884. doi:10.1016/j.socscimed.2003.11.043 PMID:15177842

Gleditsch, N. P. (2012). Whither the weather? Climate change and conflict. *Journal of Peace Research*, *49*(1), 3–9. doi:10.1177/0022343311431288

Gleick, P. H. (2014). Water, drought, climate change, and conflict in Syria. *Weather, Climate, and Society*, *6*(3), 331–340. doi:10.1175/WCAS-D-13-00059.1

Haas, J. (2009). The origins of war and ethnic violence. In J. Carman & A. Harding (Eds.), *Ancient Warfare: archaeological perspectives*. Gloucestershire: The History Press.

Hankins, C. A., Friedman, S. R., Zafar, T., & Strathdee, S. A. (2002). Transmission and prevention of HIV and sexually transmitted infections in war settings: Implications for current and future armed conflicts. *AIDS (London, England)*, *16*(17), 2245–2252. doi:10.1097/00002030-200211220-00003 PMID:12441795

Haug, G. H., Günther, D., Peterson, L. C., Sigman, D. M., Hughen, K. A., & Aeschlimann, B. (2003). Climate and the collapse of Maya civilization. *Science*, *299*(5613), 1731–1735. doi:10.1126/science.1080444 PMID:12637744

Hoegh-Guldberg, O. (2011). Coral reef ecosystems and anthropogenic climate change. *Regional Environmental Change*, *11*(0), 215–227. doi:10.1007/s10113-010-0189-2

Hoegh-Guldberg, O., Mumby, P., Hooten, A., Steneck, R., Greenfield, P., Gomez, E., & Hatziolos, M. E. et al. (2007). Coral reefs under rapid climate change and ocean acidification. *Science*, *318*(5857), 1737–1742. doi:10.1126/science.1152509 PMID:18079392

Holdstock, D. (1995). Climate, health and war. *Medicine and War, 11*(4), 145-147. doi:10.1080/07488009508409234

Homer-Dixon, T. F. (1994). Environmental scarcities and violent conflict: Evidence from cases. *International Security, 19*(1), 5–40. doi:10.2307/2539147

Hsiang, S. M., Meng, K. C., & Cane, M. A. (2011). Civil conflicts are associated with the global climate. *Nature, 476*(7361), 438-441. Doi:10.1038/nature10311

Intergovernmental Panel on Climate Change. (2013). *The physical science basis.* Geneva: Intergovernmental Panel on Climate Change.*Climatic Change*, 2013.

Intergovernmental Panel on Climate Change. (2014). *Impacts, Adaptation and Vulnerability* Geneva: Intergovernmental Panel on Climate Change.*Climatic Change*, 2014.

International Committee of the Red Cross. (2011). *Health Care in Danger: Making the case.* Geneva: International Committee of the Red Cross.

Jarvis, L., Montgomery, H., Morisetti, N., & Gilmore, I. (2011). Climate change, ill health, and conflict. *BMJ (Clinical Research Ed.)*, April. PMID:21467102

Kahl, C. H. (2006). *States, Scarcity and Civil Strife in the Developing World.* Princeton: Princeton University Press.

Kaldor, M. (2012). New and Old Wars: Organized violence in a global era (3rd ed.). Stanford: Stanford University Press.

Kelley, C. P., Mohtadi, S., Cane, M. A., Seager, R., & Kushnir, Y. (2015). Climate change in the Fertile Crescent and implications of the recent Syrian drought. *Proceedings of the National Academy of Sciences of the United States of America, 112*(11), 3241–3246. doi:10.1073/pnas.1421533112 PMID:25733898

Kennett, D. J., Breitenbach, S. F., Aquino, V. V., Asmerom, Y., Awe, J., Baldini, J. U., & Haug, G. H. et al. (2012). Development and disintegration of Maya political systems in response to climate change. *Science, 338*(6108), 788–791. doi:10.1126/science.1226299 PMID:23139330

Lape, P. V., & Chin-Yung, C. (2008). Fortification as a human response to late Holocene climate change in East Timor. *Archaeology in Oceania, 42*(4), 11-21.

Lederberg, J. (2000). Infectious history. *Science, 288*(5464), 287–293. doi:10.1126/science.288.5464.287 PMID:10777411

Marlar, R. A., Leonard, B. L., Billman, B. R., Lambert, P. M., & Marlar, J. E. (2000). Biochemical evidence of cannibalism at a prehistoric Puebloan site in southwestern Colorado. *Nature, 407*(6800), 74-78. Doi:10.1038/35024064

Martin, A. (2005). Environmental conflict between refugee and host communities. *Journal of Peace Research, 42*(3), 329–346. doi:10.1177/0022343305052015

Mazo, J. (2010). *Climate Conflict: How global warming threatens security and what to do about it.* London: Routledge, the International Institute for Strategic Studies.

McMichael, A. J. (2013). Globalization, climate change, and human health. *The New England Journal of Medicine, 368*(14), 1335–1343. doi:10.1056/NEJMra1109341 PMID:23550671

McMichael, C., Barnett, J., & McMichael, A. J. (2012). An ill wind? Climate change, migration, and health. *Environmental Health Perspectives, 120*(5), 646–654. doi:10.1289/ehp.1104375 PMID:22266739

Medact. (2008). *Rehavilitation under fire: Health care in Iraq 2003-7.* London: Medact.

Morisetti, N. (2012). Climate change and resource security. *BMJ (Clinical Research Ed.)*, March. PMID:22431658

Oxford, J. B., Sefton, A., Jackson, R., Innes, W., & Johnson, N. P. A. S. (2002). World War I may have allowed the emergence of "Spanish" influenza. *The Lancet Infectious Diseases, 2*(2), 111–114. doi:10.1016/S1473-3099(02)00185-8 PMID:11901642

Parenti, C. (2011). *Tropic of Chaos: Climate change and the new geography of violence.* New York: Nation Books.

Parker, G. (2013). *Global Crisis: War, climate change and catastrophe in the seventeenth century.* New Haven: Yale University Press.

Press, A., Bergin, A., & Gansey, E. (2013). *Heavy Weather: Climate and the Australian Defence Force.* Barton, Australia: Australian Strategic Policy Institute.

Reuveny, R. (2007). Climate change-induced migration and violent conflict. *Political Geography, 26*(6), 656-673. DOI:10.1016/j.polgeo.2007.05.001

Reuveny, R. (2008). Ecomigration and violent conflict: Case studies and public policy implications. *Human Ecology, 36*(1), 1–13. doi:10.1007/s10745-007-9142-5

Reuveny, R., & Allen, A. P. (2007). On environmental refugees and implications for the future. *Ecología Política, 33*, 21–36.

Rønnfeldt, C. F. (1997). Three Generations of Environment and Security Research. *Journal of Peace Research, 34*(4), 473–482. doi:10.1177/0022343397034004009

Salehyan, I. (2008). From climate change to conflict? No consensus yet. *Journal of Peace Research, 45*(3), 315–326. doi:10.1177/0022343308088812

Smallman-Raynor, M., & Cliff, A. D. (2001). Epidemic diffusion processes in a system of US military camps: Transfer diffusion and the spread of typhoid fever in the Spanish-American War. *Annals of the Association of American Geographers, 91*(1), 71–91. doi:10.1111/0004-5608.00234

Smallman-Raynor, M., & Cliff, A. D. (2004). The geographical spread of cholera in the Crimean War: Epidemic transmission in the camp systems of the British Army of the East, 1854-55. *Journal of Historical Geography, 30*(1), 32–69. doi:10.1016/S0305-7488(02)00084-1

Stearns, J. K. (2011). *Dancing in the Glory of Monsters: The collapse of the Congo and the great war of Africa*. New York, NY: PublicAffairs.

Sullivan, G. R., Bowman, F., Farrell, L. P., Gaffney, P. G., Kern, P. J., Lopez, T. J., & Zinni, A. C. et al. (2007). *National security and climate change*. Alexandria, VA: The CNA Corporation.

Talmon, J. L. (1980). European History as the Seedbed of the Holocaust. In J. Sonntag (Ed.), *Jewish Perspectives: 25 years of Jewish writing*. London: Secker & Warburg.

Toole, M. J., Galson, S., & Brady, W. (1993). Are war and public health compatible? *Lancet, 341*(8854), 1193–1196. doi:10.1016/0140-6736(93)91013-C PMID:8098086

Turney, C. S. M., Baillie, M., Palmer, J., & Brown, D. (2006). Holocene climatic change and past Irish societal response. *Journal of Archaeological Science, 33*(1), 34-38. doi:10.1016/j.jas.2005.05.014

UK Ministry of Defence. (2014). *Global Strategic Trends - Out to 2045*. UK Ministry of Defence.

UNHCR. (2015a). Syria regional refugee response. Retrieved from data.unhcr.org/syrianrefugees/regional/php

UNHCR. (2015b). *UNHCR Global Trends Forced Displacement in 2014*. Geneva: UNHCR.

Watts, N., Adger, W. N., Agnolucci, P., Blackstock, J., Byass, P., Cai, W., et al. (2015). Health and climate change: Policy responses to protect public health. *The Lancet.*

Welzer, H. (2012). *Climate Wars: Why people will be killed in the twenty-first century* (P. Camiller, Trans.). Cambridge: Polity Press.

Zhang, D. D., Brecke, P., Lee, H. F., He, Y.-Q., & Zhang, J. (2007). Global climate change, war, and population decline in recent human history. *Proceedings of the National Academy of Sciences of the United States of America, 104*(49), 19214–19219. doi:10.1073/pnas.0703073104 PMID:18048343

Zhang, D.D., Jim, C.Y., Lin, G.C.-S., He, Y.-Q., Wang, J.J., & Lee, H.F. (2006). Climatic change, wars and dynastic cycles in China over the last millennium. *Climatic Change, 76*(3-4), 459–477. doi:10.1007/s10584-005-9024-z

Zhang, D. D., Lee, H. F., Wang, C., Li, B., Pei, Q., Zhang, J., & An, Y. (2011). The causality analysis of climate change and large-scale human crisis. *Proceedings of the National Academy of Sciences of the United States of America, 108*(42), 17296–17301. doi:10.1073/pnas.1104268108 PMID:21969578

Zhang, D. D., Zhang, J., Lee, H. F., & He, Y.-Q. (2007). Climate change and war frequency in eastern China over the last millennium. *Human Ecology, 35*(4), 403–414. doi:10.1007/s10745-007-9115-8

Chapter 4
Waterborne Diseases and Climate Change: Impact and Implications

Maha Bouzid
University of East Anglia, UK

ABSTRACT

Waterborne diseases are caused by a multitude of pathogens and associated with a significant burden in both developed and developing countries. While the assessment of the adverse impacts of climate change on human heath from infectious diseases has mainly focused on vector-borne diseases, waterborne diseases prevalence and transmission patterns are also likely to be impacted by environmental change. This chapter will outline relevant waterborne pathogens, summarise the impact of climate change on disease transmission and explore climate change adaptation options in order to reduce the increased burden of waterborne diseases.

INTRODUCTION

Waterborne diseases are caused by bacteria, viruses and parasites that contaminate water and are transmitted to humans through consumption or contact (Leclerc, Schwartzbrod, & Dei-Cas, 2002). These pathogens are usually excreted in the faeces of infected individuals and ensure transmission of the infection to new hosts via the faecal oral route (Leclerc et al., 2002). The link between water and disease was noticed earlier on and the first publication on PubMed linking ground water

DOI: 10.4018/978-1-5225-0553-2.ch004

Figure 1. Number of publications for "water and disease" per year (1850-2015)
Source: http://www.ncbi.nlm.nih.gov/pubmed.

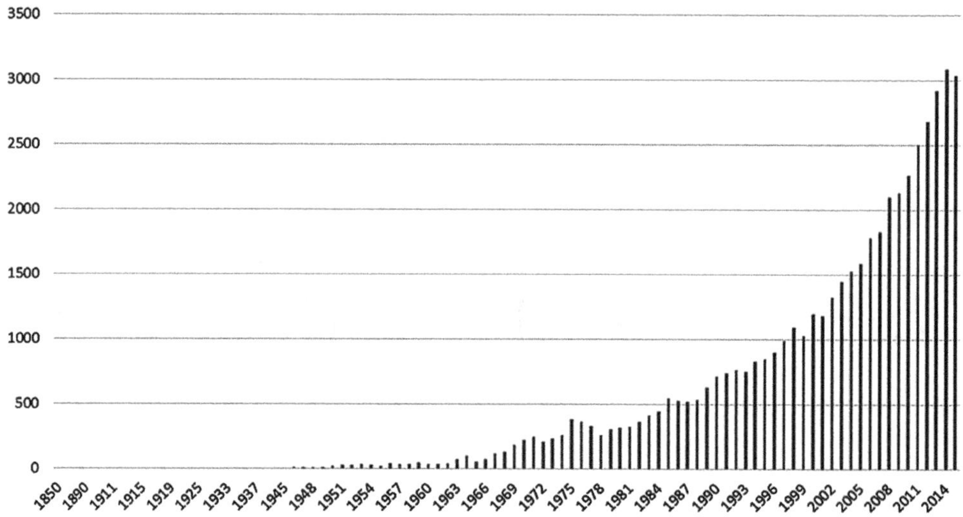

and disease dates back to 1890 (No authors listed, 1890). While some observations proved untrue (such as water transmission of smallpox), remarkably, several findings were correct and are currently backed by scientific findings, in particular the fact that droughts and heavy rain could be accompanied or followed by ill health and epidemic disease (No authors listed, 1890). The first mention of "waterborne disease" on PubMed dates back to 1895 and was associated with consumption of sewage contaminated shellfish that resulted in cholera and typhoid cases (No authors listed, 1895). The author's recommendation was to urge the local authorities to take action in order to prevent sewage contamination. This good practice recommendation is still valid today. The number of publications about water and disease has steadily increased since as shown on Figure 1 reflecting the prevalence and relevance of waterborne diseases. Indeed, some authors consider that waterborne diseases are the single largest threat to global public health (Gray, 2014).

Waterborne diseases are not confined to developing countries with poor water quality and basic sanitation. Indeed, several sizeable outbreaks occurred in developing countries (Baldursson & Karanis, 2011; Beer et al., 2015; OECD/WHO, 2003; Smith et al., 2006) and are likely to continue to occur, highlighting that water supplies could be accidently contaminated or have treatment failure. The issue in developing countries is mainly related to water availability and quality. From an international policy perspective, the need for a better understanding of the role of water in infectious diseases and improved management of water supplies were formally stated in 1996 at the Organisation for Economic Cooperation and Development workshop on "Biotechnology for Water Use and Conservation" in Cocoyoc, Mexico (P.R. Hunter, Waite, & Ronchi, 2003). In addition, improved water and sanitation was one of the Millennium Development Goals (UNICEF and World Health Organization, 2015) and is one of the post 2015 Sustainable Development Goals of the United Nations (http://www.un.org/sustainabledevelopment/water-and-sanitation/). These goals suggest that despite remarkable advances towards access to adequate water and sanitation, a large proportion of the world's poorest are still suffering from water scarcity, poor drinking water quality and inadequate sanitation and the associated high risk of infectious diseases.

Climate change refers to "a change in the state of the climate that can be identified by changes in the mean and/or the variability of its properties, and that persists for an extended period, typically decades or longer" (IPCC, 2015). Climate change is attributed directly or indirectly to human activity (UNFCC, 1992). The adverse effect of climate change on human health is no longer a debatable issue (Kim, Kabir, & Ara Jahan, 2014; McMichael & Lindgren, 2011; McMichael, Woodruff, & Hales, 2006; Semenza & Menne, 2009; Wu, Lu, Zhou, Chen, & Xu, 2016). While, the effect of climate change on vector-borne diseases has been accepted earlier on, probably, because of the obvious and well established effect of climate variables (mainly temperature and precipitation) on vector survival, biting pattern, virus multiplication and host exposure (to name just a few) (Bouzid, Colon-Gonzalez, Lung, Lake, & Hunter, 2014; Gage, Burkot, Eisen, & Hayes, 2008; A. J. McMichael et al., 2006), the effect of climate variability on waterborne diseases has been less forthcoming. This could be explained by the initial lack of knowledge of the effect of various climatic factors on the transmission of waterborne diseases and more recently limitations in quantifying these effects, the perceived lack of interest from the research, policy and stakeholder community as well as the assumption that the effect of environmental change on waterborne diseases will be negligible, at least in developed countries. In addition, it might be difficult to distinguish between climate change and other factors which can drive the change in infectious diseases (Lafferty, 2009). This chapter will summarise the most relevant waterborne pathogens, current knowledge on the relationships between weather variables and waterborne diseases as well as the impact environmental change is likely to have on these diseases.

ETIOLOGIC AGENTS OF WATERBORNE DISEASES

Several microorganisms have been classified as waterborne because of suspected or demonstrated link to water. These are excreted in the faeces of infected hosts or naturally occurring in the water (Funari, Manganelli, & Sinisi, 2012). Up until 1980, the usual culprits were typhoid, cholera and dysentery (Smith et al., 2006).

Subsequently, the list of waterborne pathogens kept expanding and several microorganisms were categorised as emerging waterborne pathogens. The up to date list of waterborne diseases and contaminants is available on the CDC website (http://www.cdc.gov/healthywater/disease/type.html). These are classified as bacterial, parasitic, viral, chemical and other (such as toxins). Most waterborne pathogens share similar characteristics ensuring their success in maintaining an ongoing transmission cycle including low infectious dose, environmental resistance and excretion in very high number in stools of infected persons, even in asymptomatic individuals (Gall, Marinas, Lu, & Shisler, 2015).

Despite significant improvements in access to drinking water and sanitation, it is estimated that 1.8 billion people still consume faecally polluted water (Onda, LoBuglio, & Bartram, 2012; World Health Organization and UNICEF, 2014). The burden is disproportionally occurring in people living in poverty, in rural areas and in developing countries (>99%) (Ashbolt, 2004; UNICEF, 2012). Consumption of contaminated water can result in gastroenteritis, with the most commonly reported symptoms are diarrhoea and vomiting. However, waterborne pathogens can cause a wide range of symptoms including fever, stomach pain, cramps, nausea, fatigue and joint pain (Andersson et al., 2014). Diarrhoea is the second cause of death in children under five years old (after pneumonia) and is associated with malnutrition and stunting (UNICEF, 2012; World Health Organization, 2013).

Bacterial Waterborne Pathogens

Bacteria were the first suspected microorganisms to cause disease generally. This was also the case for waterborne diseases. The role of faecally contaminated drinking water in cholera transmission was established as early as 1855 in London by John Snow (Snow, 1855). The source of the cholera outbreak was identified as a contaminated water pump. The index case was recognised as a baby with cholera whose faeces contaminated the local water source. Death occurred in a visitor who consumed the local water as well as in an individual who got the water transported to her dwelling. In 1893, Robert Koch was able to culture the causing bacteria *Vibrio cholerae*, isolate the bacterium from the water and stool samples and reproduce the infection (OECD/WHO, 2003), therefore, establishing the Koch's postulates defining the causative relationship between a microbe and a disease. Subsequently, a similar breakthrough happened for typhoid fever, it was established that it is transmitted via contaminated water and the causing bacterium *Salmonella typhi* was isolated (OECD/WHO, 2003). These advances highlighted the role of contaminated water in disease transmission and consequently, the need for unpolluted or treated water to maintain health, which was the start of water treatment principle. An early example of water treatment was the German city of Altona using the same water source than Hamburg (which had a ravaging cholera epidemic in 1892) but having only very few cases mainly among commuters (OECD/WHO, 2003). Altona had installed slow sand filtration, which allowed the successful elimination of cholera and typhoid causing bacteria.

Currently the list of waterborne bacterial pathogens includes *Salmonella* spp., *Shigella* spp., *Escherichia coli* O157:H7, *Campylobacter jejuni,* in addition to indigenous aquatic bacteria *Vibrio* spp., *Legionella* spp., *Aeromonas* spp., *Pseudomonas aeruginosa, Helicobacter pylori* and *Mycobacterium avium* (Leclerc et al., 2002). These bacteria cause the majority of human infections through consumption or contact. The complete list of waterborne bacterial pathogen is presented in Table 1 (World Health Organisation, 2011). Waterborne outbreaks are closely monitored in developed countries therefore resulting in good quality data and identification of the causing agent, though many mild cases are unlikely to be reported. A survey of waterborne outbreaks in Canada 1974-2001 showed that 62 out of 288 outbreaks were caused by bacteria, of which 24 outbreaks were caused by *Campylobacter,* 16 by *Salmonella* spp. and 9 by *Shigella* spp. (Schuster et al., 2005). In a case-control study of causes of diarrheal disease in infants and young children from developing countries (GEMs study), enterotoxigenic *Escherichia coli* and *Shigella* were found among the top four causes of diarrhoea, with *Aeromonas*, *Vibrio cholera* and *Campylobacter jejuni* highly prevalent in particular study sites (Kotloff et al., 2013). The study also revealed that enteropathogenic *E coli* and enterotoxigenic *E.coli* producing heat-stable toxin (ST-ETEC) are the predominant cause of death in infants aged 0-11 months (Kotloff et al., 2013). Characteristics of these waterborne bacteria including ecology, pathogenicity, virulence factors, incubation period, infectious dose and risk factors have been recently reviewed (Cabral, 2010).

Viral Waterborne Pathogens

While it is widely accepted that bacteria are major causes of diarrhoea contracted through drinking water, the role of viruses is less established (Gall et al., 2015). This is partly due to mild symptoms and self- limiting nature of the infection as well as current limitation of viral detection in clinical settings (Gall et al., 2015). Nevertheless, huge numbers of viruses are present in the human gut and many are excreted by patients and end up in the sewage system (Leclerc et al., 2002). Therefore, potentially contaminating drinking water sources in cases of sewage runoff. Waterborne viral pathogens include Adenovirus, Astrovirus, Hepatitis A, Hepatitis E, Rotavirus, Norovirus and Enterovirus (Table 1) (World Health Organisation, 2011). In addition, other viruses are suspected to be transmitted through water such as Polyomaviruses, Cytomegaloviruses, Influenza viruses and Coronaviruses, but evidence is lacking (Gall et al., 2015). These viruses are associated with a significant

Table 1. Complete list of waterborne pathogens according to the WHO guidelines for drinking-water quality, fourth edition

Waterborne Pathogens	Species	Waterborne Pathogens	Species
Bacteria	*Acinetobacter*	Protozoa	*Acanthamoeba*
	Aeromonas		*Balantidium coli*
	Bacillus		*Blastocystis*
	Burkholderia pseudomallei		*Cryptosporidium*
	Campylobacter		*Cyclospora cayetanensis*
	Enterobacter sakazakii		*Entamoeba histolytica*
	Escherichia coli pathogenic strains		*Giardia intestinalis*
	Helicobacter pylori		*Isospora belli*
	Klebsiella		*Microsporidia*
	Legionella		*Naegleria fowleri*
	Leptospira		*Toxoplasma gondii*
	Mycobacterium	Toxic Cyanobacteria, Helminths	*Dracunculus medinensis*
	Pseudomonas aeruginosa		*Fasciola* spp.
	Salmonella		Free-living nematodes
	Shigella		*Schistosoma* spp.
	Staphylococcus aureus		
	Tsukamurella		
	Vibrio		
	Yersinia		
Viruses	Adenoviruses		
	Astroviruses		
	Caliciviruses		
	Enteroviruses		
	Hepatitis A virus		
	Hepatitis E virus		
	Rotaviruses		

World Health Organisation, 2011.

disease burden globally and Rotavirus is a leading cause of severe childhood diarrhoea and mortality (> 500,000 every year) (No authors listed, 2008). In the GEMs study, Rotavirus was one of the top four causes of childhood diarrhoea in developing countries (Kotloff et al., 2013). Two Rotavirus vaccines are currently undertaking trials and initial results showed efficacy > 85% (No authors listed, 2008). Wider use of Rotavirus and Hepatitis A vaccines would significantly decrease the disease burden of these two waterborne diseases. For non-vaccine preventable viruses, removal or inactivation of infectious viruses from water sources are the only possible course of action. However, physical removal can be ineffective due to the small size of viral particles and inactivation needs to be assessed for each virus family and should be optimised to minimise toxic disinfection by-products (Gall et al., 2015).

Parasitic Waterborne Pathogens

Waterborne parasites are extremely resistant in the environment and contribute to a significant burden of diarrheal disease (Bouzid, 2014; Bouzid, Hunter, Chalmers, & Tyler, 2013). The two major waterborne parasites are *Cryptosporidium* spp. and *Giardia duodenalis* causing the majority of outbreaks worldwide (60% and 35%, respectively) (Baldursson & Karanis, 2011). *Cryptosporidium* caused several large waterborne outbreaks such as in Milwaukee (largest documented waterborne outbreak with over 403,000 cases), Ireland, UK, USA, and Sweden (Baldursson & Karanis, 2011; Chalmers, 2012; Mackenzie et al., 1994). *Cryptosporidium* and *Giardia duodenalis* cause a particular issue for water treatment companies because they are resistant to common disinfection procedures such as chlorine (Kothavade, 2012). Interestingly, outbreaks of cryptosporidiosis and giardiasis occurred in the 1990s and took the water industry by surprise because the water responsible for the outbreaks was deemed properly treated and satisfying the then current microbiological standards (Goldstein et al., 1996; Leclerc et al., 2002). This has sprung the debate about the reliability of traditional faecal indicators and despite major advancement, this debate is far from over. In addition to causing large outbreaks, *Cryptosporidium* attracted public health interest because of its chronic nature and high mortality rate in HIV/ AIDS individuals (O'Donoghue, 1995). In the GEMs study, *Cryptosporidium* was the only parasite in the top four causes of diarrhoea in children under five years old and the main cause of childhood mortality in toddlers aged 12-23 months (Kotloff et al., 2013). Other relevant waterborne parasites are *Toxoplasma gondii, Entamoeba histolytica, Cyclospora cayetanensis, Acanthamoeba spp., Microsporidia, Isospora, Blastocystis, Naegleria fowleri* and *Balantidium coli;* their prevalence, distribution and associated burden are detailed in recent reviews (Baldursson & Karanis, 2011; Bouzid, 2014; World Health Organisation, 2011).

Toxic Cyanobacteria

Cyanobacteria are photosynthetic bacteria (previously known as blue-green algae) that have ubiquitous distribution (freshwater lakes, basins, rivers, sea water, salty lakes, estuaries, soil) (World Health Organisation, 2011). Some Cyanobacteria species produce toxins that can contaminate water sources and impact on human health (Manganelli et al., 2012). Cyanobacterial toxins cause a range of adverse health outcomes including skin allergies, pneumonia, birth defects and liver damage (P. R. Hunter, 1998). Cyanobacteria are not considered infectious because they do not multiply in the human body (World Health Organisation, 2011). Human exposure occurs via consumption, contact, recreational use and haemodialysis (Manganelli et al., 2012). Cyanobacterial blooms occur when environmental conditions are favourable (nutrient enrichment and optimal temperature and light) and result in the production of high amount of toxins and associated losses of drinking, recreational and agricultural water sources (Manganelli et al., 2012).

Helminths

Helminths are worms that can be either free living or parasitic. Some helminths are transmitted via the faecal oral route when the larvae are passed in the faeces of infected hosts, contaminate the environment (including water and food) and allow transmission to new hosts. According to the WHO, drinking water is a significant mode of transmission for *Dracunculus medinensis*, *Fasciola*, nematodes and *Schistosoma* (World Health Organisation, 2011). Most of these helminths occur in inadequate water supplies mainly in developing countries. For *Schistosoma*, infection occurs when skin is in contact with fresh water containing the parasite's eggs excreted by an infected person and matured in a snail vector. The burden of Schistosomiasis is significant: 200 million infections in 75 countries and 20,000 deaths (World Health Organisation, 2011). Advances in eradication programs (for *Dracunculus*) and improvement of water and sanitation will contribute to decrease the disease burden of all waterborne and water related diseases.

WATERBORNE DISEASES, WEATHER VARIABLES, AND CLIMATE CHANGE

The link between waterborne disease and weather has been documented as early as 1850 and it was stated that "periods of low and high water…ultimately become the cause and lead to spread of the disease" and "the failure of springs and the drying-up of the rivers, such periods have always been accompanied or followed by epidemic disease" (No authors listed, 1890). These remarkable observations have subsequently been supported by research findings. In a study of waterborne disease outbreaks in the USA between 1948- 1994 (548 outbreaks were investigated), it was found that heavy rain and flooding caused contamination of water sources via sewage or agricultural runoff and was responsible for the majority of outbreaks (Curriero, Patz,

Rose, & Lele, 2001). A study of 89 drinking water outbreaks occurring in England and Wales between 1910-1999 showed a strong association between heavy rainfall in the week before the outbreak or low rainfall during the four weeks preceding the outbreak in 40% of cases (Nichols, Lane, Asgari, Verlander, & Charlett, 2009). It is therefore established that rainfall worsens the microbiological quality of drinking and bathing waters (Funari et al., 2012). Indeed, rainfall is used as a predictor of bathing waters contamination according to the European directive for the management of bathing water quality (European Parliament, 2006). In a review by Cann and colleagues using a systematic search strategy to investigate the relationships between 87 waterborne outbreaks and extreme weather events such as excessive precipitation, floods and drought, it was reported that heavy rainfall and flooding are the two events most commonly preceding the outbreaks (Cann, Thomas, Salmon, Wyn-Jones, & Kay, 2013). The authors reported that for outbreaks occurring in developed countries, *Campylobacter* spp. and *Cryptosporidium* spp. are the two major pathogens associated with contaminated drinking water supplies, while, *Leptospira* spp., *Cryptosporidium* spp., *Norovirus* and *Vibrio vulnificus* are responsible for outbreaks originating from environmental exposure (Cann et al., 2013). A recent systematic review investigated the relationship between extreme precipitation or temperature and drinking water related infection in terms of number of cases and outbreaks (Herrador et al., 2015). The Majority of outbreaks were from Asia and North America and most showed a positive association between increased precipitation or temperature and waterborne disease (Herrador et al., 2015). The link with extreme precipitation was the most influencing factor. However, the link between extremes in precipitation or temperature and waterborne disease was not demonstrated in six studies and one study reported an association between decreased precipitation and waterborne infection (Herrador et al., 2015). The cause of outbreaks was not always determined and while some studies reported on cases of gastrointestinal infections, others sought to identify the etiological agent; cholera, campylobacteriosis, typhoid fever, shigellosis, cryptosporidiosis, giardiasis, hepatitis A and paratyphoid fever were reported (Herrador et al., 2015). The multitude of pathogens involved in these outbreaks could explain the heterogeneity of associations with weather variables. The authors reported that these variations could be explained by numerous factors including the causative pathogen, geographic location, season and water supply type and concluded that their findings illustrate the complex relationship between precipitation or temperature and waterborne disease transmission (Herrador et al., 2015).

The Intergovernmental Panel on Climate Change (IPCC) projects an increase of average global temperature of 1.4- 4.8°C over the next century as well as an increase in the frequency and intensity of extreme weather events (heat waves, heavy precipitation, droughts, storms) (IPCC, 2015). Climate change is associated with adverse health impacts that would extend to waterborne diseases because of their sensitivity to weather variables, particularly precipitation and temperature (Funari et al., 2012; A. J. McMichael et al., 2006; T. McMichael, Montgomery, & Costello, 2012). The projected increases in temperature, heavy precipitation, storms and droughts are likely to disturb the global hydrological cycle and result in more waterborne infections (Patz, Grabow, & Limaye, 2014). Extreme precipitation is likely to overwhelm sewer systems, in developed and developing countries alike, thus resulting in contamination of drinking and recreational water sources and subsequent risk of infection (Patz et al., 2014). In the Great Lakes region in USA, it was estimated that extreme precipitation would increase 10-40% by the end of the century, thus resulting in increase of sewage overflow frequency by 50-120% (Patz, Vavrus, Uejio, & McLellan, 2008). In addition, climate change could interfere with the seasonality of many waterborne pathogens such as *Cryptosporidium, Giardia, Salmonella, Shigella* and *Campylobacter* (Naumova et al., 2007). This seasonality was linked to various factors such as human behaviour, prevalence of pathogens in water sources and the environment, infection in reservoirs and hosts, agricultural practices and extreme weather events (Sterk, Schijven, de Nijs, & Husman, 2013). All these factors are likely to alter under the effect of climate change and contribute to variation in disease transmission.

Climate change could affect the survival and multiplication of waterborne pathogens either positively or negatively. Projected higher temperature could result in inactivation or die off of temperature sensitive waterborne viruses, bacteria and parasites especially after long exposure due to the combined action of higher temperature and UV radiation (Gray, 2014; Hofstra, 2011). Conversely, higher water temperature could facilitate proliferation of *Vibrio* spp. in warm fresh and marine waters, therefore increasing disease transmission via drinking and recreational water as well as seafood consumption (Patz et al., 2014). Indeed, it was documented that warming of the Baltic Sea and lower salinity due to altered patterns of precipitation and runoff are the main drivers of the emergence of *Vibrio* infections in northern Europe (Baker-Austin et al., 2013). Similarly, it is thought that climate change and eutrophication (nutrient enrichment partly caused by anthropogenic pollution) are enhancing the proliferation and distribution of Cyanobacterial algal blooms, which have seen a significant increase in the last few decades (O'Neil, Davis, Burford, & Gobler, 2012). A study of the projected effect of climate change in the Netherlands estimated that by 2100 the growth rate of two toxic Cyanobacteria will double in coastal waters, with adverse impact on fish production and recreational use resulting in ecosystem devaluation (Peperzak, 2005). Therefore, climate change is likely to affect the concentration of pathogens in water sources and their fate in the environment and the subsequent infection risk (Gray, 2014; Hofstra, 2011). Pathogen concentration in surface water is directly related to disease occurrence but the associated burden differs between pathogens and is disproportionately higher in developing countries. However, for disease outbreaks to occur, the synchronised presence of various factors such as source contamination, transport to abstraction site, survival through water treatment procedures and consumption by the human host are needed (in this case for drinking water outbreaks) (Auld, MacIver, & Klaassen, 2004; Gray, 2014). Therefore, risk estimation and management at every step of water treatment is needed to reduce or eliminate the risk and accounting for climate change driven events is crucial to risk reduction.

Climate change scenarios project sea level rise in many vulnerable locations which will experience greater saltwater intrusion and increased levels of marine and estuarine bacteria such as *Vibrio cholerae* and Cyanobacteria (Lipp, Huq, & Colwell, 2002). When combined with favourable environmental conditions ensuring bacterial growth, this bacterial influx would result in higher exposure and disease risk. Conversely, freshwater input from increased runoff could result in lower salinity in the estuarine environment, which in combination with high temperature and the right amount of nutrients could favour large blooms of toxic Cyanobacteria (Funari et al., 2012). These conditions were associated with the emergence of the Cyanobacteria *Microcystis aeruginosa* in Western Australia (Robson & Hamilton, 2003).

Climate change is projected to increase droughts in mid-latitudes and semi-arid regions, some of which are already water-stressed areas (IPCC, 2015). Droughts defined as water scarcity can also affect water quality. It was emphasised that both water quantity and quality are essential for good health (Stanke, Kerac, Prudhomme, Medlock, & Murray, 2013). The effects of drought on human health include infectious diseases, malnutrition, dehydration and death. These effects are particularly dramatic in developing countries because they are associated with poor drinking water quality, basic sanitation, wastewater reuse in agriculture and multiple use of available waterbodies in order to maximise the scarce resources (Funari et al., 2012). These practices contribute to significant increase in water contamination and subsequent risk of infection. Droughts can influence the proliferation of Cyanobacteria by increasing nutrient availability due to surface water evaporation and reducing water flow; one of the largest blooms of the toxic Cyanobacteria *Anabaena circinalis* in Australia was associated with these conditions (Funari et al., 2012). Similarly, drought conditions resulted in algal blooms in Nebraska lakes, which was responsible for dogs, livestock and wildlife deaths as well as numerous reports of human infections including skin rash, lesions and gastroenteritis (Funari et al., 2012).

Droughts followed by heavy rainfall (also known as flash floods) are projected to increase due to climate change. These events are associated with extreme levels of surface water contamination because microorganisms tend to concentrate in riverbeds in dry areas and then get washed downstream when the rain falls and contaminate drinking water sources and coastal waters. In addition, these conditions favour sewage runoff into rivers and waterbodies. Such conditions were associated with outbreaks cause by *Campylobacter*, *Cryptosporidium* and *E. coli* O157:H7 (Funari et al., 2012). Although the circumstances associated with these outbreaks are well known, disease and outbreak occurrences depend on various factors including presence of viable pathogens in the environment, terrain topography, water sources, sewage location and susceptible hosts. Therefore, estimating the relationship between these variables in order to accurately predict disease occurrence is highly desirable. However, because of the complexities involved such studies are still limited, though these are mainstream in ecology and biodiversity modelling. Once these models are available, projecting the impact of climate change on waterborne diseases would be more accessible.

Although a relatively recent research field, several investigations focused on estimating the impact of climate change on waterborne diseases. Some of these focused on diarrheal diseases in general and others on specific pathogens. Most of these were summarised in recent reviews (Delpla, Jung, Baures, Clement, & Thomas, 2009; Hofstra, 2011; Sterk et al., 2013). However, these studies tend to be of qualitative nature and quantification of the impact of climate change on pathogen concentrations in relevant matrices such as surface water or coastal waters and associated risk to humans is still in its infancy (Hofstra, 2011). A study by Moors and colleagues investigating the effect of climate change on diarrhoeal disease in Northern India, taking into account changing temperature, precipitation patterns and humidity estimated 13.1% increase (0- 21% range) in diarrhoeal disease (Moors, Singh, Siderius, Balakrishnan, & Mishra, 2013).

Quantification of the effect of climate change on infectious diseases in general and waterborne diseases in particular is likely to progress in the near future as we get a grasp on environmental factors affecting pathogen survival and disease transmission, especially when combined with advanced climate change scenarios, funding of climate change and health research and involvement of stakeholders, policy makers, researchers and health officials. This would contribute to efficient adaptation strategies to mitigate the impact of climate change on human health.

ADAPTATION STRATEGIES

Adaptation is defined as "anticipating the adverse effects of climate change and taking appropriate action to prevent or minimise the damage". Adaptation options for waterborne diseases, in non-health settings, aim to reduce environmental contamination by microbial pathogens and include waste treatment, improved sanitation and wastewater management, watershed protection, environmental remediation and improved drinking water monitoring and treatment (Rose et al., 2001). For an effective adaptation strategy, a comprehensive vulnerability assessment is paramount.

This involves identifying at risk populations, events that could cause harm and opportunities to reduce the damage and protect human health (Patz et al., 2014). For example, extreme weather events (floods, storms, and heatwaves) are associated with increased risk of waterborne diseases, so it pays dividends to have early warning system in place, this includes accurate forecasting, emergency preparedness, risk communication and public education. A response action plan should also be established to reduce morbidity and mortality. This includes preventive measures such as boil water orders, advice against bathing in contaminated water and temporary bans on shellfish consumption (Rose et al., 2001). During disease outbreaks, identifying population at risk, active epidemiological surveillance, early diagnosis and adequate case management are key factors to reduce disease transmission (Confalonieri, Menezes, & de Souza, 2015).

In the study by Moors and colleagues who estimated that diarrhoeal disease is likely to increase by 13.1% in Northern India, adaptation options to overcome this increased burden were explored. These adaptation strategies were classified as reactive, preventive and national policy options. Reactive adaptation options were treatment and symptom alleviation (rehydration solution, vitamin and mineral supplementation), however, their effectiveness is dependent on awareness and care seeking behaviour (Moors et al., 2013). Preventive actions aim to reduce disease transmission and include hand hygiene, improved water quality and waste disposal and vaccination (Rotavirus) (Moors et al., 2013). Overt policy options about climate change and health were missing but several national initiatives (such as integrated water resource management, sustainable agriculture and sustainable habitat) could contribute to increase resilience and reduce the burden of diarrhoeal diseases (Moors et al., 2013). It was stated that preventive options are potentially the most effective strategy to reduce the increased burden. However, in light of the limited and slow progress in reducing diarrheal disease during the last decade, consorted action, effective implementation and policy integration are urgently needed to counterbalance the adverse effect of climate change on human health (Moors et al., 2013).

CONCLUSION

Waterborne diseases remain a public health issue in developing and developed countries. While, admittedly the nature, drivers and disease burden are vastly different, the fact that significant waterborne disease outbreaks still occur in the world most developed nations highlights that even tightly monitored and regulated water supplies can be prone to contamination. Extreme weather events are projected to increase in intensity and frequency under climate change scenarios. Heavy precipitation cause combined sewer systems to overflow and contaminate drinking water sources, which in combination with ageing sewer infrastructure and increasing population is making the problem worse. Appropriate adaptation and mitigation strategies are needed to reduce the projected increased risk of climate-sensitive diseases in the future. While climate change adaptation is still not fully integrated, especially for the health sector, increased awareness, research findings and stakeholders and policy makers' involvement are increasingly supporting the urgent need for concerted action because the cost of inaction would be too high to bear and was shown to halt economic growth and health based development targets.

REFERENCES

Andersson, T., Bjelkmar, P., Hulth, A., Lindh, J., Stenmark, S., & Widerstrom, M. (2014). Syndromic surveillance for local outbreak detection and awareness: Evaluating outbreak signals of acute gastroenteritis in telephone triage, web-based queries and over-the-counter pharmacy sales. *Epidemiology and Infection*, *142*(2), 303–313. doi:10.1017/S0950268813001088 PMID:23672877

Ashbolt, N. J. (2004). Microbial contamination of drinking water and disease outcomes in developing regions. *Toxicology*, *198*(1-3), 229–238. doi:10.1016/j.tox.2004.01.030 PMID:15138046

Auld, H., MacIver, D., & Klaassen, J. (2004). Heavy rainfall and waterborne disease outbreaks: The Walkerton example. *Journal of Toxicology and Environmental Health. Part A.*, *67*(20-22), 1879–1887. doi:10.1080/15287390490493475 PMID:15371222

Baker-Austin, C., Trinanes, J. A., Taylor, N. G. H., Hartnell, R., Siitonen, A., & Martinez-Urtaza, J. (2013). Emerging Vibrio risk at high latitudes in response to ocean warming. *Nature Climate Change*, *3*(1), 73–77. doi:10.1038/nclimate1628

Baldursson, S., & Karanis, P. (2011). Waterborne transmission of protozoan parasites: Review of worldwide outbreaks - an update 2004-2010. *Water Research*, *45*(20), 6603–6614. doi:10.1016/j.watres.2011.10.013 PMID:22048017

Beer, K. D., Gargano, J. W., Roberts, V. A., Hill, V. R., Garrison, L. E., Kutty, P. K., & Yoder, J. S. et al. (2015). Surveillance for Waterborne Disease Outbreaks Associated with Drinking Water - United States, 2011-2012. *MMWR. Morbidity and Mortality Weekly Report*, *64*(31), 842–848. doi:10.15585/mmwr.mm6431a2 PMID:26270059

Bouzid, M. (2014). Detection of Food- and Waterborne Parasites: Conventional Methods and Recent Developments. In C. A. Batt & M. L. Tortorello (Eds.), *Encyclopedia of Food Microbiology*. Elsevier Ltd. doi:10.1016/B978-0-12-384730-0.00355-4

Bouzid, M., Colon-Gonzalez, F. J., Lung, T., Lake, I. R., & Hunter, P. R. (2014). Climate change and the emergence of vector-borne diseases in Europe: Case study of dengue fever. *BMC Public Health*, *14*(1), 781. doi:10.1186/1471-2458-14-781 PMID:25149418

Bouzid, M., Hunter, P. R., Chalmers, R. M., & Tyler, K. M. (2013). Cryptosporidium pathogenicity and virulence. *Clinical Microbiology Reviews*, *26*(1), 115–134. doi:10.1128/CMR.00076-12 PMID:23297262

Cabral, J. P. (2010). Water microbiology. Bacterial pathogens and water. *International Journal of Environmental Research and Public Health*, *7*(10), 3657–3703. doi:10.3390/ijerph7103657 PMID:21139855

Cann, K. F., Thomas, D. R., Salmon, R. L., Wyn-Jones, A. P., & Kay, D. (2013). Extreme water-related weather events and waterborne disease. *Epidemiology and Infection*, *141*(4), 671–686. doi:10.1017/S0950268812001653 PMID:22877498

Chalmers, R. M. (2012). Waterborne outbreaks of cryptosporidiosis. *Annali dell'Istituto Superiore di Sanita*, *48*(4), 429–446. doi:10.4415/ANN_12_04_10 PMID:23247139

Confalonieri, U. E. C., Menezes, J. A., & de Souza, C. M. (2015). Climate change and adaptation of the health sector: The case of infectious diseases. *Virulence*, *6*(6), 550–553. doi:10.1080/21505594.2015.1023985 PMID:26177788

Curriero, F. C., Patz, J. A., Rose, J. B., & Lele, S. (2001). The association between extreme precipitation and waterborne disease outbreaks in the United States, 1948-1994. *American Journal of Public Health*, *91*(8), 1194–1199. doi:10.2105/AJPH.91.8.1194 PMID:11499103

Delpla, I., Jung, A. V., Baures, E., Clement, M., & Thomas, O. (2009). Impacts of climate change on surface water quality in relation to drinking water production. *Environment International*, *35*(8), 1225–1233. doi:10.1016/j.envint.2009.07.001 PMID:19640587

European Parliament. (2006). Directive 2006/7/EC of the European Parliament and of the Council of 15 February 2006 concerning the management of bathing water quality and repealing Directive 76/160/EEC. *Official Journal of the European Union, 2006,*64.

Funari, E., Manganelli, M., & Sinisi, L. (2012). Impact of climate change on waterborne diseases. *Annali dell'Istituto Superiore di Sanita*, *48*(4), 473–487. doi:10.4415/ANN_12_04_13 PMID:23247142

Gage, K. L., Burkot, T. R., Eisen, R. J., & Hayes, E. B. (2008). Climate and vectorborne diseases. *American Journal of Preventive Medicine*, *35*(5), 436–450. doi:10.1016/j.amepre.2008.08.030 PMID:18929970

Gall, A. M., Marinas, B. J., Lu, Y., & Shisler, J. L. (2015). Waterborne Viruses: A Barrier to Safe Drinking Water. *PLoS Pathogens*, *11*(6), e1004867. doi:10.1371/journal.ppat.1004867 PMID:26110535

Goldstein, S. T., Juranek, D. D., Ravenholt, O., Hightower, A. W., Martin, D. G., Mesnik, J. L., & Herwaldt, B. L. et al. (1996). Cryptosporidiosis: An outbreak associated with drinking water despite state-of-the-art water treatment. *Annals of Internal Medicine*, *124*(5), 459–468. doi:10.7326/0003-4819-124-5-199603010-00001 PMID:8602703

Gray, N. F. (2014). The Implications of Global Warming and Climate Change on Waterborne Diseases. In S. Percival, M. V. Yates, D. Williams, R. Chalmers, & N. Gray (Eds.), *Microbiology of Waterborne Diseases* (2nd ed., pp. 653–666). London: Academic Press, Elsevier. doi:10.1016/B978-0-12-415846-7.00036-6

Herrador, B. R. G., de Blasio, B. F., MacDonald, E., Nichols, G., Sudre, B., Vold, L., & Nygard, K. et al. (2015). Analytical studies assessing the association between extreme precipitation or temperature and drinking water-related waterborne infections: A review. *Environmental Health,* 2015. doi:10.1186/s12940-015-0014-y

Hofstra, N. (2011). Quantifying the impact of climate change on enteric waterborne pathogen concentrations in surface water. *Current Opinion in Environmental Sustainability*, *3*(6), 471–479. doi:10.1016/j.cosust.2011.10.006

Hunter, P. R. (1998). Cyanobacterial toxins and human health. *Symposium Series (Society for Applied Microbiology)*, *27*(S1), 35S–40S. doi:10.1046/j.1365-2672.1998.0840s135S.x PMID:9750360

Hunter, P. R., Waite, M., & Ronchi, E. (2003). *Drinking water and infectious disease establishing the links*. London: CRC Press IWA Publishing.

IPCC. (2015). Synthesis Report. Contribution of Working Groups I, II and III to the Fifth Assessment Report of the Intergovernmental Panel on Climate Change. Geneva, Switzerland.*Climatic Change*, 2014.

Kim, K. H., Kabir, E., & Ara Jahan, S. (2014). A review of the consequences of global climate change on human health. *J Environ Sci Health C Environ Carcinog Ecotoxicol Rev*, *32*(3), 299–318. doi:10.1080/10590501.2014.941279 PMID:25226222

Kothavade, R. J. (2012). Potential molecular tools for assessing the public health risk associated with waterborne Cryptosporidium oocysts. *Journal of Medical Microbiology*, *61*(Pt 8), 1039–1051. doi:10.1099/jmm.0.043158-0 PMID:22628454

Kotloff, K. L., Nataro, J. P., Blackwelder, W. C., Nasrin, D., Farag, T. H., Panchalingam, S., & Levine, M. M. et al. (2013). Burden and aetiology of diarrhoeal disease in infants and young children in developing countries (the Global Enteric Multicenter Study, GEMS): A prospective, case-control study. *Lancet*, *382*(9888), 209–222. doi:10.1016/S0140-6736(13)60844-2 PMID:23680352

Lafferty, K. D. (2009). The ecology of climate change and infectious diseases. *Ecology*, *90*(4), 888–900. doi:10.1890/08-0079.1 PMID:19449681

Leclerc, H., Schwartzbrod, L., & Dei-Cas, E. (2002). Microbial agents associated with waterborne diseases. *Critical Reviews in Microbiology*, *28*(4), 371–409. doi:10.1080/1040-840291046768 PMID:12546197

Lipp, E. K., Huq, A., & Colwell, R. R. (2002). Effects of global climate on infectious disease: the cholera model. *Clin Microbiol Rev, 15*(4), 757-+. doi: 10.1128/Cmr.15.4.757-770.2002

Mackenzie, W. R., Hoxie, N. J., Proctor, M. E., Gradus, M. S., Blair, K. A., Peterson, D. E., & Davis, J. P. et al. (1994). A Massive Outbreak in Milwaukee of Cryptosporidium Infection Transmitted through the Public Water-Supply. *The New England Journal of Medicine*, *331*(3), 161–167. doi:10.1056/NEJM199407213310304 PMID:7818640

Manganelli, M., Scardala, S., Stefanelli, M., Palazzo, F., Funari, E., Vichi, S., & Testai, E. et al. (2012). Emerging health issues of cyanobacterial blooms. *Annali dell'Istituto Superiore di Sanita*, *48*(4), 415–428. doi:10.4415/ANN_12_04_09 PMID:23247138

McMichael, A. J., & Lindgren, E. (2011). Climate change: Present and future risks to health, and necessary responses. *Journal of Internal Medicine*, *270*(5), 401–413. doi:10.1111/j.1365-2796.2011.02415.x PMID:21682780

McMichael, A. J., Woodruff, R. E., & Hales, S. (2006). Climate change and human health: Present and future risks. *Lancet*, *367*(9513), 859–869. doi:10.1016/S0140-6736(06)68079-3 PMID:16530580

McMichael, T., Montgomery, H., & Costello, A. (2012). Health risks, present and future, from global climate change. *BMJ (Clinical Research Ed.)*, *344*, e1359. doi:10.1136/bmj.e1359 PMID:22431661

Moors, E., Singh, T., Siderius, C., Balakrishnan, S., & Mishra, A. (2013). Climate change and waterborne diarrhoea in northern India: Impacts and adaptation strategies. *The Science of the Total Environment*, *468-469*(Suppl.), S139–S151. doi:10.1016/j.scitotenv.2013.07.021 PMID:23972324

Naumova, E. N., Jagai, J. S., Matyas, B., DeMaria, A. Jr, MacNeill, I. B., & Griffiths, J. K. (2007). Seasonality in six enterically transmitted diseases and ambient temperature. *Epidemiology and Infection*, *135*(2), 281–292. doi:10.1017/S0950268806006698 PMID:17291363

Nichols, G., Lane, C., Asgari, N., Verlander, N. Q., & Charlett, A. (2009). Rainfall and outbreaks of drinking water related disease and in England and Wales. *Journal of Water and Health*, *7*(1), 1–8. doi:10.2166/wh.2009.143 PMID:18957770

No authors listed. (1890). The Relation of Ground Water to Disease. *Science,* 16(409), 312-313. doi: 10.1126/science.ns-16.409.312

(1819). No authors listed. (1895). Shellfish and Waterborne Disease. *Br Med J*, *2*, 1175–1176.

No authors listed. (2008). Global networks for surveillance of rotavirus gastroenteritis, 2001-2008. *Wkly Epidemiol Rec,* 83(47), 421-425.

O'Donoghue, P. J. (1995). Cryptosporidium and cryptosporidiosis in man and animals. *International Journal for Parasitology*, *25*(2), 139–195. doi:10.1016/0020-7519(94)E0059-V PMID:7622324

O'Neil, J. M., Davis, T. W., Burford, M. A., & Gobler, C. J. (2012). The rise of harmful cyanobacteria blooms: The potential roles of eutrophication and climate change. *Harmful Algae*, *14*, 313–334. doi:10.1016/j.hal.2011.10.027

OECD/WHO. (2003). *Assessing Microbial Safety of Drinking Water*. OECD Publishing.

Onda, K., LoBuglio, J., & Bartram, J. (2012). Global access to safe water: Accounting for water quality and the resulting impact on MDG progress. *International Journal of Environmental Research and Public Health*, *9*(3), 880–894. doi:10.3390/ijerph9030880 PMID:22690170

Patz, J. A., Grabow, M. L., & Limaye, V. S. (2014). When It Rains, It Pours: Future Climate Extremes and Health. *Annals of Global Health*, *80*(4), 332–344. doi:10.1016/j.aogh.2014.09.007 PMID:25459335

Patz, J. A., Vavrus, S. J., Uejio, C. K., & McLellan, S. L. (2008). Climate change and waterborne disease risk in the Great Lakes region of the U.S. *American Journal of Preventive Medicine*, *35*(5), 451–458. doi:10.1016/j.amepre.2008.08.026 PMID:18929971

Peperzak, L. (2005). Future increase in harmful algal blooms in the North Sea due to climate change. *Water Science and Technology*, *51*(5), 31–36. PMID:15918356

Robson, B. J., & Hamilton, D. P. (2003). Summer flow event induces a cyanobacterial bloom in a seasonal Western Australian estuary. *Marine & Freshwater Research*, *54*(2), 139–151. doi:10.1071/MF02090

Rose, J. B., Epstein, P. R., Lipp, E. K., Sherman, B. H., Bernard, S. M., & Patz, J. A. (2001). Climate variability and change in the United States: Potential impacts on water- and foodborne diseases caused by microbiologic agents. *Environmental Health Perspectives*, *109*, 211–221. doi:10.2307/3435011 PMID:11359688

Schuster, C. J., Ellis, A. G., Robertson, W. J., Charron, D. F., Aramini, J. J., Marshall, B. J., & Medeiros, D. T. (2005). Infectious disease outbreaks related to drinking water in Canada, 1974-2001. *Canadian Journal of Public Health*, *96*(4), 254–258. PMID:16625790

Semenza, J. C., & Menne, B. (2009). Climate change and infectious diseases in Europe. *The Lancet Infectious Diseases*, *9*(6), 365–375. doi:10.1016/S1473-3099(09)70104-5 PMID:19467476

Smith, A., Reacher, M., Smerdon, W., Adak, G. K., Nichols, G., & Chalmers, R. M. (2006). Outbreaks of waterborne infectious intestinal disease in England and Wales, 1992-2003. *Epidemiology and Infection*, *134*(6), 1141–1149. doi:10.1017/S0950268806006406 PMID:16690002

Snow, J. (1855). *On the mode of communication of cholera*. London: John Churchill.

Stanke, C., Kerac, M., Prudhomme, C., Medlock, J., & Murray, V. (2013). Health effects of drought: A systematic review of the evidence. *PLoS Currents*, *5*. doi:10.1371/currents.dis.7a2cee9e980f91ad7697b570bcc4b004 PMID:23787891

Sterk, A., Schijven, J., de Nijs, T., & Husman, A. M. D. (2013). Direct and Indirect Effects of Climate Change on the Risk of Infection by Water-Transmitted Pathogens. *Environmental Science & Technology*, *47*(22), 12648–12660. doi:10.1021/es403549s PMID:24125400

UNFCC. (1992). United nation framework convention on climate change. Retrieved from: http://unfccc.int/resource/docs/convkp/conveng.pdf

UNICEF. (2012). Pneumonia and diarrhoea: Tackling the deadliest diseases for the world's poorest children. Retrieved from: www.unicef.org/eapro/Pneumonia_and_Diarrhoea_Report_2012.pdf

UNICEF and World Health Organization. (2015). Progress on sanitation and drinking water. 2015 update and MDG assessment. USA.

World Health Organisation. (2011). Guidelines for drinking-water quality, fourth edition. Retrieved from: http://www.who.int/water_sanitation_health/publications/dwq_guidelines/en/

World Health Organization. (2013). Diarrhoeal disease. Retrieved from: http://www.who.int/mediacentre/factsheets/fs330/en/

World Health Organization and UNICEF. (2014). Progress on drinking water and sanitation. Joint Monitoring Programme update 2014.

Wu, X., Lu, Y., Zhou, S., Chen, L., & Xu, B. (2016). Impact of climate change on human infectious diseases: Empirical evidence and human adaptation. *Environment International*, *86*, 14–23. doi:10.1016/j.envint.2015.09.007 PMID:26479830

Chapter 5
Environmental Change and the Emergence of Infectious Diseases:
A Regional Perspective from South America

Ulisses Confalonieri
René Rachou Research Center - Oswaldo Cruz Foundation, Brazil

Júlia Alves Menezes
René Rachou Research Center – Oswaldo Cruz Foundation, Brazil

Carina Margonari
René Rachou Research Center - Oswaldo Cruz Foundation, Brazil

DOI: 10.4018/978-1-5225-0553-2.ch005

ABSTRACT

In South America in the past decades several infectious diseases have emerged or re-emerged either as part of larger pandemics or as local processes involving autochthonous pathogens. These included arthropod-borne viral diseases, such as Dengue Fever, Chikungunya and Zika as well as viral hemorrhagic fevers, such as Hantavirus Pulmonary Syndrome, Junin, Machupo and Guanarito viruses. Parasitic disease was also important such as Malaria, endemic in the northern part of the continent, Leishmaniasis and Chagas Disease. Carrion disease, a bacterial infection originally from the Andes region, also seems to be expanding geographically. Several social and environmental processes have contributed to the emergence of these pathogens, including human migration, deforestation, road and dam building and climate shifts. Due to its high biological diversity of wildlife, arthropods and virus species in still untouched natural ecosystems in the Amazon has the greatest regional potential for the emergence of new human infections.

INTRODUCTION

South America is at the Southern part of the America Continent (10ºN; 55ºS) spanning an area of 17,819,000 million km^2, which includes 12 countries and an independent territory (French Guyana). The regional climate is predominantly hot and humid, but subtropical climate is found in mountain regions (e.g. The Andes) and temperate and polar climates occur in the southern tip of the continent, in Chile and Argentina (Canziani & Dias, 1998).

There is a high diversity of ecological areas and ecosystems such as the desert areas in northern Chile; the large plains in Venezuela and Colombia and the vast Amazon forest, which is shared by eight countries. The region has the largest freshwater system in the world due to the network formed by the Prata, Orinoco and Amazon River basins (Comisión Económica para América Latina y el Caribe [CEPAL], 2014).

The regional population is about 410 million people with about 83% living in cities. The regional life expectancy reached 73 years in 2010 (Teixeira, Paixão, & Costa, 2013). According to the Pan-American Health Organization [PAHO] (2014), safe water access reaches 89% of people in the Andes. However, this is not consistent across the region. Bolivia has sewage collection and treatment covering only 46% of households. Gross income per capita ranged from US$12,470 for Chile and Argentina to US$7,784 in Andean countries. Between 1990 and 2010, regional infant mortality rates dropped by 50%; for Brazil it was 14,6/1000 and for Chile 7,8/1000, in 2012 (PAHO, 2014; Teixeira et al., 2013).

Several endemic infectious diseases occur in South America, some with high annual incidence such as Malaria and Dengue Fever; in 2013, Brazil alone reported a total of 1,470,487 cases of Dengue Fever (PAHO, 2014). Several other endemic infections are autochthonous to the region such as Carrion Disease in the Andes and different forms of hemorrhagic infections caused by Arenavirus, such as Junin, Machupo and Guanarito hemorrhagic fever.

South America is considered vulnerable to climatic conditions, especially when combined with land use and land cover changes (Caqui, Quispe, & Zegarra, 2013; Magrin et al.; 2014). The aim of this chapter is to show the impact of environmental change and human activities on the dynamics of some infectious diseases of major importance to public health in South America.

BACKGROUND

The emergence of infectious diseases was identified as a major global health threat in the last quarter of the 20th century. Zoonotic pathogens from wild animals formed the majority of newly emerged human pathogens in the past few decades (Jones et al., 2008). Several factors have been pointed as major contributors to human infectious disease emergence: new strains of the pathogens, increased resistance to antibiotics, reduced resistance of hosts (e.g. infection by HIV), variations in human populations densities, shifts in diversity of populations of vectors and hosts, hunting of wild animals for food, deforestation and loss of biodiversity and climatic anomalies (Jones et al. 2008; Keesing et al, 2010; Pongsiri et al, 2009; Wilcox & Gubler 2005).

In general, it is acknowledged that, for disease emergence to take place, an association of different drivers is necessary. In South America, since the middle of the last century, several diseases have emerged caused by different etiological agents, from Protozoa to viral infections. Some of these were part of larger epidemics/pandemics (e.g. Cholera; Dengue Fever) but other diseases have emerged as autochthonous local processes and became endemic, such as viral haemorrhagic fevers caused by Junin and Guanarito Viruses. While some of those that have emerged or re-emerged became widely distributed (e.g. Dengue Fever), others occurred as local outbreaks with a small number of cases, restricted in time, and no more cases were reported (e.g. Sabiá Virus infections).

Field observations by scientists early in the 20th century in some regions of South America (associated with the expansion of economic activities) reported changes in the transmission of human tropical diseases as a consequence of human-induced changes to the natural environment. A classic example was the observations by Brumpt and Pedroso (1913) of outbreaks of Cutaneous Leishmaniasis in the state of São Paulo, Brazil, following the opening of a new railway to the centre of the country. The association between the natural environment and the - then poorly known- parasitic infection can be illustrated by the naming of the disease by these authors: "American Forest Leishmaniasis". These early authors have also observed that Cutaneous Leishmaniasis was more prevalent during the hottest period of the year, from November to April.

Another example of early observations of the linkage between tropical diseases and environmental factors in South America was given by Carlos Chagas, the discoverer of the American Trypanossomiasis. Also in the context of opening up of a new railway in Central Brazil, in 1905, he observed that the incidence of malaria among railroad workers was high during the summer and dropped to almost zero during the cold and dry winter (Chagas, 1907). Since these early observations, about a century ago, the environment in the South America region has undergone dramatic changes due to human encroachment on the natural ecosystems. Road buildings, deforestation for the expansion of agriculture, urban growth, and the building of dams have been the main drivers of environmental changes.

In the following section, representative groups of infectious diseases endemic to South America will be discussed in aspects of dynamic epidemiological shifts in relation to continuous changes in the physical and biological environments.

ENVIRONMENTAL CHANGES AND THEIR LINKAGES TO RELEVANT INFECTIOUS DISEASES IN SOUTH AMERICA

1. Malaria

Malaria is an endemic parasitic disease in the northern part of the South American continent, from where it extends up to the southern parts of Mexico. The disease occurs under two main forms in its various localities and sub-regions: stable transmission of endemic character and sporadic transmission of epidemic type (Cabral, Fé, Suarez-Mutis, Boia, & Carvalho-Costa, 2010).

Among the environmental changes usually associated with Malaria dynamics in South America, two process are considered most important: forest conversion due to deforestation and climatic variability in different scales such as seasonal or interannual, El Niño phenomenon, for example. El Niño corresponds to the warming phase of El Niño–Southern Oscillation (ENSO), an irregular and periodical variation in winds and sea surface temperatures over the tropical eastern Pacific Ocean. The cooling phase is known as La Niña. There has been some debate as to the relative importance of El Niño phenomenon in shaping the dynamics of this disease in specific areas of the Amazon. Several authors consider that changes in transmission patterns are more dependent on local aspects, such as failure of preventive measures or human population migration, than on large scale climatic phenomena (Cabral et al., 2010; Conn et al., 2002). However, in Colombia and Venezuela, local studies were able to demonstrate a significant influence of the El Niño on the transmission of Malaria (Delgado-Petrocelli et al., 2012; Mantilla, Oliveros, & Barnston, 2009).

In Peru, the larvae of the main Malaria vector in the Amazon region, *Anopheles darlingi*, was shown to be associated with deforestation. Sites with *A. darlingi* larvae had an average of 24.1% of forest cover, compared with 41% for sites without *A. darlingi* (Vittor et. al., 2009). In the eastern part of the Brazilian Amazon region, culicid surveys performed in a pristine national forest revealed no species of Malaria vectors, among twenty-five thousand specimens collected (Confalonieri et al., 2013; Confalonieri & Costa-Neto, 2012).

Using a geospatial model to map Malaria hotspots in the Brazilian Amazon, Valle and Lima (2014) observed that, on a large scale, forest cover and the proximity to gold mining operations were important drivers of disease risk. They have also found that areas with a longer dry season and areas with higher average rural income tended to have highest Malaria risk. The former is related to the creation of water bodies suitable to *Anopheles* sp. breeding when the rivers recede in the dry season and the latter, may be due to greater access to health services available to the high income population, which potentially increases the reporting rate. In Western Brazilian Amazon, da Silva-Nunes et al. (2008) observed, in a frontier settlement, that Malaria morbidity was strongly associated with land clearing and farming and decreased after five years of residence in these areas due to the development of immunity. A spatial clustering of Malaria was observed in the area of most recent occupation, indicating that the influx of non-immune settlers to forest fringe areas perpetuates the cycle of environmental change that favors Malaria transmission. It was also shown that the place of residence – distance of the dwelling from the forest - was important in determining risk. Using a negative binomial model of Malaria for municipalities in the Brazilian Amazon – the model associated the number of cases of Malaria to covariates as sociodemographic characteristics of the population and land cover – researchers were able to show the influence of logging, road building and forest fires on the incidence of Malaria. The overall incidence rate of Malaria was higher in the states were logging is an important activity. The influence of unpaved roads was higher in the municipalities with rates of deforestation higher than 75% (Hahn, Gangnon, Barcellos, Asner, & Patz, 2014).

2. Chagas Disease

Chagas disease, discovered in 1909 in Brazil, has been extensively studied and the knowledge about the eco-epidemiology of this disease is vast (Teixeira et al., 2001). The adaptation of the different species of disease vectors to human dwellings and man=made environments is well known. In the Amazon region, different mechanisms play a role in the increased risk of Chagas disease infection: deforestation (causing the displacement of wild animal reservoirs of the infection), adaptation of wild species of triatomine vectors to man-made structures, and the migration of susceptible hosts, both humans and domestic animals (Coura & Junqueira, 2012).

Recent studies in Panamá (Central America) have shown that anthropogenic landscape disturbance – forest fragments and peridomiciliary sites - increased vector infection with the *Trypanosoma cruzi*, the etiological agent of Chagas disease. Vector infection rates were significantly greater in deforested habitats as compared to contiguous forests. This was due to a combination of factors: although the modified environments have a low diversity of vector species they have a high abundance of mammal reservoirs of the infection and suitable conditions for the reproduction of the insect vectors and, therefore, of maintenance of the cycle of the parasite (Gottdenker, Calzada, Saldaña, & Carrol, 2011; Gottdenker, Chaves, Calzada, Saldaña, & Carrol, 2012).

Several decades ago, reports of the influence of temperature on the biology of a major Chagas disease vector – *Triatoma infestans* – found that for the same period of time, two generations of this species occurred in areas with high temperatures and only one generation in the cooler regions (Hack, 1955). Costa, Dornak, Almeida, and Peterson (2014) have modeled the distribution of *T. brasiliensis* complex of vector species according to projected temperatures and precipitation under conditions of climate change in Brazil, for the years 2020 and 2050. *Triatoma brasiliensis* is an important intra-domiciliary and peri-domiciliary transmitter of *T cruzi* in the semi-arid region of Brazil (Coura, 2015). Under scenarios of temperature increase of up to 1,72°C and a precipitation decrease of 55,6mm, the overall distribution of the species was projected to remain stable. However, the model was able to indicate a great potential for the colonization of new areas by this vector species, a fact important for the planning of epidemiological surveillance.

In Venezuela and Argentina, the study of the future distribution of the species *Rhodnius prolixus* and *T. infestans* using the ecological niche approach, has indicated a possible reduction of the area of distribution of these species under a climatic scenario for 2050 (for future climate conditions, two scenarios were considered: the representative concentration pathways (RCPs) 4.5, 6.0 and 8.5 (hereafter scenarios), which were nested to HadGEM2-ES models and bioclimatic projections for 2050) (Medone, Ceccarelli, Parham, Figuera, & Rabinovich, 2015). Cordovez, Rendon, Gonzales, and Guhl 2014 obtained similar results in Colombia. These authors have concluded that, under scenarios of increased temperature for 2035, as a result of global climate change, a reduction of the area of transmission of Chagas disease currently known in Colombia would be observed.

3. Leishmaniasis

Leishmaniasis, specially the cutaneous forms of the disease, transmitted by vectors of the genus *Lutzomyia* in South America, are known to respond to different types of environmental changes. One aspect frequently reported is the impact of agricultural encroachment in natural forests on the formation of ecological niches for the sandfly vectors of the disease.

In Brazil, coffee plantations using traditional cultivation systems were reported to facilitate the transmission of cutaneous Leishmaniasis. This occurs because coffee trees provide shadow and the decaying leaves of these trees in the soil create adequate niches for the reproduction of sandfly vectors (Alexander et al., 2001, 2009). Other types of plantations such as sugar cane, bananas and cocoa were also reported to increase the risk of Leishmaniasis transmission (Azevedo et al., 1996; Azevedo & Rangel, 1991; da Silva et al., 1999). De Araújo Pedrosa and Alencar Ximenes (2009) and Membrive et al. (2012) reported the high risk of transmission to humans in banana plantations in Brazil due to the proximity of banana trees to the dwellings of peasants.

In French Guyana, Fouque et al. (2007) have reported an expansion of Leishmaniasis transmission into areas of environmental modifications linked to gold mining, agriculture and military training. Also in the Guyana region, an increased rate of cutaneous Leishmanasis transmission was associated with deforestation (Rotureau, 2006). Although in Colombia both the population density and the number of species of sandfly vector were higher in forested areas when compared to degraded areas, in the latter the risk of Leishmaniasis transmission to human and domestic animals was higher due to the adaptation of the most important vector species to the anthropic environment.

In Brazil, shifts in the epidemiology of cutaneous Leishmaniasis were associated with building of dams (Vilela, Azevedo, Carvalho, & Rangel, 2011) and also roads and railways (Gonçalves Neto et al., 2013). Factors influencing incidence increases and the spread of *Leishmania* spp. range from habitat changes for vectors (as they frequently adapt to man-made buildings) to human population displacements associated with infrastructure projects

Climatic variations have also been associated with epidemiological changes of Leishmaniasis in the Americas (Chaves & Pascual, 2006). In Colombia, associations of El Niño and La Niña climatic phenomena and the incidence of Leishmaniasis in endemic regions of the country showed variable linkages: in some areas El Niño has had a positive correlation with cases of the disease while in others, the cases were influenced by La Niña (Cárdenas, Sandoval, Rodriguez-Morales, & Vivas, 2008). The authors have stressed the need for further studies and emphasized the role of climatic factors in the dynamics of Leishmaniasis in different areas of Colombia.

4. Bartonellosis

American Bartonellosis, also named Carrión Disease, caused by the bacterium *Bartonella bacilliformis* and transmitted by the sandfly species *Lutzomyia verrucarum*, occurs in the region of the Andes, being endemic in altitudes between 800m and 3000m. Environmental factors were reported to influence the dynamics and the spatial distribution of this disease. Chinga-Alayo, Huarcaya, Nasarre, Del Aguila, and Llanos-Cuentas (2004) have reported a positive association between disease transmission and El Niño climatic phenomenon in endemic areas of Peru. The mechanism involved is the increase in vector populations influenced by higher temperatures. Also in Peru, Zhou et al. (2001) have observed a doubling in the number of new cases of the disease during the months of El Niño, in addition to an expansion of the disease to areas not previously affected. High correlations were observed between higher incidence and increased temperatures three months before the outbreaks (Zhou et al., 2001).

Recent reports of human cases of Carrión Disease outside the well-known endemic areas, such as cases in the forest areas and on the coast, indicate a trend of geographic expansion, probably influenced by both the adaptation of the vectors to new niches and human contact during population movements for commercial activities (Maco, Maguiña, Tirado, Maco, & Vidal, 2004).

5. Viral Diseases

In South America, two major groups of endemic viral diseases are reported to be influenced by climatic shifts and environmental changes: arboviral infections and viral haemorrhagic fevers. Some of these diseases have been established in this continent for several decades such as Yellow Fever and, more recently, Dengue Fever. Others seem to be autochthonous to the region, such as the Mayaro and Oropouche fever in the Amazon (Vasconcelos et al., 2001). In the past few years, new viral introductions were reported, such as the West Nile virus (Morales et al., 2006) and Chikungunya virus, which occurred either as isolated cases or slowly expanding to become endemic (Figueiredo & Figueiredo, 2014; Carbajo & Vezzani, 2015).

The emergence of the Hantavirus Pulmonary Syndrome virus in the Americas has been associated with human disturbances of the natural systems, affecting the composition of biodiversity. This is also true for infections caused by arenaviruses, which typically occur in South America: Junin, Machupo and Guanarito (Simpson, 1978; Mills, 2006). These infections have different species of wild rodent reservoirs and those considered to be opportunistic – more adaptable to peri-domiciliary environments and presenting high fecundity and shorter life cycles – tend to predominate in the anthropic environment (Mills, 2006). Most outbreaks have been reported in anthropic environments where these species predominate, in a context of reduced diversity of potential host species (Mills, 2006). In the years 1940-1960, the emergence of rodent-borne arenavirus infections causing hemorrhagic fevers in Argentina and Bolivia, have followed a similar mechanism. Agricultural encroachment to natural pastures used for livestock production have caused a change in the local dominant species of rodents: *Calomys musculinus* in the maize crops of Argentina and *Calomys callosus* in the subsistence crops near small forested areas in Bolivia. These two species had their populations numbers greatly increased as well as adapted to the human dwellings where they have shed the virus in the urine (Johnson, 1993). Consequently, outbreaks of Machupo's disease occurred in the northern part of Bolivia during 1960 and Junin viruses have been clinically described since 1953 in Argentina.

The population dynamics of the reservoir rodents also play a role in outbreaks and in the case of the Junin virus in Argentina (Simpson, 1978), it was observed that the clustering of human cases during the harvest of maize was linked to a peak in rodent abundance. In a study of wild reservoirs of hantavirus in Paraguay, a higher rate of infection of rodent species in anthropic landscapes (e.g. agriculture) was found. When compared with areas of low infection, this high infection rate seems to be linked to habitat fragmentation, which changes the mobility of the rodents and facilitates virus transmission due to increased encounters between different rodent species, capable of increasing the probability of virus transfer from an infected animal to an uninfected member of a host population (Goodin, et. al., 2006). The role of climatic change in modulating Hantavirus infection was modeled in Argentina in regard to spatial distribution of the main reservoir rodent species. Different results were obtained according to different climatic scenarios used and for different parts of the country, but the overall distribution of the host was projected to be reduced due to changes in precipitation and temperature (Carbajo, Vera, & Gonzalez, 2009).

6. Arboviruses

The most important arboviral diseases in South America are Dengue Fever and Yellow Fever, because of the high incidence of Dengue, and the high fatality rate of Yellow Fever (Cunha & Nogueira, 2005; Tauil, Santos & Morais, 2005). Both are influenced by climatic variability and change, and Yellow Fever, which is a focal disease rooted in natural ecosystems, is also influenced by changes in land use and land cover (e.g. deforestation) (Monath & Vasconcelos, 2015; Bryant et. al., 2003; Vasconcelos et. al., 2001). Several authors have reported the influence of climate, to variable degrees, on the epidemiology of Dengue Fever in tropical America (Chowell, Cazelles, Broutin, & Munayco, 2011; Herrera-Martinez & Rodriguez-Morales, 2010; Ibarra et al., 2013; Thai & Anders, 2011). Some of these are the higher Dengue incidence during El Niño periods in Venezuela, the rainfall and minimum temperature as a predictor of *Aedes aegypti* oviposition activity in Ecuador and the influence of seasonal temperature in Dengue epidemics in Peru. Reports were also made on the influence of specific types of natural ecosystems on the production of outbreaks of Yellow Fever (Degallier et. al., 1992) and of the micrometeorology of forest on the population dynamics of mosquito's vectors of this disease (Pinto, Confalonieri, & Mascarenhas, 2009).

In regard to Dengue Fever, models have pointed out that, although climate has a role in Dengue increase in the "Southern Cone" (Argentina), other geographic factors (e.g. distance to water bodies) and demographic variables have a stronger influence on the determination of disease incidence (Carbajo, Cardo, & Vezzani, 2012). The most critical factor in the emergence/re-emergence of Dengue fever in South America is considered to be the widespread adaptation of its main vector, the mosquito *Aedes aegypti*, to artificial containers in urban settings, a fact observed in most countries of the region. This vector could also pose a major threat for the transmission of urban yellow fever in major cities of the region, a situation reported for the last time more than half a century ago (Downs, 1981). In Brazil, for example, the disease has caused major urban epidemics from the colonial period to the first decades of the 20th century when it was recorded the last urban outbreak in Rio de Janeiro, 1929; sporadic cases appeared in Acre state, in 1942 (Ferreira, Rocha, Caputto, Fonseca, & Fonseca, 2011). Since then, the successful control of the urban form of Yellow Fever, both in Brazil and in the Americas, was made possible for two reasons: the eradication of *A.aegypti* vector and the widespread use of 17D vaccine. The former occurred through the Rockefeller Foundation and the Pan American/ World Health Organization initiative which, between the years 1930 and 1950, have implemented successful programs of eradication of the vector in the Americas (Braga & Valle, 2007). The vaccine appeared in 1937, also through the Rockefeller Foundation with the involvement of Oswaldo Cruz Institute (Ferreira, Rocha, Caputto, Fonseca, & Fonseca, 2011).

Two other frequently reported arboviral infections in the region are the Oropouche and the Mayaro fever viruses. Oropouche virus infections have been epidemic in the Amazon – Brazil, Peru, Colombia – and Panama and Trinidad. Its sylvatic cycle of transmission is poorly understood and involve multiple hosts and vectors, such as sloths, monkeys and birds (Pinheiro, Travassos da Rosa, & Vasconcelos, 1998). Still, epidemics of Oropouche as an anthroponosis are well characterized, with the involvement of a single species of vector – *Culicoides paraensis* (Carpenter, Groschup, Garros, Felippe-Bauer, & Pruse, 2013). Evidence points to an adaptation of the viruses to the peri-urban habitats since their midge vector species breed in holes thumbs and roots of planted trees, and they can bite inside houses. Outbreaks are associated with periods of high rainfall intensity and can reach thousands of cases in short periods of time in the Brazilian Amazon (Pinheiro et al., 1998). In the case of Mayaro fever, outbreaks were also reported from Amazonian countries, especially Brazil, Colombia, Venezuela and Bolivia. It is considered a forest disease similar to Yellow Fever, having the same mosquito species as vector – *Haemagogus janthinomys* - as well as primates as wild reservoirs, but cases transmitted in rural areas have been also reported (Travassos da Rosa et al., 1998). In the Amazon region, the main driver for Mayaro virus occurrence is deforestation, causing limited outbreaks (Vasconcelos et al. 2001). Although few studies have addressed the role of climatic variables in Oropouche and Mayaro fever epidemiology, Vasconcelos et al. (2001) observed that increased precipitation, especially in the first half of the year for the Amazon, seems to favor the reproduction of the vectors of both diseases. In fact, other environmental factors, such as temperature and humidity, play an important role influencing mosquito competence for arboviruses and its population abundance (Aybar, Juri, Santana, Grosso, & Spinelli, 2012; Tabachnick, 2013). Given the projections of temperature rise and reduced precipitation in the Amazon region, the distribution of cases of these two diseases can reduce or increase depending on the site studied.

In regard to other less important arboviral infections in South America, the detection of West Nile virus infections from horses in Argentina was probably associated with migratory birds (Morales et al., 2006). The same mechanism was proposed for a newly emerged infection in forests of Southern Brazil, the Rocio Virus, which was detected by immunological tests in sylvatic birds captured in São Paulo state, Brazil (Lopes, Coimbra, Sacchetta, & Calisher, 1978; Ferreira et al., 1994). The Rocio virus appeared in Ribeira Valley, São Paulo state, Brazil, causing encephalitis epidemics in 1975-1976 – around 1000 cases and 10% fatality rate (Monath, 1993). A newly emerged arboviral infection in South America is Chikungunya virus, an infection also transmitted by *Aedes* mosquitoes, mainly *A. aegypti*. It is believed to have originated in Africa, where it still circulates enzootically among nonhuman primates, whereas in the Americas it was reported for the first time in 2013 (Vega-Rúa, Zouache, Girod, Failloux, & Lourenço-de-Oliveira, 2014). As of the year 2015, the South American countries Colombia, Venezuela, Brazil, Ecuador and Paraguay have reported local transmission (Carbajo & Vezzani, 2015). Since it is a disease transmitted by urban vectors, its emergence has been associated to urbanization and human migration (Figueiredo & Figueiredo, 2014). Regarding West Nile Virus, the influence of climate on its occurrence is well stablished in temperate countries, where warm winters and spring droughts are pointed out as capable of amplifying the transmission (Epstein, 2001; Paz, 2006). For Chikungunya, as for other diseases transmitted by the *Aedes* genus of mosquitoes, models projecting the distribution of *A. aegypti* vector show a potential for its advance toward South America centre, colonizing areas where the vector is absent (Campbell et al., 2015).

7. Cholera

Cholera, a gastrointestinal infection caused by the bacterium *Vibrio cholera*, has re-emerged in Peru, South America, in 1990. Although strongly influenced by socioeconomic conditions, due to its marked seasonal occurrence in endemic regions, former studies have pointed out to the influence of climatic variability in the dynamics of this disease, especially in Asia (Pascual, Bouma, & Dobson, 2002). In Peru, its emergence has been associated with the 1991-1992 El Niño climatic phenomenon (Pascual et al., 2002). It seems that the bacterium has been present in the environment months before the beginning of the outbreak in Peru and marine ecosystem changes associated with El Niño led to the propagation of *V. cholera* along the coast (Seas et al. 2000).

Recent reassessment of cholera emergence in Peru has indicated that the epidemic may have been linked to ENSO through multiple pathways, including elevated temperatures, rainfall extremes, La Niña phenomenon (cold phase of the ENSO cycle) and social vulnerability (Ramírez, Grady, & Glantz, 2013). A local study in Piura, one of the most affected districts in Peru, showed significant cholera association with sea surface temperatures in the Central Pacific Ocean and on the coast, as well as with rainfall and minimum and maximum temperatures with varying lag times (0-5 months); the strongest consistent impact was with rainfall (Ramírez, 2015).

CLIMATE AND INFECTIOUS DISEASES: ADAPTATION OPTIONS

1. Infectious Disease Control

Regardless of what models project in terms of future infectious diseases profiles, as a consequence of climactic shifts, current actions targeted to the reduction of incidence and geographical expansion of endemic infectious diseases should be considered as part of the adaptation processes. This is particularly critical for those diseases known to be affected by climate variability, such as most vector-borne and water-borne infections (Smith et al., 2014).

These control strategies vary according to disease cycles and the available technologies for their control. The following are some examples of diseases endemic to the South American region:

- **Leptospirosis:** Usually associated to urban floods following tropical storms (Dutra et al., 2015). The control strategies focus on improvement of drainage of rain water as well as on the control of rodent populations (rats), which are the main reservoirs of the pathogen;

- **Dengue Fever:** A viral disease transmitted by the urban vector species *Aedes aegypti*, a mosquito which also transmits the Zika virus and the Chikunguya virus (Ibarra et al., 2013; Musso, Nilles, & Cao-Lormeau, 2014; Roth et al., 2014; Staples & Fischer, 2014). Since there are no vaccines against these viruses, the main strategy for their control depends on the reduction of the vector populations. This can be more effective through the reduction of the breeding sites for the vector: trash in backyards, discarded tires, flowers pots and other man-made structures that accumulate rain water (Tauil, 2002);
- **Malaria:** A vector-borne disease restricted to the natural environment. The main strategy for its control is the early treatment of the infected human hosts. A secondary strategy, at the local level, is the emergency spraying of insecticides to reduce the vector population (World Health Organization [WHO], 2014);
- **Yellow Fever:** A vector-borne disease known to be associated to climate variability and forest micrometeorology (Vasconcelos et al., 2001; Bryant et al., 2003; Pinto et al., 2009). Its control depends on the preventive vaccination of the population at risk in endemic areas (Ferreira, Rocha, Caputto, Fonseca, & Fonseca, 2011).

2. Healthcare, Education, and Adaptation

Future Public Health impacts of climate change on the epidemiology of infectious diseases and on health care systems are well acknowledged (Confalonieri, Menezes & Margonari de Souza, 2015; Costello et al., 2009; Pereira & Barata, 2014; WHO, 2009). However, current health policies and strategies do not address possible changes to the epidemiological profile of the populations to be affected by climatic change. Adaptation strategies should be focused on both populations currently under climatic stress as well as those expected to be affected by future shifts to the global climate system (Ebi, Lindgren, Suk, & Semenza, 2013). The first step towards this end would be strengthening of the health care systems in order to facilitate the development of adequate responses to climatic hazards (Costello et al.; 2009).

The promotion of secure infrastructure, not affected by extreme weather, is essential for the provision of health care services on a continuous basis. Several Latin American countries have built health care installations resilient to disasters such as Peru, Colombia, Equador, Chile and Costa Rica (WHO, 2009). In the urban environment in South America events like storms, floods and landslides are responsible for a high burden of morbidity and mortality. These require comprehensive preventive actions ranging from public health monitoring and disaster preparedness, in the context of community-based risk reduction strategies (Keim, 2008).

In the municipality of Nova Friburgo, Rio de Janeiro, Brazil, heavy rains (182.8 mm/24 hours) have killed 429 persons, mostly due to trauma, in January 2011. Despite the interruption of several basic services – energy, water, transportation – an efficient health surveillance intervention was able to prevent deaths after outbreaks of dengue fever, leptospirosis and waterborne diarrhea (Porto, 2012; Pereira & Barata, 2014). This can be considered as an example of successful intervention – or reactive adaptation – in the context of climate-associated disasters causing infectious disease epidemics.

Educational approaches are also an important part of the adaptation process. Lima and Layrargues (2014), have emphasized the need for a more comprehensive understanding of the impacts of climate, associating the knowledge of climatic hazards and risks to the daily life of the population and also the need for environmental conservation as part of the adaptation strategy.

Since risks associated to climate change are socially and spatially different, a revision of the practices of governance is needed. This entails both individual adaptation strategies that respect the social and economic characteristics of the affected people, especially those more directly dependent on natural resources (Adger, 2001) and an open and participatory decision making processes involving the actors concerned with the impacts, in order to complement the necessary technical knowledge (Jacobi, 2014).] Some local initiatives have demonstrated positive results in regard to public policies targeted to climate change. In the megacity of São Paulo, Brazil, Landin and Giatti (2014) have demonstrated the existence of a positive cross sectoral dialogue in the implementation of policies targeted to climate protection, in the context of a highly complex urban area.

A survey of the population of different countries has found that middle income countries tend to favor technological solutions while developed countries are more skeptical in this regard and people believe the solutions should come from changes in habits and individual behavior (Echegaray & Afonso, 2014).

Of great importance is the need to recognize that adaptation to climate change is also a social process, involving cultural aspects, beyond the economic strategies. Adger, Barnett, Brown, Marshall and O'Brien (2013) have demonstrated that the values adopted by the human groups can influence their interpretation of the problems, their motivation and their ways to respond and adapt to climatic risks; adaptation policies that ignore cultural aspects of the population affected by climate change can fail to increase the resilience of the targeted populations.

DISCUSSION

Infectious disease emergence in South America has been driven by different factors and these depend, in a large extent, on the life cycle of the causative pathogens, their arthropod vectors and animal hosts. Also important are the environmental settings where the infections develop: processes affecting strictly urban disease cycles (e.g. Dengue fever) are quite different from those found in foci historically rooted in pristine natural ecosystems. This is due to the epidemiological complexity of the latter, which can embrace a wider variety of competent hosts and vectors as well as be more influenced by environmental and climatic factors. On the other hand, in urban cycles new human infections usually depend on the massive exposure of the susceptible human population to a high burden of pathogens in the environment or to a high density of insect vectors (e.g. mosquitoes). In the case of vector-borne diseases, emergence is usually associated with environmental shifts, such as changes in temperature or rainfall, that create new ecological niches for both adults and immatures vector species. This applies to man-made environments, such as households and buildings (e.g. in Dengue fever) and also areas with natural vegetation, as in the case of Malaria (Travassos da Rosa, Pinheiro, & Vasconcelos, 2005; Vittor et al, 2009; Carbajo et al., 2012; Hahn et al., 2014).

Another important aspect is the biodiversity richness natural ecosystems. Pathogens circulating in wildlife, independent of the human host, can "spillover" and reach humans, following changes in the animal populations or encroachment of humans into natural systems, or both (Daszak, Cunningham, & Hyatt, 2000). Therefore, areas with rich assemblages of vectors and animal pathogens have a higher potential to support shifts in hosts and emergence of new human infections. In South America, one such region is the Amazon basin, which has the largest extension of tropical forests in the world and harbors a rich biological diversity. To date, more than 180 species of arboviruses were identified in animals and insects of the Amazon and around 30 of them infect humans (Gubler, 2002). A recent review has pointed out the role of changes of the Amazon environment, such as those linked to deforestation, road building and hydroelectric dams, in driving the emergence of parasitic infections in humans (Confalonieri, Margonari, & Quintão, 2014).

The Amazon region is one of the most vulnerable part of South America. Major projects in the area are impacting the natural ecosystems and causing alterations in land cover (e.g. forest conversion to pastures, dams), which present great opportunities for infectious disease emergence (Confalonieri et al., 2014). Encroachment to areas where endemic infections are present is also a cause of concern due to the risk of disease spread. For example, the construction of roads linking the Amazon to the Pacific Coast could facilitate the emergence of Bartonellosis – a disease originally restricted to the Andes Mountains – in lowland areas (Cesario & Cesario, 2005). International travel is thought to be the main mechanism involved in the most recently introduced viral infections in South America - Zika and Chikungunya viruses (Figueiredo & Figueiredo, 2014) - a mechanism already identified in other continents (Cordel, Quatresous, Paquet, & Couturier, 2006; Weaver & Reisen, 2010).

Specifically, in relation to disease emergence in the Amazon, monitoring and surveillance strategies necessary to tackle processes leading to regional disease emergence were discussed recently (Confalonieri et. al, 2014). An interdisciplinary approach comprising control measures and clinical management combined with an integrated international surveillance is needed to manage the environmentally driven changes in the disease dynamics in the Amazon in the near future. One important strategy would be to enhance regional epidemiological surveillance, especially to sub-regions where social-environmental modifications are occurring, as is mentioned in the "Amazon Malaria Initiative" (U.S. Agency for International Development [USAID], PAHO, 2010). The epidemiological impacts of environmental change in other parts of the South American continent could also be surveyed using the same association of interdisciplinary studies (parasitology, epidemiology, ecology, land use, climatology) and epidemiological surveillance in hotspot areas.

Further investigations in specific scientific areas - parasitology, epidemiology, climatology, geoprocessing, etc – should be pursued for a broader assessment of risks of infectious disease emergence in South America. Two topics deserve special attention. First, the role of cross-immunity of the population as a "barrier" to the introduction and persistence of new viruses to the region. Herd immunity for Dengue Fever viruses, endemic in this part of the world, could prevent taxonomically close viral species (e.g. West Nile Fever Virus) from becoming endemic in the human population. Cross immunity between different flaviviruses such as those causing dengue fever and the Zika virus is well known, a fact that makes the diagnosis of past infections with these pathogens using immunological tests more complex. Second, the significance of climate warming in the emergence of pathogens in the tropics. The potential to expand the distribution of vectors and enhance the transmission potential through alterations in the life cycle (e.g. shorter gonotrophic cycles and extrinsic incubation periods) of pathogens and vectors are likely to occur in temperate climates. However, in most of the tropics, suitable temperatures and humidity for endemic disease transmission are readily available and global warming may not be a critical factor in epidemiological shifts in this region.

CONCLUSION

Infectious disease emergence in Latin America, as a consequence of environmental changes, is a complex issue. A high diversity of natural ecosystems and a vast extension of still pristine environments are important sources of microbes that circulate in wild animals and insect vectors and can potentially spill over to the human population due to encroachment of habitats following human activities to expand infrastructure and promote economic growth. Besides changes in land use practices and in land cover, climatic shifts are also expected to facilitate the emergence of pathogens in South America in the next few decades. A high degree of human population density in urban areas in this continent poses an additional risk factor for disease emergence in this region.

REFERENCES

Adger, W. N. (2001). Scales of governance and environmental justice for adaptation and mitigation of climate change. *Journal of International Development*, *13*(7), 921–931. doi:10.1002/jid.833

Adger, W. N., Barnett, J., Brown, K., Marshall, N., & O'Brien, K. (2013). Cultural dimensions of climate change impacts and adaptation. *Nature Climate Change*, *3*(2), 112–117. doi:10.1038/nclimate1666

Alexander, B., Agudelo, L. A., Navarro, F., Ruiz, F., Molina, J., Aguilera, G., & Quiñones, M. L. (2001). Phlebotomine sandflies and leishmaniasis risks in Colombian coffee plantations under two systems of cultivation. *Medical and Veterinary Entomology*, *15*(4), 364–373. doi:10.1046/j.0269-283x.2001.00322.x PMID:11776455

Alexander, B., Agudelo, L. A., Navarro, J. F., Ruiz, J. F., Molina, J., Aguilera, G., & Quiñones, M. L. et al. (2009). Relationship between coffee cultivation practices in Colombia and exposure to infection with Leishmania. *Transactions of the Royal Society of Tropical Medicine and Hygiene*, *103*(12), 1263–1268. doi:10.1016/j.trstmh.2009.04.018 PMID:19555985

Aybar, C. A. V., Juri, M. J. D., Santana, M., de Grosso, M. S. L., & Spinelli, G. R. (2012). The spatio-temporal distribution patterns of biting midges of the genus Culicoides in Salta province, Argentina. *Journal of Insect Science*, *12*(1), 145. PMID:23461794

Azevedo, A. C., & Rangel, E. F. (1991). A study of sandfly species (Diptera: Psychodidae: Phlebotominae) in a focus of cutaneous leishmaniasis in the municipality of Baturité, Ceará, Brazil. *Memorias do Instituto Oswaldo Cruz*, *86*(4), 405–410. doi:10.1590/S0074-02761991000400005 PMID:1842431

Azevedo, A. C., Vilela, M. L., Souza, N. A., Andrade-Coelho, C. A., Barbosa, A. F., Firmo, A. L., & Rangel, E. F. (1996). The sand fly fauna (Diptera: Psychodidae: Phlebotominae) of a focus of cutaneous leishmaniasis in Ilhéus, state of Bahia, Brazil. *Memorias do Instituto Oswaldo Cruz, 91*(1), 75–79. doi:10.1590/S0074-02761996000100012 PMID:8734952

Braga, I. A., & Valle, D. (2007). Aedes aegypti: Histórico do controle no Brasil. *Epidemiologia e Serviços de Saúde*, *16*(2), 113–118.

Brumpt, E., & Pedroso, A. (1913). Pesquisas epidemiológicas sobre a leishmaniose americana das florestas no estado de São Paulo. *Anais Paulistas de Medicina e Cirurgia*, *1*, 97–136.

Bryant, J., Wang, H., Cabezas, C., Ramirez, G., Watts, D., Russell, K., & Barrett, A. (2003). Enzootic transmission of yellow fever virus in Peru. *Emerging Infectious Diseases*, *9*(8), 926–933. doi:10.3201/eid0908.030075 PMID:12967489

Cabral, A. C., Fe, N. F., Suarez-Mutis, M. C., Boia, M. N., & Carvalho-Costa, F. A. (2010). Increasing incidence of malaria in the Negro River basin, Brazilian Amazon. *Transactions of the Royal Society of Tropical Medicine and Hygiene*, *104*(8), 556–562. doi:10.1016/j.trstmh.2010.03.008 PMID:20462621

Campbell, L. P., Luther, C., Moo-Llanes, D., Ramsey, J. M., Danis-Lozano, R., & Peterson, A. T. (2015). Climate change influences on global distributions of dengue and chikungunya virus vectors. *Philosophical Transactions of the Royal Society of London B: Biological Sciences*, 370(1665), 20140135.

Canziani, O. F., & Diaz, S. (1998). Latin America. In R. T. Watson, M. C. Zinyowera, & R. H. Moss (Eds.), *The Regional Impacts of Climate Change: An Assessment of Vulnerability* (pp. 187–230). Cambridge, UK: Cambridge University Press.

Caqui, K. B., Quispe, C. L. H. A., & Zegarra, R. O. (2013). Disaster risk management in the framework of the UNASUR health. In E. H. Carmo, A. Gemal, & S. Oliveira (Eds.), Health Surveillance in South America: epidemiological, sanitary and environmental (pp. 41-54). Rio de Janeiro: ISAGS.

Carbajo, A. E., Cardo, M. V., & Vezzani, D. (2012). Is temperature the main cause of dengue rise in non-endemic countries? The case of Argentina. *International Journal of Health Geographics*, *11*(1), 26. doi:10.1186/1476-072X-11-26 PMID:22768874

Carbajo, A. E., Vera, C., & González, P. L. (2009). Hantavirus reservoir *Oligoryzomys longicaudatus* spatial distribution sensitivity to climate change scenarios in Argentine Patagonia. *International Journal of Health Geographics*, *8*(1), 44. doi:10.1186/1476-072X-8-44 PMID:19607707

Carbajo, A. E., & Vezzani, D. (2015). Waiting for chikungunya fever in Argentina: Spatio-temporal risk maps. *Memorias do Instituto Oswaldo Cruz*, *110*(2), 259–262. doi:10.1590/0074-02760150005 PMID:25946252

Cárdenas, R., Sandoval, C. M., Rodriguez-Morales, A. J., & Vivas, P. (2008). Zoonoses and climate variability: The example of Leishmaniasis in southern departments of Colombia. *Animal Biodiversity and Emerging Diseases*, *1149*, 326–330.

Carpenter, S., Groschup, M. H., Garros, C., Felippe-Bauer, M. L., & Purse, B. V. (2013). Culicoides biting midges, arboviruses and public health in Europe. *Antiviral Research*, *100*(1), 102–113. doi:10.1016/j.antiviral.2013.07.020 PMID:23933421

Cesario, M., & Cesario, R. R. (2005). Infecção bacteriana rumo ao Brasil: Endêmica nos Andes, Bartonelose se alastra com abertura de estradas e degradação ambiental. *Scientific American Brasil*, *3*(34), 10–11.

Chagas, C. (1907). Adenda: Lassance. In J. R. Cançado & M. Chuster (Eds.), *Cardiopatia Chagásica* (pp. 391–413). Belo Horizonte: Fundação Carlos Chagas.

Chaves, L. F., & Pascual, M. (2006). Climate cycles and forecasts of cutaneous leishmaniasis, a nonstationary vector-borne disease. *PLoS Medicine*, *3*(8), e295. PMID:16903778

Chinga-Alayo, E., Huarcaya, E., Nasarre, C., Del Aguila, R., & Llanos-Cuentas, A. (2004). The influence of climate on the epidemiology of Bartonellosis in Ancash, Peru. *Transactions of the Royal Society of Tropical Medicine and Hygiene*, *98*(2), 116–124. doi:10.1016/S0035-9203(03)00017-8 PMID:14964812

Chowell, G., Cazelles, B., Broutin, H., & Munayco, C. V. (2011). The influence of geographic and climate factors on the timing of dengue epidemics in Perú, 1994-2008. *BMC Infectious Diseases*, *11*(1), 1. doi:10.1186/1471-2334-11-164 PMID:21651779

Comisión Económica para América Latina y el Caribe. (2014). Anuario estadístico de América Latina y el Caribe. Santiago de Chile, Chile. Retrieved from http://www.cepal.org/en/publications/37647-anuario-estadistico-de-america-latina-y-el-caribe-2014-statistical-yearbook-latin

Confalonieri, U. E., & Costa Neto, C. (2012). Diversity of mosquito vectors (Diptera: Culicidae) in Caxiuanã, Pará, Brazil. *Interdisciplinary Perspectives on Infectious Diseases*, 2012, id 741273.

Confalonieri, U. E., Margonari, C., & Quintão, A. F. (2014). Environmental change and the dynamics of parasitic diseases in the Amazon. *Acta Tropica*, *129*, 33–41. doi:10.1016/j.actatropica.2013.09.013 PMID:24056199

Confalonieri, U. E., Menezes, J. A., & Margonari de Souza, C. (2015). Climate change and adaptation of the health sector: The case of infectious diseases. *Virulence*, 6(6), 554-557. PMID:26177788

Confalonieri, U. E. C., Mascarenhas, B. M., Pinto, C. S., Santa Brigida, M., Chagas, E., & Guimarães, D. G. (2013). Distribuição Vertical e Temporal de Culicídeos (Diptera: Culicidae) em sítio na Floresta Nacional de Caxiuanã, Melgaço, Pará. In *P.L. Lisboa (Org.), Caxiuanã: paraíso ainda preservado* (pp. 117–122). Belém: Editora do Museu Paraense Emilio Goeldi.

Conn, J. E., Wilkerson, R. C., Segura, M. N. O., de Souza, R. T., Schlichting, C. D., Wirtz, R. A., & Póvoa, M. M. (2002). Emergence of a new neotropical malaria vector facilitated by human migration and changes in land use. *The American Journal of Tropical Medicine and Hygiene*, *66*(1), 18–22. PMID:12135261

Cordel, H., Quatresous, I., Paquet, C., & Couturier, E. (2006). Imported cases of chikungunya in metropolitan France, April 2005–February 2006. *Eurosurveillance*, *11*(4), E060420. PMID:16809828

Cordovez, J. M., Rendon, L. M., Gonzalez, C., & Guhl, F. (2014). Using the basic reproduction number to assess the effects of climate change in the risk of Chagas disease transmission in Colombia. *Acta Tropica*, *129*, 74–82. doi:10.1016/j.actatropica.2013.10.003 PMID:24416781

Costa, J., Dornak, L. L., Almeida, C. E., & Peterson, A. T. (2014). Distributional potential of the Triatoma brasiliensis species complex at present and under scenarios of future climate conditions. *Parasites and Vectors*, *7*(1), 238. doi:10.1186/1756-3305-7-238 PMID:24886587

Costello, A., Abbas, M., Allen, A., Ball, S., Bell, S., Bellamy, R., & Patterson, C. et al. (2009). Managing the health effects of climate change: Lancet and University College London Institute for Global Health Commission. *Lancet*, *373*(9676), 1693–1733. doi:10.1016/S0140-6736(09)60935-1 PMID:19447250

Coura, J. R. (2015). The main scenarios of Chagas disease transmission. The vectors, blood and oral transmissions - A comprehensive review. *Memorias do Instituto Oswaldo Cruz*, *110*(3), 277–282. doi:10.1590/0074-0276140362 PMID:25466622

Coura, J. R., & Junqueira, A. C. (2012). Risks of endemicity, morbidity and perspectives regarding the control of Chagas disease in the Amazon Region. *Memorias do Instituto Oswaldo Cruz*, *107*(2), 145–154. doi:10.1590/S0074-02762012000200001 PMID:22415251

Cunha, R. V., & Nogueira, R. M. R. (2005). Dengue e Dengue Hemorrágico. In J. R. Coura (Ed.), *Dinâmica das doenças infecciosas e parasitárias* (pp. 1755–1766). Rio de Janeiro: Guanabara Koogan.

da Silva, O., de Sousa, M. E., dos Santos, F. A., da Silava, P., Gazin, P., & da Silva, P. (1999). American cutaneous leishmaniasis in the sugar cane-producing area of Pernambuco, North-East Brazil. *Sante (Montrouge, France)*, *10*(2), 123–126. PMID:10960810

da Silva-Nunes, M., Codeço, C. T., Malafronte, R. S., da Silva, N. S., Juncansen, C., Muniz, P. T., & Ferreira, M. U. (2008). Malaria on the Amazonian frontier: Transmission dynamics, risk factors, spatial distribution, and prospects for control. *The American Journal of Tropical Medicine and Hygiene*, *79*(4), 624–635. PMID:18840755

Daszak, P., Cunningham, A. A., & Hyatt, A. D. (2000). Emerging infectious diseases of wildlife--threats to biodiversity and human health. *Science*, *287*(5452), 443–449. doi:10.1126/science.287.5452.443 PMID:10642539

de Araújo Pedrosa, F., & de Alencar Ximenes, R. A. (2009). Sociodemographic and environmental risk factors for American Cutaneous Leishmaniasis (ACL) in the State of Alagoas, Brazil. *The American Journal of Tropical Medicine and Hygiene*, *81*(2), 195–201. PMID:19635869

Degallier, N., Rosa, A. T. D., Hervé, J. P., Vasconcelos, P. F., Rosa, E. S., Rodrigues, S. G., & Dias, L. B. et al. (1992). A comparative study of yellow fever in Africa and South America. *Ciencia e Cultura*, *44*(2/3), 143–151.

Delgado-Petrocelli, L., Córdova, K., Camardiel, A., Aguilar, V. H., Hernández, D., & Ramos, S. (2012). Analysis of the El Niño/La Niña-southern oscillation variability and Malaria in the Estado Sucre, Venezuela. *Geospatial Health*, *6*(3), 51–57. doi:10.4081/gh.2012.122 PMID:23032283

Downs, W. G. (1981). The known and the unknown in yellow fever ecology and epidemiology. *Ecology of Disease*, *1*(2-3), 103–110. PMID:6148230

Dutra, F. R. L. S., Valadão, R. C., Confalonieri, U. E., Muller, G. V., & da Quadro, M. F. L. (2015). A influência da variabilidade da precipitação no padrão de distribuição dos casos de leptospirose em Minas Gerais, no período de 1998-2012. *Hygeia; Revista Brasileira de Geografia Médica e da Saúde*, *11*(20), 106.

Ebi, K. L., Lindgren, E., Suk, J. E., & Semenza, J. C. (2013). Adaptation to the infectious disease impacts of climate change. *Climatic Change*, *118*(2), 355–365. doi:10.1007/s10584-012-0648-5

Echegaray, F., & Afonso, M. H. F. (2014). Respostas às mudanças climáticas: Inovação tecnológica ou mudança de comportamento individual? *Estudos Avançados*, *28*(82), 155–174. doi:10.1590/S0103-40142014000300010

Epstein, P. R. (2001). West Nile virus and the climate. *Journal of Urban Health*, *78*(2), 367–371. doi:10.1093/jurban/78.2.367 PMID:11419587

Ferreira, I. B., Pereira, L. E., Rocco, I. M., Marti, A. T., Souza, L., & Iversson, L. B. (1994). Surveillance of arbovirus infections in the Atlantic forest region, State of São Paulo, Brazil: I. detection of hemagglutination-inhibition antibodies in wild birds between 1978 and 1990. *Revista do Instituto de Medicina Tropical de Sao Paulo*, *36*(3), 265–274. doi:10.1590/S0036-46651994000300011 PMID:7855491

Ferreira, K. V., Rocha, K. C., Caputto, L. Z., Fonseca, A. L. A., & Fonseca, F. L. A. (2011). Histórico da febre amarela no Brasil e a importância da vacinação antiamarílica. *Arquivos Brasileiros de Ciências da Saúde*, *36*(1), 40–47.

Figueiredo, M. L. G. D., & Figueiredo, L. T. M. (2014). Emerging alphaviruses in the Americas: Chikungunya and Mayaro. *Revista da Sociedade Brasileira de Medicina Tropical*, *47*(6), 677–683. doi:10.1590/0037-8682-0246-2014 PMID:25626645

Fouque, F., Gaborit, P., Issaly, J., Carinci, R., Gantier, J. C., Ravel, C., & Dedet, J. P. (2007). Phlebotomine sand flies (Diptera: Psychodidae) associated with changing patterns in the transmission of the human cutaneous leishmaniasis in French Guiana. *Memorias do Instituto Oswaldo Cruz*, *102*(1), 35–40. doi:10.1590/S0074-02762007000100005 PMID:17293996

Gonçalves Neto, V. S., Barros Filho, A. K. D., Santos, A. M. D., Prazeres, M. P. C. D. S., Bezerril, A. C. R., Fonseca, A. V. D. L., & Rebelo, J. M. M. (2013). An analysis of the spatiotemporal distribution of American cutaneous leishmaniasis in counties located along road and railway corridors in the State of Maranhao, Brazil. *Revista da Sociedade Brasileira de Medicina Tropical*, *46*(3), 322–328. doi:10.1590/0037-8682-0056-2012 PMID:23856875

Goodin, D. G., Koch, D. E., Owen, R. D., Chu, Y. K., Hutchinson, J. M., & Jonsson, C. B. (2006). Land cover associated with Hantavirus presence in Paraguay. *Global Ecology and Biogeography*, *15*(5), 519–527. doi:10.1111/j.1466-822X.2006.00244.x

Gottdenker, N. L., Calzada, J. E., Saldaña, A., & Carroll, C. R. (2011). Association of anthropogenic land use change and increased abundance of the Chagas disease vector Rhodnius pallescens in a rural landscape of Panama. *The American Journal of Tropical Medicine and Hygiene*, *84*(1), 70–77. doi:10.4269/ajtmh.2011.10-0041 PMID:21212205

Gottdenker, N. L., Chaves, L. F., Calzada, J. E., Saldaña, A., & Carroll, C. R. (2012). Host life history strategy, species diversity, and habitat influence Trypanosoma cruzi vector infection in changing landscapes. *PLoS Neglected Tropical Diseases*, 6(11): e1884.

Gubler, D. J. (2002). The global emergence/resurgence of arboviral diseases as public health problems. *Archives of Medical Research*, *33*(4), 330–342. doi:10.1016/S0188-4409(02)00378-8 PMID:12234522

Hack, W. H. (1955). Estudios sobre biología del *Triatoma infestans* (Klug, 1834) (Hemiptera, Reduviidae). *An Inst Med Regional*, *4*, 125–147.

Hahn, M. B., Gangnon, R. E., Barcellos, C., Asner, G. P., & Patz, J. A. (2014). Influence of deforestation, logging, and fire on Malaria in the Brazilian Amazon. *PLoS ONE*, *9*(1), e85725. doi:10.1371/journal.pone.0085725 PMID:24404206

Herrera-Martinez, A. D., & Rodriguez-Morales, A. J. (2010). Potential influence of climate variability on dengue incidence registered in a western pediatric Hospital of Venezuela. *Tropical Biomedicine*, *27*(2), 280–286. PMID:20962726

Ibarra, A. M. S., Ryan, S. J., Beltrán, E., Mejía, R., Silva, M., & Muñoz, Á. (2013). Dengue vector dynamics (*Aedes aegypti*) influenced by climate and social factors in Ecuador: Implications for targeted control. *PLoS ONE*, *8*(11), e78263. doi:10.1371/journal.pone.0078263 PMID:24324542

Jacobi, P. R. (2014). Mudanças climáticas e ensino superior: A combinação entre pesquisa e educação. *Educational Review*, *3*(spe3), 57–72. doi:10.1590/0104-4060.38107

Johnson, K. M. (1993). Emerging viruses in context: an overview of viral hemorrhagic fevers. In S. S. Morse (Ed.), *Emerging viruses* (pp. 46–47). New York: Oxford University Press.

Jones, K. E., Patel, N. G., Levy, M. A., Storeygard, A., Balk, D., Gittleman, J. L., & Daszak, P. (2008). Global trends in emerging infectious diseases. *Nature*, *451*(7181), 990–993. doi:10.1038/nature06536 PMID:18288193

Keesing, F., Belden, L. K., Daszak, P., Dobson, A., Harvell, C. D., Holt, R. D., & Ostfeld, R. S. et al. (2010). Impacts of biodiversity on the emergence and transmission of infectious diseases. *Nature*, *468*(7324), 647–652. doi:10.1038/nature09575 PMID:21124449

Keim, M. E. (2008). Building human resilience: The role of public health preparedness and response as an adaptation to climate change. *American Journal of Preventive Medicine, 35*(5), 508–516. doi:10.1016/j.amepre.2008.08.022 PMID:18929977

Landin, R., & Giatti, L. L. (2014). Política de mudança do clima no município de São Paulo, Brasil: Reflexividade e permeabilidade do Setor Saúde. *Ciencia & Saude Coletiva, 19*(10), 4149–4156. doi:10.1590/1413-812320141910.08972014 PMID:25272124

Lima, G. F. D. C., & Layrargues, P. P. (2014). Climate change, education and environment: beyond Dynamic Conservatism. *Educar em Revista*, (SPE3), 73-88.

Lopes, O. D. S., Coimbra, T. L., Sacchetta, L. D. A., & Calisher, C. H. (1978). Emergence of a new arbovirus disease in Brazil. I. Isolation and characterization of the etiologic agent, Rocio virus. *American Journal of Epidemiology, 107*(5), 444–449. PMID:665659

Maco, V., Maguiña, C., Tirado, A., Maco, C., & Vidal, J. E. (2004). Carrion's disease (*Bartonellosis bacilliformis*) confirmed by histopathology in the High Forest of Peru. *Revista do Instituto de Medicina Tropical de Sao Paulo, 46*(3), 171–174. doi:10.1590/S0036-46652004000300010 PMID:15286824

Magrin, G. O., Marengo, J. A., Boulanger, J. P., Buckeridge, M. S., Castellanos, E., Poveda, G., & Vicuña, S. (2014). Central and South America. In V. R. Barros, C. B. Field, D. J. Dokken, M. D. Mastrandrea, K. J. Mach, T. E. Bilir, & L. L. White et al. (Eds.), *Climate Change 2014: Impacts, Adaptation, and Vulnerability. Part B: Regional Aspects. Contribution of Working Group II to the Fifth Assessment Report of the Intergovernmental Panel on Climate Change* (pp. 1499–1566). Cambridge, UK: Cambridge University Press.

Mantilla, G., Oliveros, H., & Barnston, A. G. (2009). The role of ENSO in understanding changes in Colombia's annual malaria burden by region, 1960-2006. *Malaria Journal, 8*(6), 1–11. PMID:19133152

Medone, P., Ceccarelli, S., Parham, P. E., Figuera, A., & Rabinovich, J. E. (2015). The impact of climate change on the geographical distribution of two vectors of Chagas disease: implications for the force of infection. *Philosophical Transactions of the Royal Society of London B: Biological Sciences*, 370(1665), 20130560.

Membrive, N. A., Rodrigues, G., Gualda, K. P., Bernal, M. V. Z., Oliveira, D. M., Lonardoni, M. V. C., & Silveira, T. G. V. et al. (2012). Environmental and animal characteristics as factors associated with American cutaneous Leishmaniasis in rural locations with presence of dogs, Brazil. *PLoS ONE, 7*(11), e47050. doi:10.1371/journal.pone.0047050 PMID:23144799

Mills, J. N. (2006). Biodiversity loss and emerging infectious disease: An example from the rodent-borne hemorrhagic fevers. *Biodiversity*, *7*(1), 9–17. doi:10.1080/14888386.2006.9712789

Monath, T. P. (1993). Arthropod-borne viruses. In S. S. Morse (Ed.), *Emerging Viruses* (pp. 138–148). New York: Oxford University Press.

Monath, T. P., & Vasconcelos, P. F. (2015). Yellow fever. *Journal of Clinical Virology*, *64*, 160–173. doi:10.1016/j.jcv.2014.08.030 PMID:25453327

Morales, M. A., Barrandeguy, M., Fabbri, C., Garcia, J. B., Vissani, A., Trono, K., & Enría, D. et al. (2006). West Nile virus isolation from equines in Argentina, 2006. *Emerging Infectious Diseases*, *12*(10), 1559–1561. doi:10.3201/eid1210.060852 PMID:17176571

Musso, D., Nilles, E. J., & Cao-Lormeau, V. M. (2014). Rapid spread of emerging Zika virus in the Pacific area. *Clinical Microbiology and Infection*, *20*(10), O595–O596. doi:10.1111/1469-0691.12707 PMID:24909208

Pan American Health Organization. (2014). Health Situation in the Americas: Basic Indicators 2014. Retrieved from http://www.paho.org/chi/index.php?option=com_content&view=article&id=578:situacion-de-salud-en-las-americas-indicadores-basicos-de-salud-2014&catid=681:publicaciones-y-otros-recursos

Pascual, M., Bouma, M. J., & Dobson, A. P. (2002). Cholera and climate: Revisiting the quantitative evidence. *Microbes and Infection*, *4*(2), 237–245. doi:10.1016/S1286-4579(01)01533-7 PMID:11880057

Paz, S. (2006). The West Nile Virus outbreak in Israel (2000) from a new perspective: The regional impact of climate change. *International Journal of Environmental Health Research*, *16*(1), 1–13. doi:10.1080/09603120500392400 PMID:16507476

Pereira, C. A. R., & de Lima Barata, M. M. (2014). Organização dos serviços urbanos de saúde frente à mudança do clima e ao risco de desastres na América Latina. *Saúde debate*, 38(102), 624-634.

Pinheiro, F. P., Travassos Da Rosa, A. P. A., & Vasconcelos, P. F. C. (1998). An overview of Oropouche fever epidemics in Brazil and neighbour countries. In A. P. A. Travassos da Rosa, P. F. C. Vasconcelos, & J. F. S. Travassos da Rosa (Eds.), *An overview of arbovirology in Brazil and neighbouring countries* (pp. 186–192). Belém, Pará: Instituto Evandro Chagas.

Pinto, C. S., Confalonieri, U. E., & Mascarenhas, B. M. (2009). Ecology of Haemagogus sp. and Sabethes sp.(Diptera: Culicidae) in relation to the microclimates of the Caxiuanã National Forest, Pará, Brazil. *Memorias do Instituto Oswaldo Cruz, 104*(4), 592–598. doi:10.1590/S0074-02762009000400010 PMID:19722082

Pongsiri, M. J., Roman, J., Ezenwa, V. O., Goldberg, T. L., Koren, H. S., Newbold, S. C., & Salkeld, D. J. et al. (2009). Biodiversity loss affects global disease ecology. *Bioscience, 59*(11), 945–954. doi:10.1525/bio.2009.59.11.6

Porto, S. S. (2012). Saúde em desastres: a experiência de Nova Friburgo. *Nova Friburgo: Cap-Net Brasil*, 30.

Ramírez, I. J. (2015). Cholera resurgence in Piura, Peru: Examining climate associations during the 1997–1998 El Niño. *GeoJournal, 80*(1), 129–143.

Ramírez, I. J., Grady, S. C., & Glantz, M. H. (2013). Reexamining El Niño and Cholera in Peru: A climate affairs approach. *Weather, Climate, and Society, 5*(2), 148–161. doi:10.1175/WCAS-D-12-00032.1

Roth, A., Mercier, A., Lepers, C., Hoy, D., Duituturaga, S., Benyon, E., & Souares, Y. (2014). Concurrent outbreaks of dengue, chikungunya and Zika virus infections-an unprecedented epidemic wave of mosquito-borne viruses in the Pacific 2012-2014. *Eurosurveillance, 19*(41), 20929. doi:10.2807/1560-7917.ES2014.19.41.20929 PMID:25345518

Rotureau, B. (2006). Ecology of the Leishmania species in the Guianan ecoregion complex. *The American Journal of Tropical Medicine and Hygiene, 74*(1), 81–96. PMID:16407350

Seas, C., Miranda, J., Gil, A. I., Leon-Barua, R., Patz, J., Huq, A., & Sack, R. B. et al. (2000). New insights on the emergence of cholera in Latin America during 1991: The Peruvian experience. *The American Journal of Tropical Medicine and Hygiene, 62*(4), 513–517. PMID:11220769

Simpson, D. I. H. (1978). Viral haemorrhagic fevers of man. *Bulletin of the World Health Organization, 56*(6), 819. PMID:310725

Smith, K. R., Woodward, A., Campbell-Lendrum, D., Chadee, D. D., Honda, Y., Liu, Q., & Sauerborn, R. (2014). Human health: impacts, adaptation, and co-benefits. In C. B. Field, V. R. Barros, D. J. Dokken, K. J. Mach, M. D. Mastrandrea, T. E. Bilir, & L. L. White et al. (Eds.), *Climate Change 2014: Impacts, Adaptation, and Vulnerability. Part A: Global and Sectoral Aspects. Contribution of Working Group II to the Fifth Assessment Report of the Intergovernmental Panel on Climate Change* (pp. 709–754). Cambridge, UK: Cambridge University Press.

Staples, J. E., & Fischer, M. (2014). Chikungunya virus in the Americas—what a vectorborne pathogen can do. *The New England Journal of Medicine*, *371*(10), 887–889. doi:10.1056/NEJMp1407698 PMID:25184860

Tabachnick, W. J. (2013). Nature, nurture and evolution of intra-species variation in mosquito arbovirus transmission competence. *International Journal of Environmental Research and Public Health*, *10*(1), 249–277. doi:10.3390/ijerph10010249 PMID:23343982

Tauil, P. L. (2002). Aspectos críticos do controle do dengue no Brasil. *Cadernos de Saude Publica*, *18*(3), 867–871. doi:10.1590/S0102-311X2002000300035 PMID:12048612

Tauil, P. L., Santos, J. B., & Moraes, M. A. P. (2005). Febre Amarela. In J. R. Coura (Ed.), *Dinâmica das doenças infecciosas e parasitárias* (pp. 1755–1766). Rio de Janeiro: Guanabara Koogan.

Teixeira, A. R., Monteiro, P. S., Rebelo, J. M., Argañaraz, E. R., Vieira, D., Lauria-Pires, L., & Costa, J. M. et al. (2001). Emerging Chagas disease: Trophic network and cycle of transmission of *Trypanosoma cruzi* from palm trees in the Amazon. *Emerging Infectious Diseases*, *7*(1), 100–112. doi:10.3201/eid0701.070100 PMID:11266300

Teixeira, M. G., Paixão, E. S., & Costa, M. C. N. (2013). Recent epidemiological changes in South America. In E. H. Carmo, A. Gemal, Oliveira S. (Orgs), Health Surveillance in South America: epidemiological, sanitary and environmental (125-145). Rio de Janeiro: ISAGS.

Thai, K. T., & Anders, K. L. (2011). The role of climate variability and change in the transmission dynamics and geographic distribution of dengue. *Experimental Biology and Medicine*, *236*(8), 944–954. doi:10.1258/ebm.2011.010402 PMID:21737578

Travassos da Rosa, J. F. S., Pinheiro, F. P., & Vasconcelos, P. F. C. (2005). Febres hemorrágicas viróticas. In *J. R. Coura (Org.), Dinâmica das doenças infecciosas e parasitárias* (Vol. II, pp. 1741–1754). Rio de Janeiro: Guanabara-Koogan.

Travassos Da Rosa, J. F. S., Travassos Da Rosa, A. P. A., Vasconcelos, P. F. C., Pinheiro, F. P., Rodrigues, S. G., Travassos Da Rosa, E. S., . . . Cruz, A. C. R. (1998). Arboviruses isolated in the Evandro Chagas Institute, including some described for the first time in Brazilian Amazon region, their known hosts, and their pathology for man. In A. P. A. Travassos Da Rosa, P. F. C. Vasconcelos And J. F. S. Travassos Da Rosa (Eds.), An overview of arbovirology in Brazil and neighbouring countries (pp. 19-31). Belém, Pará: Instituto Evandro Chagas.

U.S. Agency for International Development. Pan American Health Organization (2010). The Amazon Malaria Initiative: Goals and Accomplishments October 2001–May 2009 (pp. 121). Washington.

Valle, D., & Lima, J. M. T. (2014). Large-scale drivers of malaria and priority areas for prevention and control in the Brazilian Amazon region using a novel multi-pathogen geospatial model. *Malaria Journal*, *13*(1), 1–13. doi:10.1186/1475-2875-13-443 PMID:25412882

Vasconcelos, P. F., Travassos da Rosa, A., Rodrigues, S. G., Travassos da Rosa, E. S., Dégallier, N., da Rosa, T., & Jorge, F. S. (2001). Inadequate management of natural ecosystem in the Brazilian Amazon region results in the emergence and reemergence of arboviruses. *Cadernos de Saude Publica*, *17*, S155–S164. doi:10.1590/S0102-311X2001000700025 PMID:11426277

Vega-Rúa, A., Zouache, K., Girod, R., Failloux, A. B., & Lourenço-de-Oliveira, R. (2014). High level of vector competence of *Aedes aegypti* and *Aedes albopictus* from ten American countries as a crucial factor in the spread of Chikungunya virus. *Journal of Virology*, *88*(11), 6294–6306. doi:10.1128/JVI.00370-14 PMID:24672026

Vilela, M. L., Azevedo, C. G., Carvalho, B. M., & Rangel, E. F. (2011). Phlebotomine Fauna (Diptera: Psychodidae) and putative vectors of Leishmaniases in impacted area by hydroelectric plant, state of Tocantins, Brazil. *PLoS ONE*, *6*(12), e27721. doi:10.1371/journal.pone.0027721 PMID:22163271

Vittor, A. Y., Pan, W., Gilman, R. H., Tielsch, J., Glass, G., Shields, T., & Patz, J. A. et al. (2009). Linking deforestation to Malaria in the Amazon: Characterization of the breeding habitat of the principal malaria vector, *Anopheles darlingi*. *The American Journal of Tropical Medicine and Hygiene*, *81*(1), 5–12. PMID:19556558

Weaver, S. C., & Reisen, W. K. (2010). Present and future arboviral threats. *Antiviral Research*, *85*(2), 328–345. doi:10.1016/j.antiviral.2009.10.008 PMID:19857523

Wilcox, B. A., & Gubler, D. J. (2005). Disease ecology and the global emergence of zoonotic pathogens. *Environmental Health and Preventive Medicine*, *10*(5), 263–272. doi:10.1007/BF02897701 PMID:21432130

World Health Organization. (2009). Para salvar vidas: hagamos que los hospitales sean seguros en las situaciones de emergencia.

World Health Organization. (2014). *WHO Guidance for Countries on Combining Indoor Residual Spraying and Long-Lasting Insecticidal Nets*. Geneva, Switzerland: World Health Organization.

Zhou, J., Lau, K. M., Laughlin, L. W., Masuoka, M., Chamberlin, J., & Lawyer, P. (2001). The effect of regional climate variability on outbreak of Bartonellosis epidemics in Peru. *Proceedings of the third Symposium on Environmental Applications: Facilitating the Use of Environmental Information*. Orlando, Florida: American Meteorological Society.

KEY TERMS AND DEFINITIONS

Anthropic Environment: New spaces formed by human actions that transforms the natural environment, such as large cities and agricultural fields.

Arboviruses: Viruses transmitted to humans by hematophagous insects and that are kept in natural or urban environment by vertebrate hosts.

Emerging Infectious Diseases: Diseases that had never been detected in humans before or which have affected a very small number of people in isolated places, such as Hantavirus Pulmonary Syndrome.

Epidemics: A higher than expected incidence of a given disease, when compared to the historical average, that occurs in a short period and in a specific geographic region.

Infectious Diseases: Diseases caused by pathogenic agents such as bacteria, protozoa and viruses that can spread from one individual to another, directly or indirectly.

Peridomiciliary: Spaces or environments surrounding human habitation, like backyard, henhouse, pigsty and barn.

Re-Emerging Infectious Diseases: Diseases that have affected large populations (worldwide or in specific countries) in the past, but were controlled and, at the present, they reappear as a major public health problem, such as Dengue Fever and Malaria.

Vector-Borne Diseases: Infections transmitted by the bite of blood-sucking arthropods such as mosquitoes, midges and bugs.

Chapter 6
Infectious Diseases and Climate Vulnerability in Morocco:
Governance and Adaptation Options

Mohamed Behnassi
Ibn Zohr University of Agadir, Morocco

Kholoud Kahime
Cadi Ayyad University of Marrakesh, Morocco

Samia Boussaa
Higher Institute of Nursing Professions and Health Techniques of Marrakesh, Morocco

Ali Boumezzough
Cadi Ayyad University of Marrakesh, Morocco

Mohammed Messouli
Cadi Ayyad University of Marrakesh, Morocco

DOI: 10.4018/978-1-5225-0553-2.ch006

ABSTRACT

Climate change is expected to affect the distribution, prevalence and life cycle of several infectious diseases. This scenario is relevant to Morocco since the country is considered by many IPCC assessments reports as a climate change hotspot with a high vulnerability to many expected impacts. Given this existing vulnerability, this chapter aims to highlight relevant vector-borne diseases, the risks of their reemergence in many vulnerable regions and the pressing need to understand their dynamics within a context marked by knowledge gaps and limited scientific evidence; underline the problematic aspects of health adaptation to climate change and the current difficulties in terms of policy and governance to manage climate-health linkages; and finally undertake an assessment of Morocco's adaptive capacity from a health perspective and formulate recommendations for effective climate-health governance and policy.

INTRODUCTION

The current realization that human beings need to be concerned about the only 'life-support system' that the Earth and its environment provide stems perhaps in part from the fact that, until fairly recently, the evolution of humankind was largely dependent on the quality of the environment and the resources it provides in terms of water, food, and favorable health conditions. These are as vital as ever, despite current levels of technology and apparent resilience in the face of often degraded environments in many parts of the world. Today, the necessary conditions for maintaining human security – i.e. water quality and quantity, food security, and health – are under threat as a result of numerous human-induced factors; among these, climate change is certainly one of the more durable aspects of anthropogenic disruptions to natural resources (Beniston 2010). It is currently recognized that with the constant increase of the planet's temperature, there are, and will be, more frequent and severe floods, droughts, storms, and heat waves; these changes to Earth's biophysical system will exacerbate and extend the rates and ranges of many diseases and overall contribute to poor health among all populations (Friel et al. 2011).

Therefore, coupled with other environmental risks, climate change poses a serious threat to the progress made towards global health and development goals in recent decades. This challenge will make efforts to ensure the conditions in which people can be healthy more difficult in a myriad of ways (Wiley 2010). Moreover, not only does climate change carry a direct and indirect health warning, it is already contributing towards widening the gap in health inequalities between people living in the developing and developed world. In other words, climate change health impacts will – perversely, unjustly and inequitably – affect those in our society who are most disadvantaged and who arguably have contributed least to it – certainly in terms of their consumption of natural resources. Thus, and according to Friel et al. (2011), without a lessening of background rates of diseases, the multiplier effects of climate change on health outcomes will greatly exacerbate existing health inequities between and within countries, which poses a major additional challenge to international development.

For Wiley (2010), climate change acts primarily as an intensifier, and to some extent a redistributor of existing threats to health. More precisely, climate change is expected to directly affect many environmental determinants of health – mainly water, air, weather, oceans, and ecosystems (Portier et al. 2010) – which may exacerbate exposure to natural disasters and disease-carrying vectors, access to safe and potable drinking water, and food security. Similarly, Boxall et al. (2009) asserts that weather and climate factors are known to affect the transmission of water and vector-borne infectious diseases. Some linkages between climate and health are fairly obvious, such as the impact of weather-related natural disasters, whereas others are less obvious and may be neglected by policymakers, such as the indirect impacts of natural disasters, especially on mental health (Wiley 2010).

Addressing the health impacts of climate change is especially challenging because both the surrounding environment and the decisions that people make influence health. In real-life situations, a host of other factors come into play in determining vulnerability including biological susceptibility, socioeconomic status, cultural competence, and the built environment. In a world of myriad 'what if' scenarios surrounding climate change, it becomes very complicated to create wise health policies for the future because of the uncertainty of predicting environmental change and human decisions. The need for sound science on which to base such policies becomes more critical than ever (Portier et al. 2010).

Within a Moroccan perspective, the main health vulnerabilities to climate change include the following: risk of reactivation of certain diseases sensitive to climate change (such as malaria, bilharzia, typhoid, leishmaniasis, dengue and cholera); possibility of re-emergence of infectious diseases, vector-borne diseases as well as diseases and deaths related to extreme weather events, especially among the most vulnerable groups; increased water and food-borne diseases; increased incidence of respiratory infections and allergies due to air pollution and sudden weather events and rising temperatures which contribute to the increased volume of ozone at ground level and precipitate the start of the pollen season; natural disasters such as floods which cause damage (affecting agriculture, livestock and real estate), deaths and illnesses suffered by victims and their families; etc.

Instead of covering all health impacts of climate change in Morocco, this chapter aims to focus on many infectious diseases that are expected to change their geographical or seasonal patterns and incidence due to climate change and variability. More specifically, climate change will likely affect the distribution, prevalence and lifecycle of several vector-borne diseases – such as Malaria and Leishmaniasis – and water-related disease – such as Schistosomiasis, and this scenario is relevant to the Moroccan context. These diseases still ravage lives in many developing and less developed countries and are likely to have a major human health burden in the coming years due to climate change and other factors. However, and despite this alarming fact, there is still limited specific scientific evidence to support policy-making processes in many countries.

In Morocco, infectious diseases are still a public health problem and the situation may be more complicated in light of climate change. Due to legal requirement to report notifiable diseases, the number of cases – both indigenous and imported – is continuously recorded for Malaria, Schistosomiasis and Leishmaniasis and showed an increase of the latter. This chapter aims to identify vector-borne diseases that are highly sensitive to climate change in Morocco, the risks of their reemergence in vulnerable regions; highlight the pressing need to understand the dynamics of these diseases, especially in a context marked by limited and uncertain scientific evidence regarding the health vulnerability to climate change; underline the problematic aspects of health adaptation to climate change and the current difficulties in terms of policy and governance action to manage climate-health linkages; and finally undertake an assessment of Morocco's adaptive capacity within a health perspective while formulating recommendations to effectively adapting the health sector to climate change and developing an effective climate-health governance and policy.

RELEVANT CLIMATE SENSITIVE INFECTIOUS DISEASES IN MOROCCO

Parasitic diseases still pose a public health challenge in Morocco. Their spread is very closely tied to risk factors – climatic, ecological, and socioeconomic – and other factors, such as urbanization, agricultural practices, beyond health authorities' traditional sphere of influence. Climate change is imposing a burden that is unevenly distributed across populations. The most vulnerable populations are in worse positions to effectively face these challenges because they have a limited ability to adapt to changing conditions. In particular, vulnerability in this context is linked to the emergence of vector-borne-diseases: Leishmaniasis, Malaria, and

Figure 1. Simplified Life cycle of Leishmaniasis

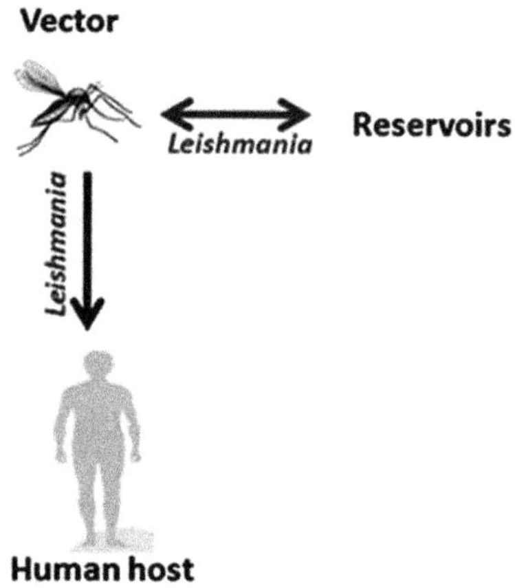

Schistosomiasis, which can be life-threatening. Therefore, there is an urgent need to understand their life cycle, habitats of reservoirs, micro niche of arthropod/insect vectors in the Moroccan context by using an eco-epidemiological approach.

Leishmaniasis

Leishmaniasis is a complex disease caused by *Leishmania* species and transmitted by a Phlebotomine sand fly (Diptera: Psychodidae) (Figure1). Two forms are known, cutaneous and visceral Leishmaniasis. The main reservoirs are dogs for Zoonotic Visceral Leishmaniasis (ZVL), rodents for Zoonotic Cutaneous Leishmaniasis (ZCL) and human for Anthroponotic Cutaneous Leishmaniasis (Figure 1) (ACL) (Rioux et al., 1982; Nejjar et al., 1998; Pratlong et al., 1991; Kahime et al., 2014).

Actually, three parasite species co-exist in Morocco:

- *Leishmania infantum*, causes mainly ZVL and transmitted by species of the subgenus *Larroussius.* It is widespread in the whole country and is more frequent in its northern part (Kahime et al., 2014, 2015a); L. *infantum* can cause Cutaneous leishmaniasis (CL) as well even if it is a rare condition with a few sporadic cases in the North of the country (especially in Sidi Kacem province) with little epidemiological data available (Kahime et al., 2015b).
- *L. major* causes ZCL with *Phlebotomus papatasi* as the vector in pre-Saharan area (Rioux et al., 1982, 1986; Kahime et al., 2014); and
- *L. tropica,* causative agent of ACL is widespread in northern and central areas, especially in semi-arid regions with *P. sergenti* as the vector (Guilvard et al., 1991; Pratlong et al., 1991; Kahime et al., 2014).

Figure 2. Simplified life cycle of Malaria

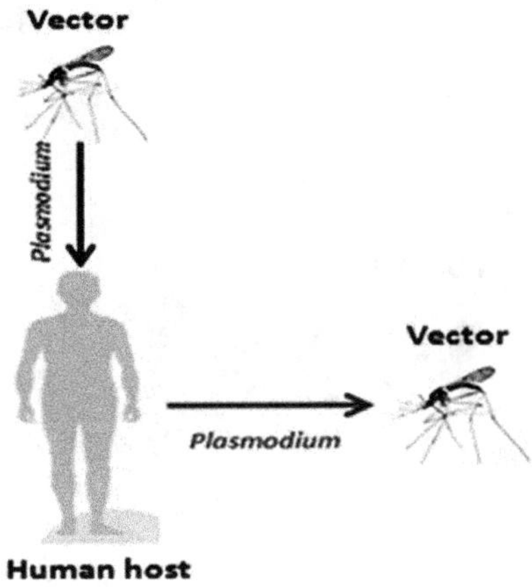

Malaria

Malaria is a mosquito-borne infectious disease caused by parasitic protozoans of the genus *Plasmodium* (*P. vivax, P. malariae, P. ovale* and *P. falciparum*) and is transmitted by female mosquito vectors of the Anopheles species. The cycle of *Plasmodium* is carried out in several stages which spread between mosquitoes and humans (Figure 2).

Malaria raged in Morocco for centuries and was an endemic disease in the majority of provinces. In 1960, a domestic program to fight the disease was launched. It allowed to control the situation after 40 years of bitter struggle. By 1999, malaria was occurring as sporadic cases of *Plasmodium vivax* in some residual foci in the north. The epidemiological assessment undertaken by the Moroccan Ministry of Health showed a shift towards the elimination of indigenous cases and the last indigenous case was reported in 2004. Morocco is now WHO certified Malaria free but imported cases are reported constantly.

In Morocco, *Anopheles labranchiae* is the main vector of malaria. In common with leishmaniasis, it is the female (vector) that bites when it needs a blood meal before laying eggs. The vectorial capacity of *A. labranchiae* is considerably high during the summer which corresponds to the rice cultivation period in northern Morocco (Faraj et al., 2008).

There are two areas with high risk of transmission: The north foci (Al Hoceima, Taounate and Chefchaoune) and the center (Beni Mellal, Khouribga and El Kelaa) (MMH, 2004; Trari and Carnevale, 2011).

Figure 3. Simplified life cycle of Schistosomiasis

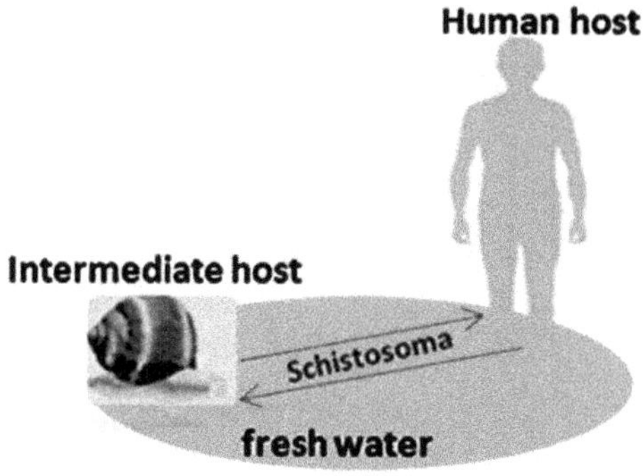

The risk of malaria resurgence in Morocco is higher because of the possible presence of parasite in human and invertebrate hosts in the last malaria foci, numerous imported cases and the presence of the main vector of Malaria in Morocco.

Schistosomiasis

Schistosomiasis is a disease caused by parasitic worms belonging to the class of trematodes and genus *Schistosoma* (S. *haematobium, S. mansoni, S. japonicum* and *S. intercalatum*). The parasite develops successively in two hosts: mollusk and human (Figure 3). This parasite is commonly found in ponds, streams and irrigation canals, housed in freshwater mollusks, and infests humans through the skin via contact with contaminated water (Figure 3).

In Morocco, the disease had spread in the oases in the south and along the southern side of the Atlas (MMH, 1980). The majority of cases have been filed in the Province of Tata, Chtouka, Taroudant, and Errachidia (WHO, 2012b). In recent years, many *S. haematobium* foci proved unstable and some even disappeared. However, creating large water supply for irrigation may lead to the onset of new foci.

Despite the existence of monitoring programs and adequate health services, cases of Malaria, Leishmaniasis and Schistosomiasis are still recorded in Morocco. As shown in Table 1, Leishmaniasis is the most reported diseaseand all cases are indigenous. In contrast, all Malaria cases are imported. For Schistosomiasis, the number of indigenous cases is rarely associated with the risk of introduction of new species, which is linked to travelers or immigrants from countries where these species are endemic.

Table 1. Annual evolution of parasitic diseases in Morocco between 2005-2014

		2005	2006	2007	2008	2009	2010	2011	2012	2013	2014
Malaria (Imported cases)		100	83	100	142	145	218	312	364	314	308
Schistosomiasis		13	4	8	2*	2*	1 *	5 (3*)	0	3	1
Leishmaniasis	ZVL	114	170	160	163	134	139	107	113	111	85
	ACL	865	1197	1938	1697	1611	2263	2100	2137	2055	2081
	ZCL	2174	2164	1352	3431	4402	6444	2219	740	537	460

MMH, 2014.
*Imported cases.

These diseases constitute a real social problem in Morocco because they disproportionately affect the poor, particularly those in vulnerable housing and environmental conditions. Loss of income and healthcare costs exacerbate the economic situation of already-disadvantaged households (Kahime et al., in press). Urban population growth, responsible for the spatial expansion of cities, creates enormous need for community facilities and equipment related to decent housing, access to clean water and sanitation and environmental preservation. The vulnerability of populations to these disease risks varies across spatial and temporal scales in response to environmental change, economic development dynamics, social capital, demographics and population structure. Understanding the interactions between human health and the environment is quite complex. In fact, the very term environment can be confusing because it is used in various acceptances. The ecosystem approach to human health is distinguished from traditional approaches by integrating all determinants of health, namely economic and environmental factors and community needs.

CLIMATE CHANGE IMPACTS ON INFECTIOUS DISEASES: THE CASE OF MOROCCO

Anthropogenic climate change may directly affect the behavior and geographical distribution of mosquito vectors and the lifecycle of the parasite, and thus change the incidence of the disease. Indirectly, climate change could also have various effects by influencing environmental factors such as vegetation and the availability of breeding sites (Martens et al., 1995).

As the climate has an important influence on environmental and socio-economic components, the juxtaposition of many of socio-economic factors e.g., poverty, economic degradation over time may convincingly increase as result to climate change. This change could, depending on the region, have adverse impacts on water resources, agriculture, biodiversity, ecosystems, food and health security.

A wide range of infectious diseases may change their geographic range, seasonality and incidence due to climate change, but there is still limited and uncertain scientific evidence about the extent of health vulnerability to climate change. For example, over the past decade, the epidemiological situation of CL has changed significantly. It is acquiring an increasingly epidemic status with geographic expansion to previously free areas and the emergence of new foci in several provinces of Morocco. 24 804 cases of zoonotic *L. major* CL and 16 852 cases of *L. tropica* anthroponotic CL were recorded between 2004 and 2013 in Morocco (Kahime et al., 2015b). ZCL acquires an increasingly epidemic status with geographic expansion into previously free areas in several provinces of South-East Morocco (Kahime et al. in press). Whereas, ACL has an extensive geographical distribution (Kahime et al., 2015b).

The leishmaniasis lifecycle, parasite-reservoir-vector, evolves depending on geographic regions and is sensitive to environmental changes that can affect the parasite, the reservoir and the vector, as well as their dynamic interaction and territorial expansion (Bounoua et al., 2013). Seasonal patterns and the correlation between the vector density and the number of cases have been documented (Feliciangeli et al., 1998). Weather and climate variables also play an important role in leishmaniasis incidence as they can constrain or exacerbate favorable conditions for the disease, such as an acceleration of the development of the parasite or synergistic changes in reservoir and vector populations that cause an explosion in the vector population. For example, increase in precipitation may increase the vegetation foliage, and thus the number and quality of breeding sites for both rodents and sandflies (Yates et al., 2002; Bounoua et al., 2013).

Distribution and seasonal activity of malaria are also sensitive to climate factors and impact the local capacity to control the disease. Malaria is one of the few climate-sensitive diseases that has been modeled by many research groups, which facilitates a thorough assessment of possible climate change effects using a multi-model inter-comparison (Caminadae et al., 2014). The distribution and population dynamics of malaria are probably more governed by abiotic than biotic factors (Southwood, 1977). Of the possible abiotic influences on the transmission cycle of malaria, temperature and rainfall are the most important; therefore, the situation could be exacerbated by climate change. Although this linkage is still uncertain, an increase in the incidence of malaria has been identified as a potential impact of climate change in South America (Lieshoutet al., 2004) and Africa (Ebi et al., 2005). Climatic factors that feed into the phenomenon could have a direct bearing on the number of malaria cases. Actually, a number of studies have reported the association between malaria cases, rainfall and temperature (Bhattarai et al., 2007). For example, a study carried out in Ethiopia revealed an association of malaria with rainfall and minimum temperature, the strength of which varied with altitude (Teklehaimanot et al., 2004). In South Africa, variations in annual cases of malaria were shown to be related to rainfall and temperature patterns (Craig et al., 2004).

According to McCreesh and Booth (2013), climate change will inevitably influence both the distribution of *Schistosoma* sp. and incidence of schistosomiasis in areas where it is currently endemic. Global warming is expected to be accompanied by perturbations in the global hydrologic cycle (Allen and Ingram, 2002), precipitation levels and pronounced changes in water availability (Allen et al., 2005). However, only a few attempts have been made to predict changes in the spatial distribution of schistosomiasis due to global warming; and the results have been conflicting (Martens et al., 1995, 1997). Thus, the nature and extent of climate change on the transmission of schistosomiasis remains poorly understood (Morgan, 2001). The lack of research in this field probably explains why climate change implications for schistosomiasis control and elimination have been largely ignored (McCreesh et al., 2015).

Snails of the genus *Bulinus* serve as the intermediate hosts of *S. haematobium* in Africa and the Eastern Mediterranean (WHO, 1990). These aquatic snails breed in waters (barrages, irrigation canals...), contaminated with infected faeces or urine from infected person or by infected wastewater, used also for irrigation. Water resources development schemes in certain areas, particularly irrigation schemes, can contribute to the introduction and spread of schistosomiasis. Consequently, intermediate hosts can determine the geographical area of schistosomiasis risk. In Morocco, the disease foci, especially in Beni Mellal, El Kelaa and Nador Provinces were linked to hydrologic programs for irrigation (barrages and irrigation canals and wells) in the 1970s (WHO, 2012b).

Potential impact of climate change on the transmission of schistosomiasis and other diseases must be juxtaposed to profound ecological, demographic, and socio-economic changes. Also, it should be mentioned that, in Morocco, there is no study that addresses the relationship between malaria or schistosomiasis and climate change.

ADAPTATION TO CLIMATE CHANGE FROM A HEALTH PERSPECTIVE: CHALLENGES AND OPPORTUNITIES

Supported by a rapidly growing body of evidence, the health implications of climate change are widely covered at international meetings and used to demand action on climate change. However, while the health sector is well involved in adaptation strategies, mitigation efforts are anticipated to largely take place outside of the sector according to Lovell (2011). Nerlander (2009) asserts that proactive health adaptation strategies are needed to protect vulnerable people from climate change impacts on their health and well-being. This is true because reducing vulnerabilities and increasing resilience in general will help populations cope with climate change health effects. For Wiley (2010), adaptation to climate change requires a set of law, policy and governance tools (such as a purpose-built legal and policy framework designed to specifically address climate risks, which may provide particular models of governance and management for adaptation to climate change (Godden et al., 2013) that are very different from what environmental policymakers have typically employed in their mitigation efforts[1]. However, to date, there has been limited research on how legal, institutional and regulatory frameworks can support and facilitate effective adaptation and responses to climate risks (Godden et al., 2013). While many factors influence resilience and the capacity to prepare for and respond to these risks, regulatory and policy frameworks remain significant drivers of collective and behavioral change.

In addition, adaptation of the health sector to climate change requires a continuous access of decision makers to relevant scientific evidence and information. Nevertheless, forecasting the impacts of climate change on human health is still limited because populations have different vulnerabilities to change and susceptibility to disease. This depends – among other factors – on hygiene practices, clothing, housing, and medical and agricultural traditions. Adaptation to the spread of disease is also determined by the economic level of a given population, the quality and coverage of medical services, and the integrity of the environments. Thus, human biological and psychological factors are primary determinants, but ecological and global systems are also involved, as are economics and access to health care, which shape the vulnerability of societies to disease. Shifts in environmental conditions that interact with the biology of disease agents can also exert profound effects. Changes in land use affect the distribution of disease carriers, such as rodents or insects, while climate influences their range and affects the timing and intensity of outbreaks. Changing social conditions, such as the growth of multimillion-inhabitant cities in the developing world and widespread ecological change, are today contributing to the spread of infectious diseases (Beniston 2010).

Adaptation to climate change is equated with the well-established concept of public health preparedness (Frumkin et al., 2008) and necessitates that traditional public health initiatives are integrated with weather warning systems, climate forecasts, and disaster preparedness. In addition, reducing vulnerabilities and increasing resilience to climate change includes strengthening health systems (Nerlander, 2009) because inadequate healthcare infrastructure contributes to the increase of vulnerability to the effects of climate change and reduces the adaptive capacity of nations (Patz and Kovats, 2002).

There are different adaptation strategies to reduce the vulnerability of human health to climate change. These strategies include measures taken before the observation of impacts (preventive measures), as well as other measures taken after the appearance of such impacts (corrective measures). In addition, adaptation may be the result of political decisions (planned adaptation), or it can be spontaneous (independent measures). In addition, adaptation in unmanaged natural systems is carried out by independent and corrective measures while the managed natural systems can take advantage from preventive measures and planned adaptation strategies. Many adaptation options may produce benefits beyond those associated with climate change; the rehabilitation and maintenance of the public health infrastructure for example is often considered the most important coping strategy, most profitable and most immediately needed.

For adaptation purposes, the World Health Organization has called on the health care sector to prepare for climate change impacts through efforts to increase resilience (CCGHC and Nova Scotia Department of Environment and Health Canada, 2013). In addition to continually engage in broader community discussions and initiatives around climate-related issues, the health care sector should constantly mainstream climate change into risk assessments and consider it when developing plans and activities. Health care and public health professionals and staff should be prepared for climate change by assessing risks from extreme weather events, readiness to manage climate-related infectious disease outbreaks or atypical cases and increasing understanding of how gradual shifts in weather can affect risk profiles. Moreover, adaptation strategies to reduce the vulnerability of human health to climate change should include public health training, monitoring, viable prevention and control programs, and more effective response systems in case of emergencies, including the setting up of multi-disciplinary health care teams to address the various health effects and implications of climate change. Furthermore, health care facilities can reduce risks of climate change through proper management of critical resources (e.g. pharmaceuticals, food, transportation, medical supplies and equipment) based on climate change considerations. A resilient health care facility is also one that commits to sustainable practices, such as water and energy conservation, promoting active transportation, and local food procurement. In investing in resilience activities in these areas, health care facilities can reduce operating costs and increase resilience in the community.

On another level, climate change will likely exacerbate already existing urban social inequities and health risks, thereby exacerbating existing urban health inequities. According to Neiderud (2015), urbanization leads to many challenges for global health and epidemiology of infectious diseases. In other terms, the rise of the new modern cities creates potential risks and challenges in the aspect of emerging infectious diseases. Risk factors in the urban environment could, for example, be the poor housing connected to inadequate water supplies, sanitation and waste management (these conditions contribute to a favorable setting for both rodents and insects which carry pathogens and soil-transmitted helminth infections[2]). Also, if buildings lack effective fuel and ventilation systems, respiratory tract infections can also be acquired. Contaminated water can spread disease, as can poor food storage and preparation, due to microbial toxins and zoonotic agents.

In addition, the new megacities can be incubators for new epidemics, and zoonotic diseases can spread in a more rapid manner and become worldwide threats. The density of inhabitants and the close contact between people in urban areas are often potential hot spots for rapid spread of infectious diseases such as severe acute respiratory syndrome (SARS) and the avian flu. Criteria for a worldwide pandemic could be met in urban centers, which could develop into a worldwide health crisis. Adequate city planning and surveillance can be powerful tools to improve global health and decrease the burden of communicable diseases, hence it must be in the mind of the governing bodies of these megacities.

Urbanization is both a cause of and potential solution to global climate change. Most population growth in the foreseeable future will occur in urban areas primarily in developing countries[3]. How this growth is managed has enormous implications for climate change given the increasing concentration and magnitude of economic production in urban localities, as well as the higher consumption practices of urbanites, especially the middle classes, compared to rural populations. There is still much to learn about the extent to which climate change affects urban health equity and what can be done effectively in different socio-political and socio-economic contexts to improve the health of urban dwelling humans and the environment. But it is clear that equity-oriented climate change adaptation means attention to the social conditions in which urban populations live – this is not just a climate change policy issue, it requires inter-sectoral action. Policies and programs in urban planning and design, workplace health and safety, and urban agriculture can help mitigate further climate change and adapt to existing impacts. If done well, these will also be good for urban health equity (Friel et al., 2011).

From a governance point of view, adaptation to climate change within a health perspective requires the consideration of many issues such as:

- Public health often depends on activities in other sectors. To ensure that climate change health effects are not overlooked, the health sector needs improved integration into strategic planning in sectors such as water, agriculture, and disaster management. This includes safeguarding the integration of health concerns into for example National Adaptation Programmes of Action (Nerlander, 2009). However, and despite emerging efforts to better integrate health and environmental objectives in international and domestic law and policy, scientists, policymakers, and advocates in these two fields remain regrettably cut off from each other (Wiley, 2010).

- Integration of health concerns into the law and governance of other related disciplines should be given high priority. Although scientific research has pointed to the interdependence of the global environment and human health, policymakers have been slow to integrate their approaches to environmental and health concerns. A robust response to climate change will require improved integration on two fronts: health concerns must be given higher priority in the response to climate change and threats associated with climate change and environmental degradation must be more adequately addressed by health law and governance (Wiley, 2010). The call for integration is being heard, but to properly heed it will require that we carefully consider our health and environmental priorities and set forth a mutually beneficial agenda for protecting both. On international level, the mitigation/adaptation response paradigm currently evolving both within and beyond the United Nations Framework Convention on Climate Change (UNFCCC) is a potentially powerful framework for thinking about the integration of health and environmental concerns more broadly[4]. It may even be a useful way of thinking about global health law and governance with respect to other upstream determinants of health (Wiley, 2010).
- Public health planning and decision making need to shift from only focusing on relatively short term risks to the projected long term impacts of climate change. It will be increasingly important to address the links between climate and health at different time- scales. Already today we need to be better at dealing with climate variability and its related health effects. Improving our capacity to prepare and respond, through using for example early warning systems and seasonal forecasts, will allow us to be better positioned to address the challenges that climate change will bring. Long-term climate projections will be increasingly important to ensure that we are prepared for risks changing over time when planning resource allocation, building infrastructure and ensuring that surveillance systems are able to detect changing patterns of disease (Nerlander, 2009).
- Research gaps to increase the understanding of climate change and health should be identified so that mitigation of and adaptation to the environmental effects of climate change in the healthiest and most efficient ways can be facilitated. Filling these gaps in many countries like Morocco also requires the strengthening of the connections between traditional researchers (clinicians, biologists, epidemiologists) and other important researchers (like meteorolo-

gists and climatologists). Additionally, expanding the understanding of the often indirect, long-term, and complex consequences of climate change for human health should be a high priority and challenging research task for both developed and developing countries.

MOROCCO'S ADAPTIVE CAPACITY AND OPTIONS NEEDED TO EFFECTIVELY MANAGE CLIMATE-HEALTH ISSUES

As shown above, it is increasingly recognized that climate change will directly and indirectly impact human health and alter the way countries should protect their vulnerable populations. Worse, the adaptive capacity varies widely from one country to another, and within countries from one region to another. Developing countries, in contrast to rich countries, lack not only expertise, but also adequate resources and effective health systems capable of managing climate-health risks.

This is also true for Morocco which is expected, in addition to current health challenges, to face many public health concerns in the future due to many factors and pressures. The data accumulated over the past decades show that the country is very concerned by climate risks and its potential effects on many sectors, including public health (MMH, 2004, 2012). These data also reveal that climate change will not affect all regions in the same way: while climate-related diseases would affect vulnerable populations in general, rural areas will suffer more due to the various effects of floods and droughts and resource depletion (the recent floods occurred in 2015 in the South of Morocco revealed the extent to which many regions are extremely vulnerable to climate-related disasters and the level of human and economic damage to be reached). This applies in particular to regions where livelihood is closely linked to natural resources (i.e. agriculture, forestry and fisheries). Malnutrition, largely as a result of poor access to food, either because of poverty or the declining agricultural productivity, can also cause various health impacts. Water supply conditions, already precarious in rural areas, could promote the development of water-borne diseases and this effect is likely to be exacerbated during periods of drought. However, changes in the distribution, extent, prevalence, incidence and seasonality of these diseases are to be expected in a context where the magnitude of change is difficult to determine with certainty.

With this prognosis, the state of the health system adds to the problem. It still suffers from many structural and organizational deficiencies which negatively affect its performance and governance. This includes: weak health infrastructure coverage; difficult access to care and health services; weak human and technical capacities with limited training and recruitment opportunities[5]; limited financing capacity of institutional actors (health costs 5.3% of GDP in Morocco, against 6.4% in Tunisia and 9% in OECD); inadequate prevention system; limited trust in the health system and professionals; low level of awareness, linked in part to a deficient educational level especially in rural remote areas, particularly about climate-sensitive diseases; difficulties related to health facilities' management methods; a drug policy relatively transparent when it comes to the determination of prices (which remain poorly harmonized with those applied in similar countries); insufficient partnership, especially between the public sector, private sector and civil society; strong centralization of decisions (which also refers to a kind of non-effective regionalization and the lack of autonomous management); and the persistence of certain health indicators below international standards (MMM, 2012).

Lemsioui (2012) highlights the insufficient consideration of natural risks in the national development plan, the lack of forecast-mapping studies of natural risks, the incomplete state of national legislation system and the absence of a framework-law on risk prevention. For the author, the cumbersome legislative system often prevents the advancement of risk management strategies of natural disasters. Meanwhile, the insufficient awareness of communities with respect to natural hazards and risk magnitude worsen the situation.

This situation may impede the implementation of effective measures to reduce the social vulnerability to climate change, especially in terms of health security. Therefore, the country, already facing many domestic problems with limited capacities to meet pressing social needs, may be more vulnerable face to climate change health impacts.

In order to foster health security in Morocco face to climate risks, adaptation options should aim to: strengthen the health system through upgrading of health facilities and the network of primary healthcare and the introduction of compulsory medical insurance; develop a robust epidemiological surveillance system; strengthen the programs against climate-sensitive diseases by promoting disease surveillance, screening and good patient management, the development of environments that promote health through sectoral programs; and strengthening intersectoral partnerships.

The universal health coverage, whose basic principle is the right to health, should also be a strategic priority. Morocco has recently engaged in this direction through the implementation of numerous initiatives – such as a medical assistance plan called RAMED[6] which aims to reduce the effects of unexpected expenses related to healthcare on the poor (ANAM, 2013) – in order to ensure adequate access to health services, while minimizing health expenses and saving the most vulnerable populations from falling into extreme poverty. Two fundamental components are given priority: access to required health services and financial protection to prevent the degradation of public health from leading vulnerable people to poverty. However, these initiatives are only a first step which must be strengthened through additional funding, outreach to all vulnerable social groups, and the implementation of appropriate governance measures.

In addition, populations need to be regularly sensitized and informed about climate-related health risks and its consequences. Training and awareness programs as well as capacity building for prevention and action against infectious diseases should be an integral part of adaptation options to be considered.

Insufficient or inadequate information on climate and health issues may prevent the adoption of a genuine strategic perspective. Thus, improving knowledge on climate-health issues (including studies analyzing and modeling the evolution of climate parameters and their impact on related notifiable infectious diseases) is needed to better understand the future and prioritize adaptation options. Also, allowing policymakers continuous access to viable and high quality information is crucial to good decision making. To do this, exchange between scientists, experts and decision makers should be institutionalized (via for instance setting up of a competence center for climate and health research, committees of experts, observatories, etc.). In addition, the government should encourage the development of a national expertise and strengthening the links between researchers, decision makers and professionals concerned by climate-health issues.

There is also the need to develop a strategic framework to manage climate change and health issues. Such a framework should enable a continuous exchange between research and policy spheres. The underlying approach consists of guiding all relevant actors during policy making and implementation and reciprocally help define research needs and priorities based on lessons learned from practices.

Given the diverse aspects of climate change-health issues and associated uncertainties, it is still difficult to analyze and investigate these issues. Therefore, caution is recommended in any work aiming to forecast and interpreting risks, especially that human health is in permanent interaction with ecosystems whose biotic and abiotic elements are affected by climate change (ONERC, 2007).

The establishment and improvement of early warning systems, including seasonal climate forecasting, have also become essential to support decision and policy-making processes. Similarly, it is useful to have a surveillance mechanism that can monitor extreme weather events and climate-related epidemics which may affect human health and safety.

Also, the adaptation strategy in the health sector should be guided by the equity imperative, which requires the involvement of all relevant stakeholders in engaged processes. Such involvement will ensure the inclusion of stakeholders' expectations in a spirit of solidarity and complementarity.

The management of climate-related health impacts requires the development of a 'multilevel governance' which offers a more flexible system and promotes the efficiency, consistency and economies of scale. In this governance framework, the health policy also includes action on socio-economic and environmental determinants of health (access to financial resources, nutrition, healthy environment, education, etc.). However, Morocco should gather many conditions to ensure the efficiency of this governance framework such as: Reforming its legal and policy framework related to health and environment; fostering the human rights approach within the health and environment areas; developing regional health and climate plans and strategies; mainstreaming health and climate change in social and human science research and teaching; bridging the gap between health/climate expertise and policy makers; promoting the full involvement of civil society actors in health and climate policy making and implementation processes, etc.

In addition to these, adaptation options to climate change in Morocco from a health perspective should focus on the following key priorities: The allocation of sufficient financial and human resources according to public health needs while ensuring that these resources are distributed fairly in a spirit of regionalization; the establishment of adequate, operational and sufficient health infrastructure, covering the whole territory and ensuring a balance between regions and urban and rural areas; the development of a good health governance by adopting effective management methods, reforming health institutions, strengthening the network of primary healthcare, strengthening the compulsory medical coverage and the development of health professionals' capacities. To support the implementation of such option, Morocco may develop health-focused projects to be submitted to the UNFCCC for funding.

CONCLUSION

This chapter has attempted to demonstrate how infectious diseases in Morocco – mainly Leishmaniasis, Malaria, and Schistosomiasis – are expected to change their geographical or seasonal patterns and incidence due to climate change and variability. Given the gravity of these vector-borne-diseases, the analysis highlights the pressing need to understand its dynamics, especially in a context marked by limited and uncertain scientific evidence regarding the health vulnerability to climate change in Morocco.

In addition to this, the analysis underlines the problematic aspects of adaptation to climate change from a health perspective and the current difficulties in terms of policy and governance action to manage climate-health linkages.

Finally, and by reference to the Moroccan context, an assessment of the national adaptive capacity within the health sector has been undertaken. Relevant recommendations to effectively adapt to climate change and develop effective climate-health governance and policy have been formulated.

REFERENCES

Allen, M. R., & Ingram, W. J. (2002). Constraints on future changes in climate and the hydrologic cycle. *Nature*, *419*(6903), 224–232. doi:10.1038/nature01092 PMID:12226677

Amal, L. (2012). *La prévention et la gestion des risques au Maroc, Ministère de l'Énergie, des Mines, de l'Eau et de l'Environnement*. Rabat: Département de l'Environnement.

Barkia, H., Barkia, A., Nhammi, H., & Belghyti, D. (2011). La schistosomiase au Maroc: de sa découverte à l'après-élimination. *La Revue de Santé de la Méditerranée orientale*, 17(3).

Beniston, M. (2010). Climate change and its impacts: Growing stress factors for human societies. *International Review of the Red Cross*, *92*(879), 557–568. doi:10.1017/S1816383110000342

Bhattarai, A., Abdullah, S., Kachur, P., Martensson, A., Abbas, A., Khatib, R., & Bjorkman, A. et al. (2007). Impact of Artemisinin-Based Combination Therapy and Insecticide-Treated Nets on Malaria Burden in Zanzibar. *PLoS Medicine*, *4*, e309. doi:10.1371/journal.pmed.0040309 PMID:17988171

Bounoua, L., Kahime, K., Houti, L., Blakey, T., Ebi, K. L., Zhang, P., & Boumezzough, A. et al. (2013). Linking climate to incidence of zoonotic cutaneous leishmaniasis (*L. major*) in Pre-Saharan North Africa. *International Journal of Environmental Research and Public Health*, *10*(8), 3172–3191. doi:10.3390/ijerph10083172 PMID:23912199

Boxall Alistair B.A. et al. (2009). Impacts of Climate Change on Indirect Human Exposure to Pathogens and Chemicals from Agriculture. *Environmental Health Perspectives*, 117(4), 508-514.

Canadian Coalition for Green Health Care and the Nova Scotia Department of Environment and Health Canada. (2013), *Health Care Facility Climate Change Resiliency Toolkit*. Retrieved from http://www.greenhealthcare.ca/climateresilienthealthcare/CCGHC-HealthCareFacilityResiliencyChecklist.pdf

Craig, M. H., Kleinschmidt, I., Le Sueur, D., & Sharp, B. L. (2004). Exploring 30 Years of Malaria Case Data in KwaZulu-Natal, South Africa: Part II. The Impact of Non-Climatic Factors. *Tropical Medicine & International Health*, *9*(12), 1258–1266. doi:10.1111/j.1365-3156.2004.01341.x PMID:15598257

Cyril, C., Sari, K., Joacim, R., Adrian, M., Tompkins, A. M., Andrew, P., & Simon, J. L. et al. (2014). Impact of climate change on global malaria distribution. 3286–3291. *Proceedings of the National Academy of Sciences of the United States of America, 11*(9), 3286–3291. doi:10.1073/pnas.1302089111 PMID:24596427

Ebi, K. L., Hartman, J., Chan, N., McConnell, K. J., Schlesinger, M., & Weyant, J. (2005). Climate Suitability for Sta ble Malaria Transmission in Zimbabwe under Different Climate Change Scenarios. *Climatic Change, 73*(3), 375–393. doi:10.1007/s10584-005-6875-2

Faraj, C., Ouahabi, S., Adlaoui, E., Boccolini, D., Romi, R., & El Aouad, R. (2008). Risque de reemergence du paludisme au Maroc étude de la capacité vectorielle d'*Anopheles labranchiae* dans une zone rizicole au nord du pays. *Parasite (Paris, France), 15*(4), 605–610. doi:10.1051/parasite/2008154605 PMID:19202769

Feliciangeli, M. D., & Rabinovich, J. E. (1998). Abundance of *Lutzomyia ovallesi* but not *Lu. gomezi* (Diptera: Psychodidae) correlated with cutaneous leishmaniasis incidence in north-central Venezuela. *Medical and Veterinary Entomology, 12*(2), 121–131. doi:10.1046/j.1365-2915.1998.00072.x PMID:9622364

Friel, S., Bowen, K., Campbell-Lendrum, D., Frumkin, H., McMichael, A. J., & Rasanathan, K. (2011). Climate Change, Noncommunicable Diseases, and Development: The Relationships and Common Policy Opportunities. *Annual Review of Public Health, 32*(1), 133–147. doi:10.1146/annurev-publhealth-071910-140612 PMID:21091194

Friel, S., Trevor, H., Tord, K., Gordon, M., Monge, P., & Joyashree R. (2011). Urban Health Inequities and the Added Pressure of Climate Change: An Action-Oriented Research Agenda. *Journal of Urban Health*, 88(5), 886-895.

Frumkin, H., Hess, J., Luber, G., Malilay, J., & McGeehin, M. (2008). Climate change: The public health response. *American Journal of Public Health, 98*(3), 435–445. doi:10.2105/AJPH.2007.119362 PMID:18235058

Godden Lee et al. (2013), *Governance arrangements for climate change adaptation and natural disaster risk management - An initial analysis of legal and regulatory issues*, Victorian Centre for Climate Change Adaptation Research, available at: <http://www.vcccar.org.au/sites/default/files/publications/VCCCAR%20report%20Governance%20Arrangements%20for%20web%20060913.pdf

Guilvard, E., Rioux, J. P., Gallego, M., Pratlong, F., Mahjour, J., Martinezorlega, E., & Martini, A. et al. (1991). *Leishmania tropica* au Maroc III- Rôle de *Phlebotomus sergenti*. A propos de 89 isolats1. *Annales de Parasitologie Humaine et Comparee, 66*, 96–99. PMID:1776784

Kahime, K., Bounoua, B., Messouli, M., Boussaa, S., & Boumezzough, A. (in press). Evaluation of Eco-Adaptation Strategies of Health to Climate change: Case of Zoonotic Cutaneous Leishmaniasis (ZCL) as Vulnerability Indicator in Pre-Saharan region of Morocco. In M. Behnassi & K. Mcglade (Eds.), *Environmental Change and Human Security in the Middle East and Africa*. Netherlands: Springer.

Kahime, K., Boussaa, S., Bounoua, L., Fouad, O., Messouli, M., & Boumezzough, A. (2014). Leishmaniasis in Morocco: Diseases and vectors. *Asian Pacific Journal of Tropical Disease*, *4*, 530–534. doi:10.1016/S2222-1808(14)60671-X

Kahime, K., Boussaa, S., Laamrani-El Idrissi, A., Nhammi, H., and Boumezzough, A. (2015b). Epidemiological study on acute cutaneous leishmaniasis in Morocco. *Journal of Acute Diseases*. doi:10.1016/j.joad.2015.08.004.

Kahime, K., Boussaa, S., Ouanaimi, F., & Boumezzough, A. (2015a). Entomological Assessment of Human Visceral Leishmaniasis Risk in area with sporadic cases, Morocco. *Acta Tropica*, *48*, 58–65. doi:10.1016/j.actatropica.2015.04.006 PMID:25930188

Lieshout, V. M., Kovats, R. S., Livermore, M. T. J., & Martens, P. (2004). Climate change and malaria: Analysis of the SRES climate and socio-economic scenarios. *Global Environmental Change*, *14*(1), 87–99. doi:10.1016/j.gloenvcha.2003.10.009

Lovell, S. A. (2011). Health governance and the impact of climate change on Pacific small island developing states. *IHDP Update Issue*, *1*, 50–55.

Martens, W. J. M., Jetten, T. H., & Focks, D. A. (1997). Sensitivity of malaria, schistosomiasis and dengue to global warming. *Climatic Change*, *35*(2), 145–156. doi:10.1023/A:1005365413932

Martens, W. J. M., Jetten, T. H., Rotmans, J., & Niessen, L. W. (1995). Climate change and vector-borne diseases: A global modeling per-spective. *Global Environmental Change*, *5*(3), 195–209. doi:10.1016/0959-3780(95)00051-O

McCreesh, N., & Booth, M. (2013). Challenges in predicting the effects of climate change on *Schistosoma mansoni* and *Schistosoma haematobium* transmission potential. *Trends in Parasitology*, *29*(11), 548–555. doi:10.1016/j.pt.2013.08.007 PMID:24064438

Morgan, J. A. T., Dejong, R. J., Snyder, S. D., Mkoji, G. M., & Loker, E. S. (2001). *Schistosoma mansoni* and *Biomphalaria*: Past history and future trends. *Parasitology*, *123*(7 Suppl.), 211–228. doi:10.1017/S0031182001007703 PMID:11769285

Moroccan Ministry of Health (MMH). (1980). *Revue bibliographique des publications et travaux sur la bilharziose au Maroc* (Vol. 1 et 2).

Moroccan Ministry of Health (MMH). (2004). *Direction de l'épidémiologie et de lutte contre les maladies (1966–2004) Données épidémiologiques des maladies sous surveillance. Bulletin épidémiologique*. Maroc: Ministère de la Santé.

Moroccan Ministry of Health (MMH). (2012). La Stratégie Sectorielle "Santé" 2012-2016, http://srvweb.sante.gov.ma/Docs/Documents/secteur%20santé.pdf

Moroccan Ministry of Health (MMH). (2014). A report on progress of control programs against parasitic diseases. Directorate of Epidemiology and Disease Control, Ministry of Health, Rabat, Morocco. Retrieved from http://www.sante.gov.ma/departements/delm/index-delm.htm

Agence Nationale de l'Assurance Maladie (ANAM) (2013), Feuille de Route 2014-2018. Retrieved from http://www.anam.ma/upload/document/feuillederoute.pdf

Neiderud Carl-Johan. (2015). How urbanization affects the epidemiology of emerging infectious diseases. *Infection Ecology and Epidemiology*, 5. Retrieved from http://www.infectionecologyandepidemiology.net/index.php/iee/article/view/27060 doi:10.3402/iee.v5.27060

Nejjar, R., Lemrani, M., Malki, A., Ibrahimy, S., Amarouch, H., & Benslimane, A. (1998). Canine leishmaniasis due to Leishmania infantum MON-1 in northern Morocco. *Parasite (Paris, France)*, *5*(4), 325–330. doi:10.1051/parasite/1998054325 PMID:9879555

Nerlander, L. (2009). Climate Change and Health, Commission on Climate Change and Development. Retrieved from http://www.pacificdisaster.net/pdnadmin/data/original/CCCD_2009_CC_health.pdf

Nicky, M., Grigory, N., & Mark, B. (2015). Predicting the effects of climate change on *Schistosoma mansoni* transmission in eastern Africa. *Parasites & Vectors*, *8*(4). doi:10.1186/s13071-014-0617-0 PMID:25558917

ONERC. (2007). *Changements climatiques et risques sanitaires en France*. Rapport de l'Observatoire national sur les effets du réchauffement climatique, Retrieved from http://www.developpement-durable.gouv.fr/IMG/ecologie/pdf/Rapport_ONERC_version_site_27-09-07_-_1.67Mo.pdf

Patz, J., & Kovats, R. (2002). Hotspots in climate change and human health. *British Medical Journal*, November, 325. PMID:12424173

Portier, C. J., Thigpen Tart, K., Carter, S. R., Dilworth, C. H., Grambsch, A. E., Gohlke, J., . . . Whung, P.-Y. (2010). A Human Health Perspective On Climate Change: A Report Outlining the Research Needs on the Human Health Effects of Climate Change. Research Triangle Park, NC: Environmental Health Perspectives/ National Institute of Environmental Health Sciences.; Retrieved from https://www.niehs.nih.gov/health/materials/a_human_health_perspective_on_climate_change_full_report_508.pdf

Pratlong, F., Rioux, J. A., Dereure, J., Mahjour, J., Gallego, M., Guilvard, E., & Saddiki, A. et al. (1991). *Leishmania tropica* au Maroc. IV. Diversité isozymique intrafocale. *Annales de Parasitologie Humaine et Comparee*, *66*, 100–104. PMID:1776781

Rioux, J. A., Lanotte, G., Petter, F., Dereure, J., Akalay, O., Pratlong, F., & Velez, G. ... Serres, E. 1986. Les leishmanioses cutanées du bassin méditerranéen occidental: de l'identification enzymatique à l'analyse éco-épidemiologique, l'exemple de trois 'foyers', tunisien, marocain et français. In Leishmania. Taxonomie et Phylogenése. Applications Écoépidémiologiques, ed. Rioux, J.A. Montpellier, France: Institut Méditerranéen d'Études Épidémiologiques et Écologiques, 365-395.

Rioux, J. A., Petter, F., Akalay, O., Lanotte, G., Ouazani, A., Seguignes, M., & Mohcine, A. (1982). *Meriones shawi* (Duvernoy, 1842) (Rodentia, Gerbillidae), réservoir de *Leishmania major* Yakimoff et Shokhor, 1914 dans le Sud Marocain. *Comptes Rendus de l'Académie des Sciences*, *294*, 515–517.

Royaume du Maroc. (2010). *Seconde communication nationale à la CCNUCC*. Secrétariat de l'État Chargé de l'Eau et de l'Environnement.

Southwood, T. R. E. (1977). Habitat, the templet for ecological strategies? *Journal of Animal Ecology*, *46*(2), 337–365. doi:10.2307/3817

Teklehaimanot, H. D., Schwatrz, J., Teklehaimanot, A., & Lipsitch, M. (2004). Alert Threshold Algorithms and Malaria Epidemic Detection. *Emerging Infectious Diseases*, *10*(7), 1220–1226. doi:10.3201/eid1007.030722 PMID:15324541

Trari, B., & Carnevale, P. (2011). De la préélimination à l'élimination du paludisme au Maroc. Quels risques pour l'avenir? *Bulletin de la Société de Pathologie Exotique*, *104*(4), 291–295. doi:10.1007/s13149-011-0156-2 PMID:21638201

Wiley, L.F. (2010). Mitigation/ adaptation and health: Health policymaking in the global response to climate change and implications for other upstream determinants. *Journal of Law, Medicine and Ethics*, 38(3), 629.

World Health Organization. (1990). Health education in the control of schistosomiasis (Ch. 8). Freshwater snails. Intermediate hosts of schistosomiasis and foodborne trematode infections, 337p.

World Health Organization. (2012a). *Accelerating work to overcome the global impact of neglected tropical diseases: a roadmap for implementation.* Geneva, Switzerland: WHO.

World Health Organization. (2012b). *Élimination de la schistosomiase au Maroc, une réalité et un succès après trois décennies de lutte.*

Yates, T. L., Mills, J. N., Parmenter, C. A., Ksiazek, T. G., Parmenter, R. R., Vande Castle, J. R., & Peters, C. R. (2002). The ecology and evolutionary history of an emergent disease: Hantavirus pulmonary syndrome. *Bioscience*, *52*(11), 989–998. doi:10.1641/0006-3568(2002)052[0989:TEAEHO]2.0.CO;2

ENDNOTES

1 Climate change mitigation strategies are also synonymous with health improvement, whether through improved housing, active transport, changes to patterns of food consumption or economic localization.

2 Soil-transmitted helminth infections are among the most common infections worldwide and affect the poorest and most deprived communities. They are transmitted by eggs present in human faeces which in turn contaminate soil in areas where sanitation is poor. The main species that infect people are the roundworm (*Ascaris lumbricoides*), the whipworm (*Trichuris trichiura*) and hookworms (*Necator americanus* and *Ancylostoma duodenale*). For more information: http://www.who.int/mediacentre/factsheets/fs366/en/

3 About a century ago, only 20% of the world's population lived in cities, and in the least developed countries the percentage was only 5%. Approximately half of the world's population now live in these urban centers. The two inhabited continents, which currently are the least urbanized, are Asia and Africa, with respectively 48 and 40% of the population living in cities. These percentages are expected to rise dramatically by the year 2050 to 64 and 56% respectively (Neiderud, 2015).

4 The Paris Agreement recently adopted by the COP21 recognizes the links between climate change and health concerns: It recognizes the social, economic and environmental value of voluntary mitigation actions and their co-benefits for adaptation, health and sustainable development; and acknowledges that

Parties should, when taking action to address climate change, respect, promote and consider their respective obligations on human rights, the right to health… (FCCC/CP/2015/L.9/Rev.1).

5 The remote geography of many rural areas in Morocco also creates additional challenges such as the retention and training of medical staff, the collection of health system data, and the provision of infrastructure.

6 RAMED: Régime d'Assistance Médicale.

Chapter 7
Emergence of the Ebola Virus Disease in West Africa

Samir Dervisevic
Norfolk and Norwich University Hospital and University of East Anglia, UK

ABSTRACT

This chapter gives an overview on the recent outbreak of Ebola Virus Disease in West Africa which has lasted for over seventeen months. The Ebola virus has been implicated as a causative agent of viral haemorrhagic fever occurring in Central Africa over the last thirty-nine years. However, the Ebola virus has not previously been recognised as an endemic virus causing outbreaks of viral illness in West Africa. The start of what was to become the largest Ebola virus disease (EVD) outbreak in known history was first reported to the World Health Organization (WHO) on the 23rd of March 2014 and since then it has transformed into an unprecedented and severe epidemic affecting the three countries of West Africa (Guinea, Liberia and Sierra Leone). The emergence of this lethal virus in a setting of profound poverty, a dysfunctional public-health and a weak government infrastructure alarmed the wider world and caused dread from an uncontrollable spread.

DOI: 10.4018/978-1-5225-0553-2.ch007

INTRODUCTION

An emerging infectious agent is defined as a causative agent of a new or previously unrecognised infectious disease in a defined population. The Ebola virus has been implicated as a causative agent of viral haemorrhagic fever occurring in Central Africa over the last thirty-nine years (Baize et al., 2014). People living in West Africa have certainly been familiar with outbreaks of another zoonotic virus causing an acute viral illness: Lassa virus. However, the Ebola virus has not previously been recognised as an endemic virus causing outbreaks of viral illness in West Africa.

The start of what was to become the largest Ebola virus disease (EVD) outbreak in known history was first reported to the World Health Organization (WHO) on the 23rd of March 2014 and since then it has transformed into an unprecedented and severe epidemic affecting the three countries of West Africa (Guinea, Liberia and Sierra Leone) and lasting for seventeen months (Baize et al., 2014). The name of the causative agent, Ebola, has for a long time instilled fear and anxiety, not just in the endemic areas of Africa, hardest hit by this virus, but also around the world. Since the start of the epidemic, people's awareness of the impact of the Ebola virus infection became even more acute as it became obvious that the virus could be spread by air travel. The emergence of this lethal virus in a setting of profound poverty, a dysfunctional public-health system and a weak government infrastructure alarmed the wider world and caused dread from an uncontrollable spread. Mankind witnessed a multitude of human losses, orphaned and abandoned children, uncollected bodies lying on the roads, infected patients wandering outside medical compounds searching for food, decimated medical and nursing staff and social disruption. The emergence of the Ebola virus disease in the low-income countries of West Africa has demonstrated the limitation of the unsupported local efforts, largely reactive, to contain the epidemic. Yet, the initial world response could be summarised in one word: delay.

Unlike previous outbreaks of Ebola in Africa, this epidemic had spread along the roads, reaching villages, towns and the crowded slums of the capitals of all three countries (Baize et al., 2014). Cases of the Ebola virus disease were reported in other countries outside the affected area and with the global movement of people and workforce, there was a real possibility of global spread. During the first half of 2015, the epidemic had spread much faster than the response from the international community. Once the effective surveillance and development of containment systems were put in place, the number of new cases started to drop. Towards the end of 2015, the number of new cases became very low; however even after declaring all three countries Ebola free, sporadic cases still occurred demonstrating the unpredictability of the virus.

The WHO statistics by the 14th of February 2016 showed that there have been 28,639 registered cases of Ebola virus disease and 11,316 deaths (WHO, 2015). The outbreak of the Ebola virus disease in West Africa has claimed more lives than all previous Ebola outbreaks combined (Centers for Disease Control and Prevention, 2014). The majority of cases were recorded in the three West African countries, Guinea, Liberia and Sierra Leone. However, a limited number of exported cases was confirmed in seven other countries worldwide including: Nigeria, Mali, Senegal, the United States of America, Italy, Spain and the United Kingdom.

The ease with which the Ebola virus disease entered and spread within the human population in the affected area illustrated infrastructure weaknesses of the health care systems of several developing countries as well as the fragility and the lack of preparedness for immediate response of the world healthcare systems. The majority of medical and scientific staff that have been involved with the epidemic, as well as many politicians now support the view that international aid came too late and was for the most part ineffective. The three countries hardest hit by the outbreak (Guinea, Liberia, and Sierra Leone) have sustained severe economic consequences which are likely to last for many years to come (Baize et al., 2014). The potential of this epidemic to become global highlighted the need for joint collaborative action in order to improve monitoring of emerging viral pathogens, their rapid diagnosis including efforts to control their spread and the necessity to invest in new treatments and prophylaxis. It also emphasised the risks associated with hunting/consuming 'bushmeat' especially that these practices have the potential to cause epidemics and pandemics.

Ebola Virus

Ebola viruses are a group of enveloped, non-segmented, negative-strand ribonucleic acid containing viruses belonging to the family of *Filoviridae* which are the causative agents of one of the most lethal viral haemorrhagic fevers in humans and nonhuman primates. Members of the family include the single species Marburg virus, Ebola virus with five distinctive species and the single species Cuevavirus, affecting *Miniopterus Schreibersii* bats in southern Europe (Negredo et al., 2011) whilst its pathogenicity in humans is still unknown. The majority of studies involving a description of filoviruses have concentrated on the more infamous Ebola species.

Ebola species include Zaire Ebolavirus, Sudan Ebolavirus, Bundibuguo Ebolavirus and Taï Forest Ebolavirus, which are all pathogenic in humans and nonhuman primates and are endemic in Africa (Feldmann, 2014), as well as Reston Ebolavirus which appears to be pathogenic in nonhuman primates but not in humans and is endemic in the Philippines (Barrette et al., 2009). Although the outbreaks caused by filoviruses amongst human population generate intense media coverage, it is worth mentioning that the same viruses have on number of occasions severely reduced populations of endangered great ape species in the Congo basin of Central Africa without much media interest.

Since the first discovery of the Filoviruses in 1967 (Siegert, Shu, Slenczka, Peters, & Muller, 1967), a total of forty-one outbreaks affecting humans have been recognised (Centers for Disease Control and Prevention, 2014). Twenty-nine of these outbreaks were caused by the Ebola virus disease and twelve by Marburg virus disease, all occurred in sub-Saharan Africa. The Ebola virus was first described in 1976 during the sudden explosive outbreaks in rural remote villages in the Democratic Republic of Congo and South Sudan (World Health Organization, 1976a, 1976b). In addition to these two countries, several subsequent epidemics have been reported in Gabon, the Republic of Congo and Uganda. Since 2000, the outbreaks of Zaire Ebola virus have been documented almost every year in the Democratic Republic of Congo, Gabon and the Republic of Congo. The Taï Forest Ebola virus involved only a single human infection which resulted from contact with an infected chimpanzee in the Ivory Coast (Formenty et al., 1999). Outbreaks of the Sudan Ebola virus have been recorded in South Sudan in 2000, 2001 and 2004 and in Uganda in 2011 and 2012 (Towner et al., 2004).

The current outbreak of EVD is believed to have been caused by a variant of the virus which diverged from a common zoonotic ancestor virus in Central Africa in 2004 and was ten years later transmitted from its natural reservoir by a single spill over event in Guinea, crossed into Sierra Leone and resulted in an intense transmission between humans (Gire et al., 2014). The first case was believed to be an eighteen-month-old boy who felt ill on the 26th of December 2013 in a village of the Guéckédou Prefecture in the south of Guinea, probably after being infected by a fruit bat or some other wild animal which introduced the virus in the area. By the time he died, three other members of his family had already been infected and all succumbed to illness in January 2014. These deaths triggered a chain of events and on the 10th of March 2014 the local hospitals in Guéckédou and Macenta alerted the Ministry of Health of Guinea and Médecins sans Frontières working in the area about clusters of cases presenting with fever, vomiting, severe diarrhoea and a high mortality rate.

Blood samples collected from these patients were sent to the biosafety level 4 laboratories in Lyon, France and Hamburg, Germany, where the causative agent was confirmed to be Ebola. The first suspected cases of EVD in neighbouring Sierra Leone were recorded soon after the onset of the outbreak in Guinea as people in the area freely moved across the borders. The Ministry of Health of Sierra Leone announced the first confirmed case on the 25th of May 2014 and in the weeks after this announcement, there were hundreds more confirmed cases and several hundreds of unconfirmed ones as the virus reached the critical mass for wider dissemination.

Between June and October 2014, over 4600 people died (WHO, 2015). Cross-border markets had been shut down, farmers had fled affected areas, leaving fruit and crops un-harvested and rotting in the fields. Borders closures, trade and travel restrictions and local restrictions of goods and movement of people halted commercial activities in many areas and slowed disease transmission. The national governments of the three countries affected by the EVD reported smaller numbers of cases to the WHO. However, this did not contain the outbreak. Nigeria reported the first out of two imported case of EVD on the 25th of July 2014 and a week later two American aid workers tested positive for EVD and were flown back to the USA. On the 8th of August 2014, after several months of delay, the WHO declared the epidemic of EVD a Public Health Emergency of International Concern.

Clinical presentation of the Ebola virus disease is often nonspecific and usually include an incubation period of two to twenty-one days following which the affected individual develops fever, headache, weakness, muscle pain, red eyes and blood clotting disorder (disseminated intravascular coagulation) which is manifested by bleeding from the gastrointestinal system and occasionally external bleeding (Feldmann, Sanchez, & Geisbert, 2013). End-stage disease is often manifested with shock, hypotension and multiple organ failure. Filoviruses infect their human hosts through mucosal surfaces or damaged skin as well as by parenteral introduction (needles already used on person infected with these viruses). Infection with EVD causes a high case fatality rate of 30-90%, depending on the virus species (World Health Organization, 1976b).

Burial rituals and handling of dead bodies are particularly high risk activities as the concentration of virus is highest before and following death (Baize et al., 2014). The common practice included washing dead bodies by hand before burials. Relatives of the infected people would also come to bring food and stay close to them. Both practices have contributed to high exposure and transmission rates. Being sick was a stigma and the families would hide their sick, and even deceased people, so that they would not be taken away to hospitals. The repercussion of this practice was that those infected would get weaker, and having received no medical help would become a continuous source of infection, subsequently increasing the risk of transmission to non-infected members of their families. In the later stage of the outbreak, the survivors would often be shunned by the community or even expelled. Social customs and traditional beliefs have therefore greatly contributed to an unabated spread of the EVD.

During the current outbreak which started in 2014, the live virus has been detected in the semen of survivors of the disease several months after disease remission (Mackay & Arden, 2015), the implication of which was the possibility of disease transmission after the persons have been declared cured. Ebola virus genetic material (RNA) has been detectable in semen (day 272), aqueous humour (day 63), sweat (day 40), urine (day 30), vaginal secretions (day 33), conjunctival fluid (day 22), faeces (day 19) and breast milk (day 17) after the onset of disease (Chughtai, Barnes, & Macintyre, 2016), demonstrating that the virus can persist in several different body fluids even after being cleared from the blood. Taking into consideration the high case fatality of EVD and uncertainties surrounding transmission characteristics, Ebola survivors should be considered potentially infectious for a period of time after immediate clinical recovery and medical teams looking after these people and the immediate family contacts should be informed about the risk of disease transmission.

Natural Reservoir of Ebola Viruses

In the case of the Ebolavirus, a driver of emergence is a human contact with unknown or suspected natural host of this virus and the establishment of a virus-host relationship in terms of virus adaptation to a new host. This greatly depends on the probability of productive encounters as well as the concentration of the viruses and hosts in the same area (Barrette et al., 2009). Living in a close proximity of each other or overlapping habitats may facilitate interspecies transmission.

Our current understanding is that humans are end hosts highly susceptible to infection with Ebola and Marburg species, apart from the Reston Ebola virus, whilst nonhuman primates are end hosts susceptible to all filoviruses (Barrette et al., 2009). However, neither humans nor nonhuman primates are considered to be reservoirs of these viruses. Following every major outbreak of the Ebola virus disease, significant efforts were undertaken in order to identify the potential animal reservoirs and arthropod vectors. The ecological studies conducted in Gabon and the Republic of Congo between 2003 and 2006 have implicated three species of fruit bats as reservoirs and vectors for transmission of the Zaire Ebola virus in Africa (Leroy et al., 2009). Virus sequences were identified in *Hypsignatus monstrosus, Epomops frangueti and Myonycteris torquata* (Leroy et al., 2005). Unlike end hosts, bats infected with Ebola virus remain healthy and are capable of shedding the virus for weeks (Leroy et al., 2005).

Bats as a Reservoir of Zoonotic Viruses

An ancient species, bats represent the second largest, most widely distributed mammalian order on the planet that have been around for more than 65 million years, long before the appearance of *Homo sapiens* (Teeling et al., 2005). They inhabit all continents with the exception of Antarctica. Over the last two decades, bats have become a focus of interest due to the recognition that they are reservoirs of several zoonotic viruses which, although not pathogenic in their own species, except for

Lyssaviruses, can cause significant morbidity and mortality in humans (Jones, Purvis, MacLarnon, Bininda-Emonds, & Simmons, 2002). Known examples of viruses found in bats, in addition to filoviruses causing high-profile haemorrhagic fever such as Ebola and Marburg, include Severe Acute Respiratory Syndrome coronavirus (SARS CoV), Nipah, Hendra, rabies and rabies-related lyssaviruses (Teeling et al., 2005). The features making bats an ideal reservoir of viruses, include their immune system which is capable of repairing damaged DNA quickly and the fact that they are the only mammalian species capable of flying. Certainly, because bats were among the earliest mammalian species to develop, it is quite possible that their innate and acquired immune responses have developed important qualitative or quantitative differences from those of other mammals.

The bats (Chiroptera), comprising of over 930 species, belong to the second largest mammalian order, making up 20% of all mammalian species (Jones et al., 2002). The order is further subdivided into two suborders, the Megachiroptera, a single family of flying foxes or fruit bats, which is found only in tropical Africa, and the Microchiroptera, containing 16 families, present across the globe (Jones et al., 2002). The latter suborder consists mainly of insectivorous as well as fruit eating and three species of sanguivorous bats, a feature not found in other mammals. The viruses found in Megachiroptera in addition to Ebola, Marburg, SARS CoV, Nipah, Hendra, rabies and rabies-related lyssaviruses, include several members of the families of alphaviruses, flaviviruses, bunyaviruses and paramyxoviruses (Drexler et al., 2012).

Bats are the only mammalian species capable of flying and the members of Microchiroptera suborder also have the unique capability to echolocate. The majority of bats are nocturnal and only few species are diurnal. Their roosts in the natural environment include forests and caves. However, they also live in man-made structures such as house-roofs, bridges, mines and ruins, bringing them into close proximity with humans and animal livestock. This contributes to the transmission of bat viruses from bats to animals and humans. Expansion of the human population and the consequential increase in acquiring new agricultural land, as well as deforestation and intensification of livestock production, are reducing the bats' natural habitat and forcing the species to adapt to new niches closer to areas of human habitation (Wong, Lau, Woo, & Yuen, 2007). Scientists have pointed out that deforestation and drastic changes in forest ecosystems in tropical Africa are the commonality between several outbreaks of Ebola (Muyembe-Tamfum et al., 2012). Substantial areas of forest have over the years been sold off to logging companies and the amount of forest that remains is at risk of disappearing. In Liberia, for example, agricultural expansion has resulted in the destruction of 20 percent of the country's dense forests since 1979, according to a government report (Bayol & Chevalier, 2004). In the area affected in the current outbreak, there are several big mines. Mining and shipping iron ore require putting new infrastructure in place, which could result in further destruction of the forests. These changes are of particular importance for the spread of zoonotic diseases as they allow exposure of human hosts to a rich pool of pathogen diversity ('pathogen pool hypothesis') (K. A. Murray & Daszak, 2013).

There are around 60 types of viruses (59 being RNA viruses) which have been isolated from the bat species (Wong et al., 2007). In view of the changes in bat habitats bringing these mammals in ever closer proximity with humans, there is a significant potential for the generation of emerging viral infections in the human population. Yet, less than 2% of human pathogens have bats as natural reservoirs (Dobson, 2005). Despite the fact that bats are natural reservoirs for several viruses, there is little evidence that they succumb to the same viruses, except from Lyssaviruses which cause morbidity and mortality in this species (Baker, Schountz, & Wang, 2013). The low pathogenicity of the majority of viruses in bats has led to suggestions that during the bat-virus co-evolution, bats developed an effective mechanism for controlling viral replication (Zhang et al., 2013). The key to controlling viruses is likely to be the bat's immune system. The B cells of the mammalian immune system, which play a significant role in the defence against viral infections, are produced in the bone marrow. However, being the only flying mammals, bats do have hollow bones and may compensate by overproducing B calls in their pelvis and legs (Zhang et al., 2013). Little is known about the second part of the bat's adaptive immune system, the T cells. However, there have been genetic changes in their innate immune system that evolved concomitantly with a high metabolic rate during the flight, causing the need to continuously repair the resulting DNA damage, which is a feature that differs from other mammalian species (Zhang et al., 2013). Some authors have suggested that through increased metabolism and higher body temperature that bats develop during flight, they can increase the rate of their immune responses, potentially representing an evolutionary adjuvant to their immune system (O'Shea et al., 2014). These features, according to the authors, enable bats to tolerate a greater diversity of viruses which may have a high potential for virulence when they spill over to other mammals.

Bats could be an ideal reservoir of emerging viruses. They have come to prominence over the last two decades, most notably due to the impact of high profile viral infections, such as SARS and viral haemorrhagic fevers such as Ebola and Marburg, causing high mortality and instilling fear of death. Humans and bats, both mammalian species, are used to some extent, to share habitats. Recent human activities, including population explosion, intensification of animal farming and agriculture as well as deforestation, urbanisation and eco-tourism, have increased the contact between bats and humans and bats and domestic animals. As demonstrated in the examples of the Nipah and Hendra viruses (Halpin, Young, Field, & Mackenzie, 1999; K. Murray et al., 1995; Paton et al., 1999), domesticated animals have indeed become amplifying hosts, in whom viruses multiply rapidly to high levels, providing an important source of further infection. The likelihood is that human exposure to bats will intensify in the future and that we might face several other newly emerging viral diseases originating in bats.

Understanding the ecology of viruses carried by bats and the subsequent triggers of an outbreak will help in our efforts to contain new emerging viral diseases. However, our understanding of the diversity of viral infections in bats, the mechanism of virus shedding, the bat immune system and its control of viral infections is limited. It is becoming very important to understand the mechanism by which bats effectively co-exist with their viral pathogens, including viral receptors and immune responses. Such understanding is of paramount importance in devising strategies to combat emerging viral infections and associated public health measures. Prevention strategies such as distance creation between bat habitats, fruiting trees and farms with domesticated animals would be highly desirable.

Bats are an ancient species highly capable of adapting to new circumstances. Any attempt to reduce their number in order to reduce zoonotic transmission is counterproductive as they perform several important biological functions such as insect control and pollination of fruiting trees. Moreover, the likelihood is that human activity in the future will overlap with bat habitats. Areas of high risk could include newly acquired farming land via the destruction of forests, new human dwellings in proximity of forested areas, and mining and logging activity in areas in which bats reside. Instead, improved surveillance of bat species and the viruses they carry will become important as well as our understanding of the infection dynamics in the bat hosts and their potential for spillover to humans and domestic animals.

The current epidemic of EVD in West Africa probably started long before the small boy in a village of the Guéckédou Prefecture became infected, which would be difficult to investigate and confirm. It has been noticed that Equatorial African fruit bats migrate within the area, thus movement of Ebola virus through bat colonies from Central Africa into West Africa would be possible. It is thought that more surveillance and use of sensitive molecular testing for the presence of viruses causing haemorrhagic fevers, combined with proper case management and earlier involvement of the international community could have potentially attenuated the epidemic.

Changes in Human Behaviour that Have Potentially Influenced the Emergence of Ebola

In order to understand the emergence of new zoonotic viruses such as the Ebola virus we need to consider changes in human behaviour that may have impacted the human-wildlife interactions and resulted in pressures on wildlife populations. It is beyond dispute that a steady increase in the human population in terms of numbers and density have resulted in more frequent exposures to different wildlife species (Paton et al., 1999). A comprehensive literature review has identified that almost three quarters of emerging infectious diseases are caused by a range of different zoonotic pathogens (Taylor, Latham, & Woolhouse, 2001) to which humans get exposed.

Amongst the human activities related to the increase in human population is bushmeat hunting and selective extraction of high-value timber species (deforestation). Hunting for food is an old human practice that carries a significant risk of exposure to animal pathogens including several viruses. Hunting mammals, particularly nonhuman primates which are phylogenetically close to humans and bats, carries a particular risk of transmission of animal pathogens. The demand for bushmeat in West and central Africa is four times greater than that in the Amazon Basin in South America (Fa, Juste, Del Val, & Castroviejo, 1995). 60% of mammal species in the Congo Basin have been exploited unsustainably and a total of 4.5 million tons of bushmeat extracted annually (Fa, Peres, & Meeuwig, 2001). Bearing in mind poverty in parts of West Africa, weak and fragile infrastructure following recent wars in Liberia and Sierra Leone and lack of resources one can understand that consumption of bushmeat has played an important dietary role among the poorest in these societies. Bushmeat hunting is widespread in some communities in West Africa and bat bushmeat seems to function as subsistence and a source of animal protein. The Straw-coloured Fruit Bat, *Eidolon helvum*, which can harbour Ebola virus, is widely hunted and eaten in the areas affected by EVD. Shooting, netting and handling live bats means coming into contact with bat blood and indeed, people involved in hunting, butchering and consuming bats could potentially get infected by the virus through bites, scratches, spillage of bat bodily fluids, tissues and excrement. Scientists linked nearly all human outbreaks in Gabon and the Democratic Republic of Congo to handling of dead animals by villagers or hunters (Richardson, 2012).

On the other hand, the process of deforestation played a significant role. Interior forest -dependent bat species lose their habitat and have to search for new niches and food sources, which adds to the already prevalent bat populations adapted to living in the human-modified landscape. For example, Straw-coloured bats have been observed in man-altered environments, including city parks (IUCN, 2008). Furthermore, deforestation results in a concentration of bat species in the remaining forest patches, where the risk of disease transmission between bat colonies, other wildlife and humans entering these patches increases significantly. In addition, road building may have facilitated both of these relevant activities by reducing the costs of extracting timber and allowing much easier access to wildlife hunters.

Possible Climatic Triggers of Ebola Outbreaks

Meteorological and climatic factors may have been associated with several infectious diseases and could have influenced disease transmission pathways (Anyamba et al., 2009). Regional weather patterns influence floral characteristics and surface water attributes within an area and indeed, distribution of these resources may influence species fitness and distribution, population density, migration patterns as well as contact probabilities between infected and susceptible hosts within a single species or between different species. There has been little analysis of the meteorological conditions associated with outbreaks of Ebola virus. However, one study by Pinzon and colleagues examined eight outbreaks of Ebola in the period from 1994 and 2002 and found an association with drier-than-normal conditions at the end of the rainy season (Pinzon et al., 2004). Certainly, hydrologic changes could interfere with forest fruit production and foraging behaviour of several animal species. Therefore, climate change could influence transmission pattern and disease risk of bat transmitted viral infections.

CONCLUSION

The emergence of the Ebola virus in West Africa has demonstrated that a variant of this virus has crossed the species barrier and initiated a successful transmission between humans. This has been the largest Ebola virus disease (EVD) outbreak in known history, lasting seventeen months and still causing sporadic cases, with the potential to perpetuate the outbreak. So far, EVD has remained a local epidemic affecting the populations of Guinea, Liberia and Sierra Leone which are highly interconnected. Only few imported cases were introduced into the other parts of Africa, Europe and the United States of America via air and road travel. In terms of lessons learnt, the recent Ebola epidemic has demonstrated the limited ability of the world's public health system to rapidly respond to an emerging viral infectious pathogen and taught us that constant vigilance and improved understanding of such pathogens is paramount. Furthermore, in order to tackle similar epidemics in the future, a joint coordinated outbreak response will be required in order to mobilise available necessary resources including medical personnel and supplies, diagnostic technology, food and water. Governments and organisations outside the affected outbreak regions will need to get actively involved and provide assistance and support to frontline countries. Information collection and rapid communication will need to be developed in order to allow rapid identification of the infectious agents and collaboration in order to contain the spread of disease. Finally, community-based surveillance should be in place in order to rapidly identify pathogens such as Ebola virus as well as to detect sickness and death before outbreaks go out of control. This will require an international partnership involving governments in endemic areas as well as in developed countries, international agencies and NGOs.

REFERENCES

Anyamba, A., Chretien, J. P., Small, J., Tucker, C. J., Formenty, P. B., Richardson, J. H., & Linthicum, K. J. et al. (2009). Prediction of a Rift Valley fever outbreak. *Proceedings of the National Academy of Sciences of the United States of America*, *106*(3), 955–959. doi:10.1073/pnas.0806490106 PMID:19144928

Baize, S., Pannetier, D., Oestereich, L., Rieger, T., Koivogui, L., Magassouba, N., & Gunther, S. et al. (2014). Emergence of Zaire Ebola virus disease in Guinea. *The New England Journal of Medicine*, *371*(15), 1418–1425. doi:10.1056/NEJMoa1404505 PMID:24738640

Baker, M. L., Schountz, T., & Wang, L. F. (2013). Antiviral immune responses of bats: A review. *Zoonoses and Public Health*, *60*(1), 104–116. doi:10.1111/j.1863-2378.2012.01528.x PMID:23302292

Barrette, R. W., Metwally, S. A., Rowland, J. M., Xu, L., Zaki, S. R., Nichol, S. T., & McIntosh, M. T. et al. (2009). Discovery of swine as a host for the Reston ebolavirus. *Science*, *325*(5937), 204–206. doi:10.1126/science.1172705 PMID:19590002

Bayol, F., & Chevalier, J. F. (2004). Current state of the forest cover in Liberia.

Centers for Disease Control and Prevention. (2014). Outbreaks Chronology: Ebola Virus Disease. Retrieved from http://www.cdc.gov/vhf/ebola/outbreaks/history/chronology.html

Chughtai, A. A., Barnes, M., & Macintyre, C. R. (2016). Persistence of Ebola virus in various body fluids during convalescence: Evidence and implications for disease transmission and control. *Epidemiology and Infection*, 144(8), 1652-1660. doi:10.1017/S0950268816000054 PMID:26808232

Dobson, A. P. (2005). Virology. What links bats to emerging infectious diseases? *Science*, *310*(5748), 628–629. doi:10.1126/science.1120872 PMID:16254175

Drexler, J. F., Corman, V. M., Muller, M. A., Maganga, G. D., Vallo, P., Binger, T., & Drosten, C. et al. (2012). Bats host major mammalian paramyxoviruses. *Nature Communications*, *3*, 796. doi:10.1038/ncomms1796 PMID:22531181

Fa, J. E., Juste, J., Del Val, J. P., & Castroviejo, J. (1995). Impact of Market Hunting on Mammal Species in Equatorial Guinea. *Conservation Biology*, *9*(5), 1107–1115. doi:10.1046/j.1523-1739.1995.951107.x

Fa, J. E., Peres, C. A., & Meeuwig, J. (2001). Bushmeat exploitation in tropical forest: An intercontinental comparison. *Conservation Biology*, *16*(1), 232–237. doi:10.1046/j.1523-1739.2002.00275.x

Feldmann, H. (2014). Ebola--a growing threat? *The New England Journal of Medicine*, *371*(15), 1375–1378. doi:10.1056/NEJMp1405314 PMID:24805988

Feldmann, H., Sanchez, A., & Geisbert, T. W. (2013). Filoviridae: Marburg and Ebola viruses. In D. M. Knipe & P. M. Howley (Eds.), *Fields Virology* (6th ed., pp. 923–956). Philadelphia: Lippincott Williams & Wilkins.

Formenty, P., Hatz, C., Le Guenno, B., Stoll, A., Rogenmoser, P., & Widmer, A. (1999). Human infection due to Ebola virus, subtype Cote d'Ivoire: Clinical and biologic presentation. *The Journal of Infectious Diseases*, *179*(Suppl. 1), S48–S53. doi:10.1086/514285 PMID:9988164

Gire, S. K., Goba, A., Andersen, K. G., Sealfon, R. S., Park, D. J., Kanneh, L., & Sabeti, P. C. et al. (2014). Genomic surveillance elucidates Ebola virus origin and transmission during the 2014 outbreak. *Science*, *345*(6202), 1369–1372. doi:10.1126/science.1259657 PMID:25214632

Halpin, K., Young, P. L., Field, H., & Mackenzie, J. S. (1999). Newly discovered viruses of flying foxes. *Veterinary Microbiology*, *68*(1-2), 83–87. Retrieved from http://www.ncbi.nlm.nih.gov/pubmed/10501164 doi:10.1016/S0378-1135(99)00063-2 PMID:10501164

IUCN. (2008). The IUCN Red List of Threatened Species (Version 2014.2). Retrieved from http://www.iucnredlist.org

Jones, K. E., Purvis, A., MacLarnon, A., Bininda-Emonds, O. R., & Simmons, N. B. (2002). A phylogenetic supertree of the bats (Mammalia: Chiroptera). *Biological Reviews of the Cambridge Philosophical Society*, *77*(2), 223–259. Retrieved from http://www.ncbi.nlm.nih.gov/pubmed/12056748 doi:10.1017/S1464793101005899 PMID:12056748

Leroy, E. M., Epelboin, A., Mondonge, V., Pourrut, X., Gonzalez, J. P., Muyembe-Tamfum, J. J., & Formenty, P. (2009). Human Ebola outbreak resulting from direct exposure to fruit bats in Luebo, Democratic Republic of Congo, 2007. *Vector Borne and Zoonotic Diseases (Larchmont, N.Y.)*, *9*(6), 723–728. doi:10.1089/vbz.2008.0167 PMID:19323614

Leroy, E. M., Kumulungui, B., Pourrut, X., Rouquet, P., Hassanin, A., Yaba, P., & Swanepoel, R. et al. (2005). Fruit bats as reservoirs of Ebola virus. *Nature*, *438*(7068), 575–576. doi:10.1038/438575a PMID:16319873

Mackay, I. M., & Arden, K. E. (2015). Ebola virus in the semen of convalescent men. *The Lancet Infectious Diseases*, *15*(2), 149–150. doi:10.1016/S1473-3099(14)71033-3 PMID:25467652

Murray, K., Rogers, R., Selvey, L., Selleck, P., Hyatt, A., Gould, A., & Westbury, H. et al. (1995). A novel morbillivirus pneumonia of horses and its transmission to humans. *Emerging Infectious Diseases*, *1*(1), 31–33. doi:10.3201/eid0101.950107 PMID:8903153

Murray, K. A., & Daszak, P. (2013). Human ecology in pathogenic landscapes: Two hypotheses on how land use change drives viral emergence. *Curr. Opin. Virol.*, *3*(1), 79–83. doi:10.1016/j.coviro.2013.01.006 PMID:23415415

Muyembe-Tamfum, J. J., Mulangu, S., Masumu, J., Kayembe, J. M., Kemp, A., & Paweska, J. T. (2012). Ebola virus outbreaks in Africa: Past and present. *The Onderstepoort Journal of Veterinary Research*, *79*(2), 451. doi:10.4102/ojvr.v79i2.451 PMID:23327370

Negredo, A., Palacios, G., Vazquez-Moron, S., Gonzalez, F., Dopazo, H., Molero, F., & Tenorio, A. et al. (2011). Discovery of an ebolavirus-like filovirus in Europe. *PLoS Pathogens*, *7*(10), e1002304. doi:10.1371/journal.ppat.1002304 PMID:22039362

O'Shea, T. J., Cryan, P. M., Cunningham, A. A., Fooks, A. R., Hayman, D. T., Luis, A. D., & Wood, J. L. et al. (2014). Bat flight and zoonotic viruses. *Emerging Infectious Diseases*, *20*(5), 741–745. doi:10.3201/eid2005.130539 PMID:24750692

Paton, N. I., Leo, Y. S., Zaki, S. R., Auchus, A. P., Lee, K. E., Ling, A. E., & Ksiazek, T. G. (1999). Outbreak of Nipah-virus infection among abattoir workers in Singapore. *Lancet*, *354*(9186), 1253–1256. doi:10.1016/S0140-6736(99)04379-2 PMID:10520634

Pinzon, J. E., Wilson, J. M., Tucker, C. J., Arthur, R., Jahrling, P. B., & Formenty, P. (2004). Trigger events: Enviroclimatic coupling of Ebola hemorrhagic fever outbreaks. *The American Journal of Tropical Medicine and Hygiene*, *71*(5), 664–674. Retrieved from http://www.ncbi.nlm.nih.gov/pubmed/15569802 PMID:15569802

Richardson, J. (2012). Deadly African Ebola Virus Linked to Bushmeat. *Food Safety News*, September 17.

Siegert, R., Shu, H. L., Slenczka, W., Peters, D., & Muller, G. (1967). Zur Ätiologie einer unbekannten, von Affen ausgegangenen menschlichen Infektionskrankheit. *Deutsche Medizinische Wochenschrift*, *92*(51), 2341–2343. doi:10.1055/s-0028-1106144 PMID:4294540

Taylor, L. H., Latham, S. M., & Woolhouse, M. E. (2001). Risk factors for human disease emergence. *Philosophical Transactions of the Royal Society of London. Series B, Biological Sciences*, *356*(1411), 983–989. doi:10.1098/rstb.2001.0888 PMID:11516376

Teeling, E. C., Springer, M. S., Madsen, O., Bates, P., O'Brien, S. J., & Murphy, W. J. (2005). A molecular phylogeny for bats illuminates biogeography and the fossil record. *Science*, *307*(5709), 580–584. doi:10.1126/science.1105113 PMID:15681385

Towner, J. S., Rollin, P. E., Bausch, D. G., Sanchez, A., Crary, S. M., Vincent, M., & Nichol, S. T. (2004). Rapid diagnosis of Ebola hemorrhagic fever by reverse transcription-PCR in an outbreak setting and assessment of patient viral load as a predictor of outcome. *Journal of Virology*, *78*(8), 4330–4341. Retrieved from http://www.ncbi.nlm.nih.gov/pubmed/15047846 doi:10.1128/JVI.78.8.4330-4341.2004 PMID:15047846

WHO. (2015). S., Lau, S., Woo, P., & Yuen, K. Y. (2007). Bats as a continuing source of emerging infections in humans. *Reviews in Medical Virology*, *17*(2), 67–91. Retrieved from http://www.who.int/csr/disease/ebola/enWong doi:10.1002/rmv.520

World Health Organization. (1976b). Ebola haemorrhagic fever in Zaire. *Bulletin of the World Health Organization*, *56*, 271–293. PMID:307456

World Heath Organization. (1976a). Ebola haemorrhagic fever in Sudan. Report of a WHO/International Study Team. *Bulletin of the World Health Organization*, *56*, 247–270. PMID:307455

Zhang, G., Cowled, C., Shi, Z., Huang, Z., Bishop-Lilly, K. A., Fang, X., & Wang, J. (2013). Comparative analysis of bat genomes provides insight into the evolution of flight and immunity. *Science*, *339*(6118), 456–460. doi:10.1126/science.1230835 PMID:23258410

Chapter 8
Rift Valley Fever and the Changing Environment:
A Case Study in East Africa

Johanna Lindahl
Swedish University of Agricultural Sciences, Sweden & International Livestock Research Institute, Kenya

Bernard Bett
International Livestock Research Institute, Kenya

Timothy Robinson
International Livestock Research Institute, Kenya

Delia Grace
International Livestock Research Institute, Kenya

ABSTRACT

Rift Valley fever is a severe disease affecting both humans and animals. The Rift Valley fever virus can be transmitted by body fluids, and the most common way for humans to get infected is from animals. The virus is also vector-borne and can be transmitted by many species of mosquitoes. As with other vector-borne diseases, the epidemiology may vary in response to environmental changes. Here the effects of climate and land use changes on Rift Valley fever, as well as on other vector-borne diseases, are discussed. The effect of irrigation in East Africa on inter-epidemic transmission of RVF is discussed in greater detail, followed by recommendations for future research and actions.

DOI: 10.4018/978-1-5225-0553-2.ch008

INTRODUCTION

The last century has seen a period of ecological change, unprecedented in recent times, with dramatic reductions in pristine habitats, ecosystem services and biodiversity along with equally dramatic increases in numbers of people and of domestic animals. Most land use practices change the environment and are due to human influence, often related to feeding the growing population, such as increased irrigation and land deforestation to allow crop cultivation. Environmental change is ubiquitous, and new diseases continue to emerge, while concerns about the impact of environmental change on human health grow. However, while it seems intuitive that degradation of ecosystems will increase the risk of disease, detailed knowledge on how specific environmental changes in a given context affect disease transmission is often lacking. As a consequence, decision makers may inadvertently increase health risks.

This chapter uses the example of Rift Valley fever (RVF) to explore relations between disease dynamics and environmental change. Two main drivers, climate and land use changes, can be identified. We focus on irrigation as a type of land use change. In addition, RVF is characterized, describing the knowns and unknowns about its epidemiology and recapitulating the history and impact of RVF outbreaks in east Africa. Next we discuss relations between RVF and climate and land use change. Findings from an ongoing study, which shows that an irrigated area can support endemic transmission of RVF, without any signs of outbreaks, are also presented and the relevance of this is discussed.

IMPORTANCE AND DRIVERS OF EMERGING DISEASE

Emerging infectious diseases in both animals and humans cause major economic and health burdens in every part of the world. On average, a new human disease appears every four months and around 75% of emerging diseases are zoonotic (Jones et al., 2008). Most originate from wildlife, and the study of disease emergence has a strong focus on wildlife. However, economically important emerging diseases often involve domestic animals. For example, between 1997 and 2009, six major emerging diseases have together cost at least 80 billion USD: the Nipah virus outbreak in Malaysia, West Nile fever in the USA, severe acute respiratory syndrome (SARS, starting in Asia), highly pathogenic avian influenza (HPAI, starting in Asia), bovine spongiform encephalopathy (BSE, starting in the UK) and RVF in East Africa (World Bank, 2012). In all of these, livestock or animals farmed for human consumption provided either a reservoir or a bridge to transmit the disease to people. Later outbreaks of emerging infectious diseases, such as Middle East respiratory syndrome (MERS) and the Ebola outbreak in West Africa, were also caused by viruses with an animal reservoir; in the case of MERS livestock (camels) are an amplifying host, and in the case of Ebola a livestock interface has been suspected (Atherstone, Smith, Ochungo, Roesel, & Grace, 2015; Wong et al., 2015; Yuen, 2015).

The burden of infectious diseases is not uniform, and in low-income countries a high proportion of disease stems from zoonotic diseases and diseases recently emerged from animals (Grace, Gilbert, Randolph, & Kang'ethe, 2012). In Africa, diseases affect poor people disproportionally and further contribute to their poverty in a vicious circle. In particular, zoonotic diseases have the potential to harm both the livelihoods and health of those depending on livestock. Africa is also the continent where more than half of all outbreaks of emerging infectious diseases verified by WHO between 1996 and 2009 occurred, and where the time lags between outbreak detection and public alerts are the longest (Chan et al., 2010).

Moreover, demographic growth is predicted to remain high in Africa, with the continent's population predicted to reach 4 billion in 2100 (from 1 billion in 2014) (Gerland et al., 2014). This rapid population growth is likely to drive equally rapid changes in ecosystems, including crop expansion into marginal areas, irrigation, deforestation, urban sprawl, road building, mining and bush meat harvesting (Grace & Bett, 2014). Depending on how these changes affect the number of susceptible animals and humans, their risks of exposure and the infectiousness of the infected individuals, they may either increase or decrease disease incidence (Lindahl & Grace, 2015). Land-use change often drives disease and has been estimated to be responsible for more than 20% of disease emergence on the island continent of Australia (McFarlane, Sleigh, & McMichael, 2013). In addition, the risks of land use associated diseases are often exacerbated by change and variation in climate changes and the poor adaptive capabilities of communities.

Influence of Environmental Change on Arthropods and Arthropod-Borne Diseases

The dependence of arthropod vectors on the surrounding environment makes them susceptible to change. Arthropods are exothermic, and thus their life cycles, as well as possible pathogen replication within them, are strongly affected by temperature (Khasnis & Nettleman, 2005; Kramer & Ebel, 2003). Vectors are also dependent on suitable habitats for breeding, determined by factors such as soil, land cover and use, temperature, moisture, rainfall patterns and vegetation. When climate changes, the consequence may be that a vector changes its potential distribution and seasonal pattern of activity (Githeko, Lindsay, Confalonieri, & Patz, 2000; Patz, Campbell-Lendrum, Holloway, & Foley, 2005; Russell, 1998). Climate change may also result in changes in precipitation with varying effects on vectors (Githeko et al., 2000). For example, drastically increased precipitation may flush out larval habitats (Impoinvil et al., 2011; Khasnis & Nettleman, 2005; Murty, Rao, & Arunachalam, 2010), but the long term effect might be establishment of more breeding grounds.

The Effect of Climate and Indirect Consequences

Climate changes have direct and indirect impacts on vector-borne disease transmission. Changes in temperature can change vector competence, and thus cause mosquito species (that are minor or insignificant vectors) to become more important (Mellor & Leake, 2000; Tabachnick, 2013). Whether or not, and how efficiently, an arthropod can act as a vector depends on many factors. Temperature affects the longevity of the vector and the rate of virus multiplication within it (Kramer & Ebel, 2003; Tabachnick, 2013). In addition, humidity and wind could influence the longevity and dispersal of the vector, and precipitation may determine the spatial distribution and suitability of breeding grounds.

So how will vector-borne diseases be affected when climate changes? The complex nature of vector-borne disease transmission makes predictions difficult. It is important to understand the impact of climate change on environmental factors that may influence vector dynamics (Tabachnick, 2010), such as changes in land use, population movements and interactions. Increased temperatures may increase the need for irrigation or dams, thus creating vector habitats, and movement and interactions between humans and different animal species bring new vectors into contact with new hosts. The transmission from the vectors depends on the vector capacity (Cohuet, Harris, Robert, & Fontenille, 2010), defined as the number of potentially infective bites an individual is exposed to during one day from one particular vector species (Black & Moore, 2005).

Extrinsic and intrinsic factors needed to be estimated for the calculation of vector capacity:

- The probability that a vector feeds on a specific host, which depends on the feeding frequency and the proportion of meals taken from that host.
- The vector density in relation to the host density.
- The probability that the vector survives each day.
- The vector competence, which is the proportion of vectors taking an infected meal that become infected and can transmit the infection to another host.
- The duration of the vector's life after the incubation period of the pathogen in the vector.

It is clearly an over-simplification, therefore, to equate increase in vector-borne transmission with increase in temperature. Temperature affects the intrinsic factors of the arthropods, such as development rate and longevity (and thus vector abundance), pathogen incubation period within the vector (extrinsic incubation period (EIP)), behaviour and reproduction. Increased temperatures speed processes up within the arthropods and thus decrease both the incubation period in the vectors and their longevity (Kramer & Ebel, 2003; Tabachnick, 2013). In general, all factors contributing to increased vector abundance, shorter incubation periods, and increased vector survival are likely to increase the disease transmission. However, factors contributing to an increased host density and proportion of suitable hosts could also result in increased disease transmission, and these factors may also be dependent on environmental changes, such as increased droughts making farmers keep more goats instead of relying on crop production.

The indirect effects of climate change on ecosystems are thus also important. Decreased surface water availability may cause large numbers of potential hosts to aggregate closer to the water sources which could intensify transmission, and reduced availability of water bodies could negatively impact the predators that feed on mosquito larvae (Epstein, 2001). In addition, even when an increase in disease incidence is observed after climate change, there are many possible confounding factors such as droughts reducing crop yield and impacting the nutritional status of a community, interrupted infrastructure, which may for example make vaccination campaigns or health care impossible, or decreased social and health services. All these events may contribute to the subsequent increase in disease incidence (Lindahl & Grace, 2015).

Due to these complexities, there are uncertainties as to how different vector-borne diseases may be affected by different aspects of climate change. Each vector-pathogen system has a different range of tolerated temperatures, with optimal transmission occurring within a tighter range (Mellor & Leake, 2000). Since minimum temperatures are predicted to change more than maximum temperatures, it has been suggested that the effects on vector-borne disease transmission are more likely to go from sub-optimal to more optimal, than from optimal to temperatures higher than those ideal for transmission (Ostfeld, 2009). As an example, the lower temperature allowing malaria transmission is estimated to be around 14-18°C, whereas the upper limit is around 35-40°C. When changes occur, it is likely to affect disease transmission in borderline regions more than in areas of core suitability (Githeko, Lindsay et al. 2000), although temperature changes that could make transmission more optimal in the local setting are likely to have an impact. Cold temperatures affect the duration of the mosquito's breeding season and some species are able to continue breeding at lower temperatures (Mogi, 1996). Increases in temperature could lead to a life span shorter than the incubation period, or higher mosquito mortality, therefore, limiting disease transmission. However, the capacity of mosquitoes to find suitable microhabitats to avoid heat is an efficient adaptation strategy.

However, it is not certain that the increased potential for disease transmission caused by climate change will be observable. If the transmission of vector-borne diseases is already high, it may not be possible to observe small changes in transmission rates, and this could be the case especially in tropical low-income countries (Reiter, 2001). The opposite scenario may occur in developed countries, where socioeconomic factors may prohibit increased spread of diseases, in spite of climate becoming more permissive. Mosquitoes capable of transmitting malaria are present as far north as Scandinavia, but because of improved living standards and healthcare, malaria disappeared from Europe in the 20th century (Hulden & Hulden, 2009), and is unlikely to return.

The Effect of Land Use Changes

Climate changes, though an important contributor to mosquito habitat suitability, is not the only factor responsible for distribution changes of vector-borne diseases. Indeed, most of the new introductions of invasive mosquito species have not been related directly to climate change (Gubler et al., 2001). Globalization, increased travels and trade provide excellent means for transportation of adult mosquitoes, as well as their eggs. In spite of these risks, travels and trade are highly promoted and unlikely to be reduced (Lindahl & Grace, 2015).

Vectors have different requirements for their breeding grounds, and these are often, but not always, dependent on meteorological factors. One of the major vectors for RVF in West Africa, *Aedes vexans*, breeds in seasonal ponds, which vary in size with the rainfall, since the mosquito eggs also require a period of draught before the rains in order to hatch (Vignolles et al., 2009). Some of the factors influencing the availability of suitable breeding grounds are, however, affected by anthropogenic activities, such as irrigation, agricultural intensification, urbanization, and deforestation. In addition, climate and land use changes interact with each other as well as other drivers. For example, climate induced deterioration of agricultural conditions may increase the rural-urban migration; it may initiate schemes for irrigation, or cause people to switch agricultural practices. Furthermore, large-scale changes in land use such as deforestation may contribute to climate change.

Irrigation may be an especially potent driver of vector-borne disease in Africa given current forecasts for large-scale expansion. Irrigation has been associated with a plethora of diseases including malaria, schistosomiasis, and RVF (Keiser et al., 2005; Steinmann, Keiser, Bos, Tanner, & Utzinger, 2006). Africa has an enormous potential for irrigation: just 4% of the total cultivated area is irrigated compared to 37% in Asia and 14% in Latin America (You, 2008). It is expected that heavy investments in irrigation are likely to occur over the coming decades and smallholder irrigation could cover 30-40 million ha, benefitting 400 million people (Keiser et al., 2005; Xie, You, Wielgosz, & Ringler, 2014). However, it is possible that this may expose them to additional health risks.

RIFT VALLEY FEVER: AN EMERGING DISEASE

Rift Valley fever (RVF) was identified in the 1930s in the Rift Valley region in Kenya, although probable outbreaks could be traced as far back as 1912 (Daubney, Hudson, & Garnham, 1931; Davies, 2010; Findlay, 1932). Analyses of the viral genome suggests that the original emergence of the virus, proposed to have evolved rather recently, might be due to early land use changes in the shape of colonial agricultural development with introduction of large scale ruminant farming (Pepin, Bouloy, Bird, Kemp, & Paweska, 2010).

The virus was not known to be zoonotic until human cases were described in the 1951 South African outbreak (Davies, 2010), and the full zoonotic potential was realized only in 1977 when an outbreak in Egypt resulted in about 600 human deaths, among some 200,000 human cases (Nanyingi et al., 2015; Soumaré et al., 2012). Rift Valley fever is considered one of the emerging infectious diseases that is having the most severe impacts in the African continent, and is present in most of sub-Saharan Africa, with extensive outbreaks occurring in Kenya, Tanzania, Madagascar, Mauretania, Senegal, Namibia, Mozambique, South Africa, Zimbabwe, Zambia and Sudan (Figure 1), but additional countries have also experienced small outbreaks, or there has been evidence of virus circulation (Nanyingi et al., 2015; Swanepoel & Coetzer, 2004). With outbreaks extending outside sub-Saharan Africa and occurring as far north as Egypt and the Arabian Peninsula, it is feared that the virus has the potential to spread into Europe and to other parts of the world (Chevalier, Pépin, Plée, & Lancelot, 2010; Moutailler, Krida, Schaffner, Vazeille, & Failloux, 2008; Turell, Sardelis, O'Guinn, & Dohm, 2002; Turell et al., 2008). Within Africa, RVF strains seem to be circulating over large distances, with samples in Madagascar shown to be similar to both North and East African strains and to the strains causing outbreaks in the Arabian Peninsula (Sall et al., 1998; Shoemaker et al., 2002). Phylogenetic studies suggest that RVF virus has been introduced into West Africa from other parts of Africa on five different occasions (Soumaré et al., 2012). Trade and transport of infected animals has been suggested to be one of the mechanisms that allowed virus dispersal.

Figure 1. Map over the countries in Africa which have experienced extensive outbreaks of Rift Valley fever (grey); the insert of Kenya shows the study areas for the case study on Rift Valley fever, with dark grey for the pastoral areas in Ijara district, and light grey for the areas close to the irrigation scheme in Tana River district. *Swanepoel & Coetzer, 2004.*

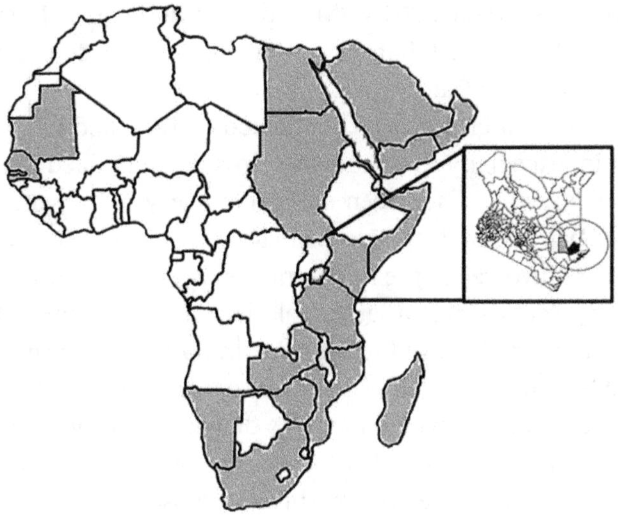

The RVF virus belongs to the *Bunyaviridae* family, in the genus Phlebovirus. This is a large viral family with more than 300 disease-causing viruses which have tripartite single-stranded RNA. Although the viruses often are transmitted by vectors such as mosquitoes or ticks, they may also be transmitted directly by body fluids, especially fetal materials that may carry high loads of virus. The RVF virus can be transmitted by numerous mosquitoes from at least 6 genera and over 30 species (Pepin et al., 2010; Walter & Barr, 2011). However, even though many mosquitoes may harbour the virus, they may not all be important for the disease epidemiology, and different vectors are important in different parts of Africa. In East Africa, it is believed that the primary vectors maintaining the virus and starting outbreaks are floodwater breeding *Aedes* species, whereas secondary vectors fuelling the outbreak are likely to be *Culex* species (Anyamba et al., 2010; Sang et al., 2010; Vilaly et al., 2013).

The virus has a predilection for the liver but spreads to many parts of the body after a host is infected (Pepin, Bouloy, Bird, Kemp, & Paweska, 2010). Clinical cases in humans are usually characterized by hemorrhagic fever and encephalitis, whereas animals abort or have hemorrhages and necrotic hepatitis (Elliott, 1997). Case fatality in young livestock can be high, and although it is difficult to assess accurately how many humans were actually infected in an outbreak, as many as

47% of confirmed and probable cases may succumb to the disease (Mohamed et al., 2010). In addition, the virus frequently causes abortion storms (Pepin et al., 2010). In the earlier outbreaks, it was reported that indigenous livestock were unaffected (Davies & Martin, 2006); however, breed has not been reported as one of the risk factors for animal infection, summarized by recent reviews (Nanyingi et al., 2015; Nicholas, Jacobsen, & Waters, 2014). Moreover, if indigenous breeds are as likely as exotic breeds to be infected, but not to develop disease, seroprevalence studies are unable to detect such a trait.

Although much research has been conducted on RVF and its vectors, some uncertainties remain, especially as to how the virus is maintained during the periods between epidemics, which have been occurring every 5-15 years in East Africa. The vectors for RVF are divided into primary and secondary vectors, where the primary vectors are involved in the initial transmission of the virus to host animals, and the secondary vectors play a larger role later in the outbreak by propagating it. *Aedes* mosquitoes are believed to be most important for disease transmission, especially those breeding in flood water. In East Africa, these mosquitoes breed in depressions termed dambos, where the eggs diapause during the dry periods, and hatch during subsequent flooding that follows prolonged rain (Logan et al., 1991). Transovarial transmission has been shown in many mosquito species, and adult reared from larvae collected in naturally flooded dambos have been shown to contain the virus (Linthicum, Davies, Kairo, & Bailey, 1985). It is therefore believed that the virus may survive in the eggs for years while waiting to hatch, and that this could explain how the virus persists during the inter-epidemic periods. It is also possible that immature stages of mosquitoes can be infected by virus present in the breeding habitats, and it has been shown that larvae of *Culex pipiens*, *Aedes mcintoshi* and *A. circumluteolus* can become infected experimentally and subsequently transfer the virus when feeding as adults (Turell, Linthicum, & Beaman, 1990).

Another explanation is that wildlife reservoirs maintain the virus in an endemic sylvatic cycle. A number of studies have shown that wild animals naturally seroconvert to RVF (Britch et al., 2013; Evans et al., 2008; Olive, Goodman, & Reynes, 2012). Sudden increases in rainfall could lead to more mosquitoes hatching, and the increased number of vectors cause spillover onto livestock and humans. Secondary vectors, such as other *Culex* species also multiply after the rains, which promotes the outbreak. It is even possible that the virus is maintained by a low level circulation among domestic animals.

The epidemiology of RVF is further complicated because body fluids of infected animals (including fetal membranes) are infectious and this is the main cause of infection in humans, although they may still get infected from mosquitoes (Anyamba et al., 2009). Therefore, certain occupation groups such as herders, farmers,

Figure 2. Epidemiology of Rift Valley fever

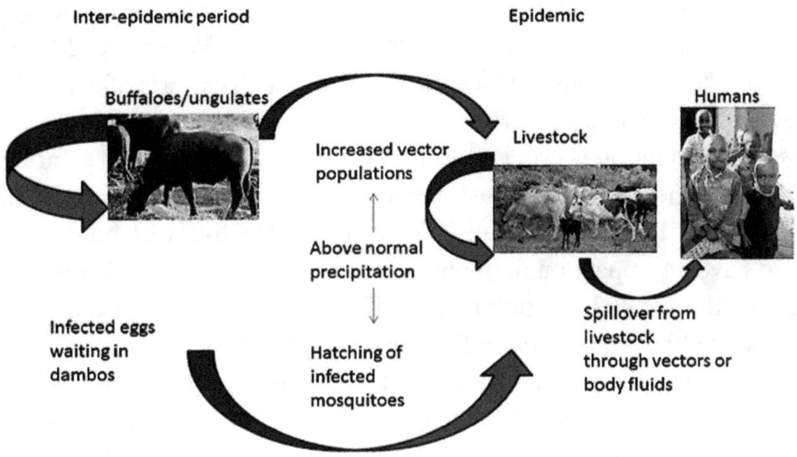

slaughterhouse workers and veterinarians are at greater risk of infection. Similar contacts may occur during care or slaughtering of infected animals or possibly from ingestion of raw milk.

Whereas RVF is mostly known for the severe outbreaks, it is believed to be endemic and circulating between the outbreaks. Molecular epidemiological estimates suggest that the virus may not only persist but also expand undetected during these periods (Pepin et al., 2010). Possible ways in which RVF could be maintained during epidemiological cycles are shown in Figure 2.

Rift Valley Fever Outbreaks in East Africa and Their Impacts

Whereas RVF outbreaks have been associated with semi-arid or irrigated areas in West and North Africa, outbreaks in East Africa often occur in plateau grasslands (Martin et al., 2008). In 1997, Kenya experienced heavy rains, and this was followed by an outbreak of RVF, which may have infected as many as 27 000 humans (Woods et al., 2002). In this outbreak, the main risk factors for human infections were sheltering animals in the home and contact with sheep blood or body fluids. Farmers reported losing up to 70% of their sheep and goats, and as much as 95% of young lambs (Woods et al., 2002). The next major outbreak started in 2006 in Kenya and 2007 in Tanzania, continuing with Somalia and Sudan. While the Tanzania outbreak had a total number 511 suspected cases, it is unknown how many were really infected and it is believed that only severe cases were detected (Mohamed et al., 2010). For the outbreak in Sudan, up to 75,000 humans may have been exposed (Hassan, Ahlm, Sang, & Evander, 2011). It was estimated that more than 900 people died in the outbreak in the four countries affected (Nanyingi et al., 2015). Based on interviews, it was estimated that more than 400,000 livestock died in Kenya during the outbreak, with large economic consequences (Rich & Wanyoike, 2010). Also in Tanzania the economic consequences were severe, with exports of cattle decreased by more than 50% (Sindato, Karimuribo, & Mboera, 2012).

Livestock production is crucial for the livelihood in the arid and semi-arid areas of East Africa, where many are pastoralists, and an outbreak of a severe infectious disease such as RVF can be devastating for the individual livestock holders; however, it will also have an impact on national economy (Anyamba, Linthicum, & Tucker, 2001). After the 2006-2007 outbreak in Kenya, pastoralists ranked RVF as the disease with the highest impact on livestock-derived livelihoods (Jost et al., 2010). This outbreak was estimated to have caused large economic losses to the region. The World Bank (2012) estimated that the entire outbreak in East Africa caused losses of 30 million USD, similar to estimates by Rich and Wanyoike (2010).

Rift Valley Fever, Climate, and Land Use Change

RVF epidemics occur when environmental, ecological and climatic factors are conducive. Many studies have demonstrated the association with excessive rainfall, the El Nino Southern Oscillation (ENSO), changes in normalized difference vegetation index (NDVI) and presence of water bodies (Métras et al., 2011; Tourre, Lacaux, Vignolles, & Lafaye, 2009; Anyamba, Linthicum, & Tucker, 2001). However, the association with weather, which is very clear in East Africa where outbreaks occur after periods of above normal rainfall (in intervals of around 5-15 years) (Anyamba, Linthicum, & Tucker, 2001; Davies, Linthicum, & James, 1985), is not as evident in West Africa (Chevalier, Rocque, & Baldet, 2004).

There have been several attempts to model how climate change impacts RVF; either using knowledge-driven spatial models, or mathematical dynamic transmission models (Métras et al., 2011). The results of some models are being made available online, such as the model by the healthy future project and University of Liverpool (http://www.healthyfutures.eu/ (Morper-Busch, Kienberger, & Hagenlocher, 2015)). The modelling commonly takes into account mosquito vectorial capacity with biting frequencies and life cycle dependent on temperature, and breeding dependent on precipitation. However, there is a need to incorporate other environmental factors, such as land use, human and animal densities, water storage and irrigation, in such models so as to predict RVF transmission more accurately. Although climate is a very important driver of vector dynamics, more factors may be of importance and we lack a complete understanding of the life cycles of the vectors (Tabachnick, 2010). The consequences of this may be suboptimal prediction models and inaccurate decisions taken in the face of outbreaks.

Methods including remote-sensing have also proved very beneficial for studying spatial dynamics of RVF, and the influence of climate and environmental changes, and through this it has been possible to link the dynamics of *Aedes vexans* with rainfall and changes in water bodies (Tourre et al., 2009). Work in West Africa also shows the possibility of including these techniques in creating risk maps and early warning systems (Vignolles, Tourre, Mora, Imanache, & Lafaye, 2010).

Land use changes may have similar effects to climate change, by providing increased mosquito breeding grounds, and changing the availability and movements of different hosts. Increased water storage and utilization along rivers, as well as development of dams and irrigation systems, enlarge the areas that could be suitable for mosquito development, which can increase the risk of RVF transmission. A few outbreaks have been associated with anthropogenic changes to water bodies, such as the 2000-2001 outbreak in Yemen (Abdo-Salem et al., 2006), and the 1987 epidemic in the Senegal River basin (Thonnon et al., 1999). However, it must also be remembered that the purpose of providing a community with irrigation or water reservoirs often is to improve the livelihood of the inhabitants, which may lead to socioeconomic improvements, including increased living standards, improved nutritional status and access to health care and prophylactic measures. These socioeconomic factors may reduce the exposure of humans to pathogens, and make them more resistant to disease, which is likely to reduce the risks of disease spread (Lindahl & Grace, 2015). In addition, agriculture increase, following irrigation, makes the communities more resilient to the devastating effect of a Rift Valley fever outbreak in cattle, by providing more diversified means of income and food sources (Figure 3).

Circulation of RVF Virus in Irrigated Areas during the Inter-Epidemic Period: A Case Study from Kenya

Kenya has considerable potential to increase irrigation (You, Xie, Wood-Sichra, Guo, & Wang, 2014) and there are many ongoing and planned irrigation schemes. The Northeast part of Kenya has been the focus of the last outbreaks of RVF, and the first case in the 2006 outbreak was detected in Garissa (Sang et al., 2010). The land bordering the Tana River is semi-arid with increasing irrigation. This area, including Garissa, and Tana River counties, is a hot spot for RVF outbreaks (Martin et al., 2008), and is also high in biodiversity with many endangered species threatened by land use change (Medley, 1993, 2009). Figure 1 is showing the location of the study areas within Kenya.

In this case study conducted in 2013-14, the aim was to study the circulation of RVF in an irrigated area by investigating seroprevalence of RVF in humans and animals; perform an entomological survey and interview actors in the animal value chains.

Figure 3. How environmental changes can affect Rift valley fever transmission

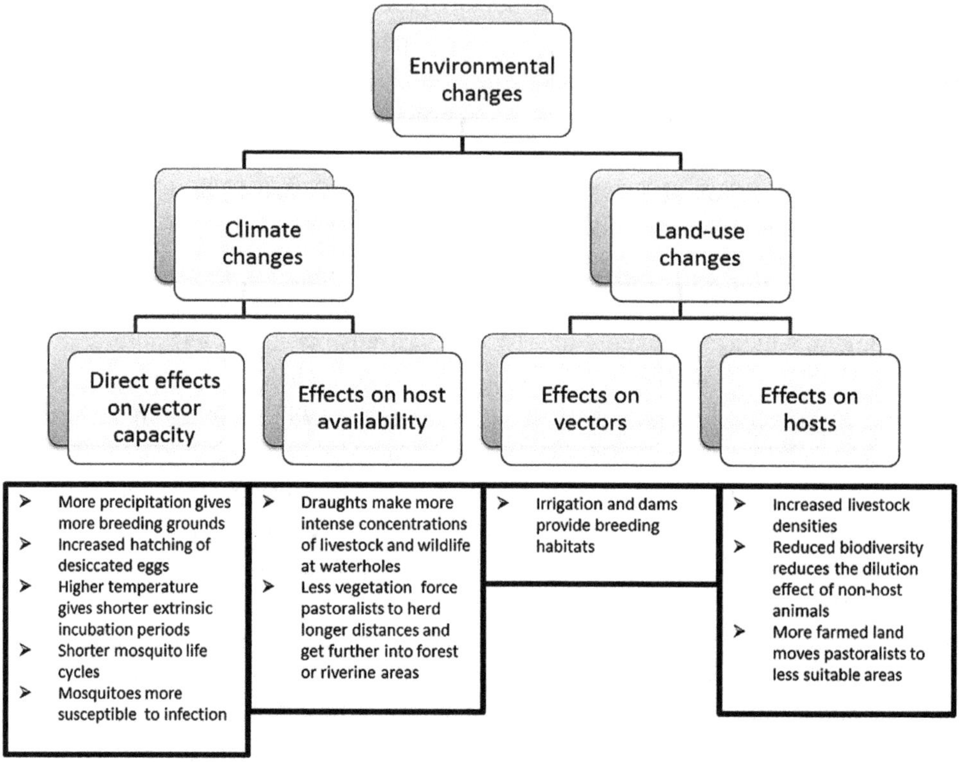

Serum samples from humans, aged 5-90 years old, were collected in villages both from irrigated and non-irrigated, pastoral areas. In total, 303 people were sampled within the irrigation schemes in Bura and Hola in Tana River County, and 728 in the pastoral areas in Ijara district in Garissa County. Serum was tested using a commercial competitive enzyme-linked immunosorbent assay (ELISA) (Biological Diagnostic Supplies Limited, UK) according to manufacturer's instruction, but with a modified conjugate dilution (1:12 000) and adjusted incubation time. The original protocol was modified after consultation with the manufacturer to allow controls to fall within the fixed range before analyses. This ELISA does not differentiate IgG and IgM antibodies (Kortekaas et al., 2013), and therefore it is not possible to tell if infections causing the seropositivity are old or new. Overall 21.5% (95% confidence interval 19-24%) were seropositive, with no significant difference between the areas. Significantly more adults (over 18 years) were seropositive ($p<0.001$), and significantly more men than female ($p=0.04$).

Table 1. Rift Valley fever seroprevalence in humans and ruminants in Northeastern Kenya

	Humans	Ruminants
Overall seropositivity	21.48%	25.59%
Young*	3.71%	12.31%
Adults	30.69%	30.22%
Male	24.88%	14.81%
Female	19.22%	28.80%

*Young ruminants include weaners and younger animals. Young humans include people <18 years.

Ruminants (sheep, goats, and cattle) were sampled in randomly selected villages within the two irrigation schemes located in Hola and Bura, in Tana River County. A total of 1997 animals were sampled, and tested with the same ELISA as the human samples. The overall seroprevalence was 25.6% (95% confidence interval 23.7-27.6%), with no difference between the two irrigation schemes, nor between the seroprevalences in goats, sheep and cattle. Juveniles had a significantly ($p<0.001$) lower seroprevalence rates than adults, 12.3% compared to 30.2%. The seroprevalence among adult ruminants may be confounded by previous vaccination campaigns in sheep and goats, campaigns were known to have taken place but there were no records of animals vaccinated. However, similar prevalence pattern was observed in unvaccinated cows: 36.4% in adults, significantly higher than in calves (19.1%, $p<0.001$). The last outbreak of RVF occurred in 2006-2007, therefore, juveniles could not have been infected during the outbreak. The level of prevalence detected is supportive of local endemic transmission (Table 1).

In the same study, entomological surveys were implemented with Kenyan Medical Research Institute (KEMRI) as has been described previously (Sang et al., 2010). Preliminary observations demonstrate that irrigated farms (specifically drainage canals) support the development of the primary vectors of RVF, mainly *Aedes mcintoshi* and *Aedes ochraceus*. These mosquitoes are believed to develop in floodwater after periods of prolonged heavy rainfall (Linthicum, Davies, Bailey, & Kairo, 1984) and it has been shown that they are infected with RVF in the same area (Sang et al., 2010). The irrigated areas had significantly higher numbers of both primary (*Aedes* spp) and secondary (*Culex* spp) vectors (Sang et al. unpublished data). This emerging evidence shows that irrigated fields can mimic breeding conditions established in flooded areas and there is a need to integrate vector management activities with irrigation, especially in areas that are conducive for RVF transmission.

Interviews and focus groups discussions among livestock value chain actors in Tana River during the same time showed that knowledge about RVF was high in some groups, such as animal traders and slaughterhouse workers, where 88 and 90% respectively knew about the disease, whereas only 40% of milk traders knew about it. Out of 43 interviewed value chain actors, only two said they would report dead livestock to veterinarian, whereas nine reported that they would slaughter the animal. It was also common to slaughter sick animals, and the main reason for this was meat consumption (49%) (Nyokabi, 2015).

There was a higher seroprevalence rate in female animals, which may warrant further studies. Since RVF may cause reproductive failures and abortions in females, and aborted materials contain high viral loads (Pepin et al., 2010), this might indicate that the females are infected through contact with this infected material and not only through mosquito bites. This also indicates that even endemic RVF transmission could have economic impacts, and this might go undetected in an environment where there are many other pathogens causing abortions and reproductive failure. In humans, men are more exposed to infections. A higher prevalence in male herders has been observed in previous studies, and risk factors shown to increase the risk of seropositivity include contact with livestock, handling sick animals, fetal materials, or activities connected to slaughtering (Nanyingi et al., 2015).

It was observed that seroconversions must have occurred in the inter-epidemic period, since some positive animals were too young to have been infected in the last known outbreak 2006-07. Seroconversion in livestock during inter-epidemic periods has been observed before. Infected vectors have been encountered during inter-epidemic periods in Kenya, as well as seroconversions in both wildlife and livestock (Linthicum et al., 1985). This has also been found in other countries; in a recent study in Tanzania, 8% of the small ruminants were seropositive, including animals borne after the last outbreak (Wensman et al., 2015). Awareness is often decreasing in inter-epidemic periods (Martin et al., 2008), although we found varying levels of awareness among different risk categories.

Overall, this study showed that an irrigated area can have an endemic transmission of RVF, without any sign of outbreaks, but that seroprevalence is not necessarily higher in these areas compared to non-irrigated ones, in spite of significantly higher number of mosquitoes. This indicates that although it may be evident that irrigation increases the abundance of vectors, other factors also play a role in determining if infection will occur. Higher number of vectors are however likely to contribute to increased risk of outbreaks, but would need to be studied more intently during an epidemic period. Future work should include looking into how factors such as irrigation, agricultural and pastoral behaviour changes, animal and human densities, and socioeconomic parameters, including education and living standards impact on RVF disease transmission and outbreak occurrence.

SOLUTIONS AND RECOMMENDATIONS

As shown, both climate and land use changes can affect the epidemiology of RVF. The case of irrigation is particularly relevant in sub-Saharan Africa, which has low levels of irrigation and high levels of population growth, and many believe large investments in irrigation are necessary for sufficient food supply. Irrigation schemes are often constructed in dry/hot areas where rates of vector development are likely to be rapid, if water sources provide suitable breeding sites. This increased risk should be considered and monitored when designing new irrigation schemes or planning new dams. In addition, the effect of climate change on the introduction of new vectors and pathogens should also be considered in order to use the most suitable technologies and be prepared for public health consequences.

There is no single optimal solution for dealing with the risk of RVF outbreaks. Depending on the risks in the region, vaccination campaigns before predicted outbreaks may be the most cost efficient option, but require good forecasting models, and for policy makers to be well informed about the disease. In addition, it is important that stakeholders and policy makers are informed about the risks of RVF transmission between epidemics and RVF emergence. Other strategies include better diagnostics, case management and farming practices to reduce RVF outbreak risks.

Importantly, there is a lack of reliable guidelines, tools or procedures to steer emergency response measures in the event of an outbreak. In the past, the losses associated with outbreaks have in part been a result of insufficient capacity offered by decision makers, who lacked accurate information about efficient preventive and response measures. More recently, there have been promising developments in planning for RVF control in East Africa. These include:

- **Risk Maps and Early Warning Systems:** Known disease hotspots have been geo-referenced and analysed together with detailed data on precipitation, rainfall, soil types, land use and vegetation to produce RVF risk maps (Martin et al., 2008; Sindato et al., 2014). These maps can be used to target surveillance efforts, and also in trade negotiations. This could put an end to the blanket trade bans and their disastrous effects on people's livelihoods. The maps have also been combined with socio-economic variables, which define vulnerability in order to determine how best to respond to an outbreak in local communities.
- **Contingency Plans and Standard Methods and Procedures:** These are key to mounting early, effective action in the face of an emergency. These are pre-prepared plans and operational protocols to create uniformity in animal disease detection, diagnostic and control procedure across a region. The

Government of Kenya produced and endorsed a contingency plan for RVF in 2010 which was updated in 2014. In addition, a standards methods and procedures manual has been produced by AU-IBAR (2014).

- **Decision Support Tool:** Policy makers face a dilemma in responding to RVF. Too early response and they may cause unnecessary panic. Too late and impacts will be more severe. A decision support tool has been developed comprising a list of the sequence of events related to increasing and decreasing the risk of an RVF epizootic and the appropriate actions that should match these events (Consultative Group for RVF Decision Support, 2010).
- **Vaccination Guidance:** This is often lacking among policy makers. Model-based analyses can offer guidance on preventative action by indicating the proportion of animals that need to be vaccinated to prevent spread. In addition, model-based analyses demonstrate that vaccinating strategically over a five-year period could reduce the thresholds needed to stop an epidemic. Such periodic vaccinations could be done annually or biannually, depending on logistical/economic considerations and the type of vaccine available.
- **Impact Assessment Tools:** A system dynamics model that combines biological and socio-economic processes involved in Rift Valley fever transmission has been developed as an integrated framework for evaluating the impacts of the disease along livestock value chains in north eastern Kenya.
- **Disease Scenarios:** These are plausible stories, not projections, and illustrate what might happen, based on given assumptions in the chosen settings. They can be used to identify and discuss a range of possible futures and used to create action to build resilience. Scenarios have been developed for RVF along with some other emerging diseases in Africa (http://www.diseasescenarios.org/scenarios.html), and others have looked at how agriculture and disease risks in general will develop in different future scenarios (Oborn et al., 2013).

FUTURE RESEARCH DIRECTIONS

As mentioned above, there is a lack of evidence regarding RVF epidemiology and factors that maintain RVF between outbreaks. Such knowledge is necessary for efficient future interventions. Future research should focus on filling the gaps identified here and reported in the literature:

- The importance of different vectors, including their vector competencies, their breeding requirements and the risks of transovarial transmission.
- The inter-epidemic transmission, including the role of wildlife and domestic animals, and the risks for humans.

- The role of socioeconomic factors in improving resilience of vulnerable populations during outbreaks.
- The impacts of land use change and climate change on disease epidemiology.

CONCLUSION

It is difficult to predict exactly how environmental changes will affect the incidence of vector-borne diseases, and it must be remembered that other changes, not directly influencing the vectors directly, may play very substantial roles. RVF as a disease may be directly influenced by environmental and climatic changes, and increasing irrigation may be one cause of increased vector populations. However, this may not be directly translated into higher prevalence of disease, and it is important to further study the other factors that may contribute to this. In order to increase the preparedness for RVF outbreaks, it is important to bridge the knowledge gaps and to improve surveillance systems and decisions support tools that can aid stake holders.

Whereas social insecurity, political unrest and poverty may drive the emergence of disease, improved housing, education, nutrition and health care services will render people less exposed to mosquito bites and more resistant to disease. Increased irrigation and agricultural outputs may help in achieving this. Improving socio-economic conditions is therefore likely to increase people's resilience to the negative effects of environmental change with regard to vector-borne diseases.

REFERENCES

Abdo-Salem, S., Gerbier, G., Bonnet, P., Al-Qadasi, M., Tran, A., Thiry, E., & Roger, F. et al. (2006). Descriptive and spatial epidemiology of Rift valley fever outbreak in Yemen 2000-2001. *Annals of the New York Academy of Sciences*, *1081*(1), 240–242. doi:10.1196/annals.1373.028 PMID:17135517

Anyamba, A., Chretien, J.-P., Small, J., Tucker, C. J., Formenty, P. B., & Richardson, J. H., … Linthicum, K. J. (2009). Prediction of a Rift Valley fever outbreak. *Proceedings of the National Academy of Sciences of the United States of America*, 106(3), 955–9. doi:doi:10.1073/pnas.0806490106 doi:10.1073/pnas.0806490106

Anyamba, A., Linthicum, K. J., Small, J., Britch, S. C., Pak, E., de La Rocque, S., & Swanepoel, R. et al. (2010). Prediction, assessment of the Rift Valley fever activity in East and Southern Africa 2006-2008 and possible vector control strategies. *The American Journal of Tropical Medicine and Hygiene*, *83*(2 Suppl.), 43–51. doi:10.4269/ajtmh.2010.09-0289 PMID:20682905

Anyamba, A., Linthicum, K. J., & Tucker, C. J. (2001). Climate-disease connections: Rift Valley fever in Kenya. *Cadernos de Saude Publica, 17*, S133–S140. doi:10.1590/S0102-311X2001000700022 PMID:11426274

Atherstone, C., Smith, E., Ochungo, P., Roesel, K., & Grace, D. (2015). Assessing the Potential Role of Pigs in the Epidemiology of Ebola Virus in Uganda. *Transboundary and Emerging Diseases*, n/a. doi:10.1111/tbed.12394 PMID:26310206

AU-IBAR. (2014). Standard Methods and Procedures (SMPs) for Control of Rift Valley Fever (RVF) in the Greater Horn of Africa. Nairobi.

Black, W. C. I. V., & Moore, C. G. (2005). Population biology as a tool to study vector-borne diseases. In W. C. Marquardt, B. C. Kondratieff, C. G. Moore, J. E. Freier, H. H. Hagedorn, W. C. I. V Black, … S. Higgs (Eds.), Biology of disease vectors (2nd ed., pp. 187–206). San Diego: Elsevier Academic Press.

Britch, S. C., Binepal, Y. S., Ruder, M. G., Kariithi, H. M., Linthicum, K. J., Anyamba, A., & Wilson, W. C. et al. (2013). Rift Valley fever risk map model and seroprevalence in selected wild ungulates and camels from Kenya. *PLoS ONE, 8*(6), e66626. doi:10.1371/journal.pone.0066626 PMID:23840512

Chan, E. H., Brewer, T. F., Madoff, L. C., Pollack, M. P., Sonricker, A. L., & Keller, M., … Brownstein, J. S. (2010). Global capacity for emerging infectious disease detection. *Proceedings of the National Academy of Sciences* (Vol. 107, pp. 21701–21706). doi:doi:10.1073/pnas.1006219107 doi:10.1073/pnas.1006219107

Chevalier, V., Pépin, M., Plée, L., & Lancelot, R. (2010). Rift Valley fever--a threat for Europe? *Euro Surveillance: Bulletin Européen Sur Les Maladies Transmissibles. European Communicable Disease Bulletin, 15.*

Cohuet, A., Harris, C., Robert, V., & Fontenille, D. (2010). Evolutionary forces on Anopheles: What makes a malaria vector? *Trends in Parasitology, 26*(3), 130–136. doi:10.1016/j.pt.2009.12.001 PMID:20056485

Consultative Group for RVF Decision Support. (2010). Decision-support tool for prevention and control of Rift Valley fever epizootics in the Greater Horn of Africa. *The American Journal of Tropical Medicine and Hygiene, 83*(2 Suppl.), 75–85. PMID:20682910

Daubney, R., Hudson, J. R., & Garnham, P. C. (1931). Enzootic hepatitis or rift valley fever. An undescribed virus disease of sheep cattle and man from east africa. *The Journal of Pathology and Bacteriology, 34*(4), 545–579. doi:10.1002/path.1700340418

Davies, F. G. (2010). The historical and recent impact of Rift Valley fever in Africa. *The American Journal of Tropical Medicine and Hygiene, 83*(2 Suppl.), 73–74. doi:10.4269/ajtmh.2010.83s2a02 PMID:20682909

Davies, F. G., & Martin, V. (2006). Recognizing Rift Valley Fever. *Veterinaria Italiana, 42*(1), 31–53. PMID:20429078

Elliott, R. M. (1997). Emerging viruses: The Bunyaviridae. *Molecular Medicine (Cambridge, Mass.)*, 3. PMID:9323708

Epstein, P. (2001). West Nile virus and the climate. *Journal of Urban Health, 78*(2), 367–371. doi:10.1093/jurban/78.2.367 PMID:11419587

Evans, A., Gakuya, F., Paweska, J. T., Rostal, M., Akoolo, L., Van Vuren, P. J., & Feikin, D. R. et al. (2008). Prevalence of antibodies against Rift Valley fever virus in Kenyan wildlife. *Epidemiology and Infection, 136*(09), 1261–1269. doi:10.1017/S0950268807009806 PMID:17988425

Findlay, G. M. (1932). Rift valley fever or enzootic hepatitis. *Transactions of the Royal Society of Tropical Medicine and Hygiene, 25*(4), 229–IN11. doi:10.1016/S0035-9203(32)90042-X

Gerland, P., Raftery, A. E., ev ikova, H., Li, N., Gu, D., Spoorenberg, T., & Wilmoth, J. et al. (2014). World population stabilization unlikely this century. *Science, 346*(6206), 234–237. doi:10.1126/science.1257469 PMID:25301627

Githeko, A. K., Lindsay, S. W., Confalonieri, U. E., & Patz, J. A. (2000). Climate change and vector-borne diseases: A regional analysis. *Bulletin of the World Health Organization, 78*, 1136–1147. PMID:11019462

Grace, D., & Bett, B. (2014). Zoonotic diseases and their drivers in Africa. In C. D. Butler (Ed.), Climate Change and Global Health (pp. 228–236). Boston: CABI. Retrieved from https://books.google.com/books?hl=en&lr=&id=qVDNBAAAQBAJ&pgis=1 doi:doi:10.1079/9781780642659.0228 doi:10.1079/9781780642659.0228

Grace, D., Gilbert, J., Randolph, T., & Kang'ethe, E. (2012). The multiple burdens of zoonotic disease and an Ecohealth approach to their assessment. *Tropical Animal Health and Production, 44*(S1), S67–S73. doi:10.1007/s11250-012-0209-y PMID:22886445

Gubler, D. J., Reiter, P., Ebi, K. L., Yap, W., Nasci, R., & Patz, J. A. (2001). Climate variability and change in the United States: Potential impacts on vector- and rodent-borne diseases. *Environmental Health Perspectives, 109*, 223–233. doi:10.2307/3435012 PMID:11359689

Hassan, O. A., Ahlm, C., Sang, R., & Evander, M. (2011). The 2007 Rift Valley fever outbreak in Sudan. *PLoS Neglected Tropical Diseases*, *5*(9), e1229. doi:10.1371/journal.pntd.0001229 PMID:21980543

Hulden, L., & Hulden, L. (2009). The decline of malaria in Finland--the impact of the vector and social variables. *Malaria Journal*, *8*(1), 94. doi:10.1186/1475-2875-8-94 PMID:19422702

Impoinvil, D. E., Solomon, T., Schluter, W. W., Rayamajhi, A., Bichha, R. P., Shakya, G., & Baylis, M. et al. (2011). The Spatial Heterogeneity between Japanese Encephalitis Incidence Distribution and Environmental Variables in Nepal. *PLoS ONE*, 6(7). PMID:21811573

Jones, K. E., Patel, N. G., Levy, M. A., Storeygard, A., Balk, D., Gittleman, J. L., & Daszak, P. (2008). Global trends in emerging infectious diseases. *Nature*, *451*(7181), 990–993. doi:10.1038/nature06536 PMID:18288193

Jost, C. C., Nzietchueng, S., Kihu, S., Bett, B., Njogu, G., Swai, E. S., & Mariner, J. C. (2010). Epidemiological assessment of the Rift Valley fever outbreak in Kenya and Tanzania in 2006 and 2007. *The American Journal of Tropical Medicine and Hygiene*, *83*(2 Suppl.), 65–72. doi:10.4269/ajtmh.2010.09-0290 PMID:20682908

Keiser, J., De Castro, M. C., Maltese, M. F., Bos, R., Tanner, M., Singer, B. H., & Utzinger, J. (2005). Effect of irrigation and large dams on the burden of malaria on a global and regional scale. *The American Journal of Tropical Medicine and Hygiene*, *72*(4), 392–406. PMID:15827275

Khasnis, A. A., & Nettleman, M. D. (2005). Global warming and infectious disease. *Archives of Medical Research*, *36*(6), 689–696. doi:10.1016/j.arcmed.2005.03.041 PMID:16216650

Kortekaas, J., Kant, J., Vloet, R., Cêtre-Sossah, C., Marianneau, P., Lacote, S., & Brun, A. et al. (2013). European ring trial to evaluate ELISAs for the diagnosis of infection with Rift Valley fever virus. *Journal of Virological Methods*, *187*(1), 177–181. doi:10.1016/j.jviromet.2012.09.016 PMID:23022325

Kramer, L. D., & Ebel, G. D. (2003). Dynamics of flavivirus infection in mosquitoes. In T. J. Chambers & T. P. Monath (Eds.), *Flaviviruses: Pathogenesis and Immunity* (Vol. 60, pp. 187–232). San Diego: Academic Press Inc.

Lindahl, J. F., & Grace, D. (2015). The consequences of human actions on risks for infectious diseases: a review. *Infection Ecology & Epidemiology, 5.*

Linthicum, J. J., Davies, F. G., Bailey, C. L., & Kairo, A. (1984). Mosquito species encountered in a flooded grassland dambo in Kenya. *Mosquito News*, *44*(2), 228–232.

Linthicum, K. J., Davies, F. G., Kairo, A., & Bailey, C. L. (1985). Rift Valley fever virus (family Bunyaviridae, genus Phlebovirus). Isolations from Diptera collected during an inter-epizootic period in Kenya. *The Journal of Hygiene*, *95*(1), 197–209. doi:10.1017/S0022172400062434 PMID:2862206

Logan, T. M., Linthicum, K. J., Thande, P. C., Wagateh, J. N., Nelson, G. O., & Roberts, C. R. (1991). Egg hatching of Aedes mosquitoes during successive floodings in a Rift Valley fever endemic area in Kenya. *Journal of the American Mosquito Control Association*, *7*(1), 109–112. PMID:2045800

Martin, V., Chevalier, V., Ceccato, P. N., Anyamba, A., Simone, L., De, , & Lubroth, J., … Domenech, J. (. (2008). The impact of climate change on the epidemiology and control of Rift Valley fever. *Revue Scientifique et Technique (International Office of Epizootics)*, *27*(2), 413–426. doi:10.20506/rst.27.2.1802 PMID:18819669

McFarlane, R. A., Sleigh, A. C., & McMichael, A. J. (2013). Land-Use Change and Emerging Infectious Disease on an Island Continent. *International Journal of Environmental Research and Public Health*, *10*(7), 2699–2719. doi:10.3390/ijerph10072699 PMID:23812027

Medley, K. E. (1993). Primate Conservation along the Tana River, Kenya: An Examination of the Forest Habitat. *Conservation Biology*, *7*(1), 109–121. doi:10.1046/j.1523-1739.1993.07010109.x

Medley, K. E. (2009). Patterns of forest diversity along the Tana River, Kenya. *Journal of Tropical Ecology*, *8*(04), 353. doi:10.1017/S0266467400006684

Mellor, P. S., & Leake, C. J. (2000). Climatic and geographic influences on arboviral infections and vectors. *Revue Scientifique Et Technique De L Office International Des Epizooties*, *19*, 41–54.

Métras, R., Collins, L. M., White, R. G., Alonso, S., Chevalier, V., Thuranira-McKeever, C., & Pfeiffer, D. U. (2011). Rift Valley fever epidemiology, surveillance, and control: What have models contributed? *Vector Borne and Zoonotic Diseases (Larchmont, N.Y.)*, *11*(6), 761–771. doi:10.1089/vbz.2010.0200 PMID:21548763

Mogi, M. (1996). Overwintering strategies of mosquitoes (Diptera: Culicidae) on warmer islands may predict impact of global warming on Kyushu, Japan. *Journal of Medical Entomology*, *33*(3), 438–444. doi:10.1093/jmedent/33.3.438 PMID:8667392

Mohamed, M., Mosha, F., Mghamba, J., Zaki, S. R., Shieh, W.-J., Paweska, J., & Njenga, M. K. et al. (2010). Epidemiologic and clinical aspects of a Rift Valley fever outbreak in humans in Tanzania, 2007. *The American Journal of Tropical Medicine and Hygiene*, *83*(2 Suppl.), 22–27. doi:10.4269/ajtmh.2010.09-0318 PMID:20682902

Morper-Busch, L., Kienberger, S., & Hagenlocher, M. (2015). Healthy Futures Atlas: An Open-source WebGIS to Support Infectious Disease Intervention Planning in Eastern Africa. *Journal for Geographic Information Science*, 1, 460–463.

Moutailler, S., Krida, G., Schaffner, F., Vazeille, M., & Failloux, A.-B. (2008). Potential vectors of Rift Valley fever virus in the Mediterranean region. *Vector Borne and Zoonotic Diseases (Larchmont, N.Y.)*, *8*(6), 749–753. doi:10.1089/vbz.2008.0009 PMID:18620510

Murty, U. S., Rao, M. S., & Arunachalam, N. (2010). The effects of climatic factors on the distribution and abundance of Japanese encephalitis vectors in Kurnool district of Andhra Pradesh, India. *Journal of Vector Borne Diseases*, *47*, 26–32. PMID:20231770

Nanyingi, M. O., Munyua, P., Kiama, S. G., Muchemi, G. M., Thumbi, S. M., Bitek, A. O., … Njenga, M. K. (2015). A systematic review of Rift Valley Fever epidemiology 1931–2014. *Infection Ecology & Epidemiology, 5*.

Nicholas, D. E., Jacobsen, K. H., & Waters, N. M. (2014). Risk factors associated with human Rift Valley fever infection: Systematic review and meta-analysis. *Tropical Medicine & International Health : TM & IH*, *19*(12), 1420–1429. doi:10.1111/tmi.12385 PMID:25252137

Nyokabi, S. (2015). *Biosecurity Measures in Meat and Milk Value Chains: A Study in Bura sub-county, Kenya*. Germany: Hohenheim University.

Oborn, I., Bengtsson, J., Hedenus, F., Rydhmer, L., Stenström, M., Vrede, K., & Magnusson, U. et al. (2013). Scenario development as a basis for formulating a research program on future agriculture: A methodological approach. *Ambio*, *42*(7), 823–839. doi:10.1007/s13280-013-0417-3 PMID:23836311

Olive, M.-M., Goodman, S. M., & Reynes, J.-M. (2012). The role of wild mammals in the maintenance of Rift Valley fever virus. *Journal of Wildlife Diseases*, *48*(2), 241–266. doi:10.7589/0090-3558-48.2.241 PMID:22493102

Ostfeld, R. S. (2009). Climate change and the distribution and intensity of infectious diseases. *Ecology*, *90*(4), 903–905. doi:10.1890/08-0659.1 PMID:19449683

Patz, J. A., Campbell-Lendrum, D., Holloway, T., & Foley, J. A. (2005). Impact of regional climate change on human health. *Nature*, *438*(7066), 310–317. doi:10.1038/nature04188 PMID:16292302

Pepin, M., Bouloy, M., Bird, B. H., Kemp, A., & Paweska, J. (2010). Rift Valley fever virus (Bunyaviridae: Phlebovirus): an update on pathogenesis, molecular epidemiology, vectors, diagnostics and prevention. *Veterinary Research*, *41*(6), 61. doi:10.1051/vetres/2010033 PMID:21188836

Reiter, P. (2001). Climate change and mosquito-borne disease. *Environmental Health Perspectives*, *109*(S1), 141–161. doi:10.1289/ehp.01109s1141 PMID:11250812

Rich, K. M., & Wanyoike, F. (2010). An assessment of the regional and national socio-economic impacts of the 2007 Rift Valley fever outbreak in Kenya. *The American Journal of Tropical Medicine and Hygiene*, *83*(2 Suppl.), 52–57. doi:10.4269/ajtmh.2010.09-0291 PMID:20682906

Russell, R. C. (1998). Mosquito-borne arboviruses in Australia: The current scene and implications of climate change for human health. *International Journal for Parasitology*, *28*(6), 955–969. doi:10.1016/S0020-7519(98)00053-8 PMID:9673874

Sall, A., Zanotto, P. de A., Vialat, P., Sene, O., Bouloy, M., Sall, A. A., & Sene, O. K. et al. (1596–1597). M. B. (1998). Origin of 1997-98 Rift Valley fever outbreak in East Africa. *Lancet*, *352*(9140).

Sang, R., Kioko, E., Lutomiah, J., Warigia, M., Ochieng, C., O'Guinn, M., & Richardson, J. et al. (2010). Rift Valley fever virus epidemic in Kenya, 2006/2007: The entomologic investigations. *The American Journal of Tropical Medicine and Hygiene*, *83*(2 Suppl.), 28–37. doi:10.4269/ajtmh.2010.09-0319 PMID:20682903

Shoemaker, T., Boulianne, C., Vincent, M. J., Pezzanite, L., Al-Qahtani, M. M., Al-Mazrou, Y., & Ksiazek, T. G. et al. (2002). Genetic analysis of viruses associated with emergence of Rift Valley fever in Saudi Arabia and Yemen, 2000-01. *Emerging Infectious Diseases*, 8(12). PMID:12498657

Sindato, C., Karimuribo, E., & Mboera, E. G. (2012). The epidemiology and socio-economic impact of Rift Valley fever epidemics in Tanzania: A review. *The Onderstepoort Journal of Veterinary Research*, *79*(2), E1. doi:10.4102/ojvr.v79i2.467

Sindato, C., Karimuribo, E. D., Pfeiffer, D. U., Mboera, L. E. G., Kivaria, F., Dautu, G., & Paweska, J. T. et al. (2014). Spatial and temporal pattern of Rift Valley fever outbreaks in Tanzania; 1930 to 2007. *PLoS ONE*, *9*(2), e88897. doi:10.1371/journal.pone.0088897 PMID:24586433

Soumaré, P. O. L., Freire, C. C. M., Faye, O., Diallo, M., de Oliveira, J. V. C., Zanotto, P. M. A., & Sall, A. A. (2012). Phylogeography of Rift Valley Fever virus in Africa reveals multiple introductions in Senegal and Mauritania. *PLoS ONE*, *7*(4), e35216. doi:10.1371/journal.pone.0035216 PMID:22539961

Steinmann, P., Keiser, J., Bos, R., Tanner, M., & Utzinger, J. (2006). Schistosomiasis and water resources development: Systematic review, meta-analysis, and estimates of people at risk. *The Lancet Infectious Diseases*, *6*(7), 411–425. doi:10.1016/S1473-3099(06)70521-7 PMID:16790382

Swanepoel, R., & Coetzer, J. A. W. (2004). Rift Valley fever. In J. A. W. Coetzer & R. C. Tustin (Eds.), *Infectious Diseases of Livestock* (2nd ed., pp. 1037–1070). Cape Town: Oxford University Press.

Tabachnick, W. J. (2010). Challenges in predicting climate and environmental effects on vector-borne disease episystems in a changing world. *The Journal of Experimental Biology*, *213*(6), 946–954.

Tabachnick, W. J. (2013). Nature, nurture and evolution of intra-species variation in mosquito arbovirus transmission competence. *International Journal of Environmental Research and Public Health*, *10*(1), 249–277. doi:10.3390/ijerph10010249 PMID:23343982

Thonnon, J., Picquet, M., Thiongane, Y., Lo, M., Sylla, R., & Vercruysse, J. (1999). Rift valley fever surveillance in the lower Senegal river basin: Update 10 years after the epidemic. *Tropical Medicine & International Health*, *4*(8), 580–585. doi:10.1046/j.1365-3156.1999.00437.x PMID:10499082

Tourre, Y. M., Lacaux, J.-P., Vignolles, C., & Lafaye, M. (2009). Climate impacts on environmental risks evaluated from space: a conceptual approach to the case of Rift Valley Fever in Senegal. *Global Health Action, 2*.

Turell, M. J., Dohm, D. J., Mores, C. N., Terracina, L., Wallette, D. L. Jr, Hribar, L. J., & Blow, J. A. et al. (2008). Potential for North American Mosquitoes to Transmit Rift Valley Fever Virus 1. *Journal of the American Mosquito Control Association*, *24*(4), 502–507. doi:10.2987/08-5791.1 PMID:19181056

Turell, M. J., Linthicum, K. J., & Beaman, J. R. (1990). Transmission of Rift Valley fever virus by adult mosquitoes after ingestion of virus as larvae. *The American Journal of Tropical Medicine and Hygiene*, *43*(6), 677–680. PMID:2267972

Turell, M. J., Sardelis, M. R., O'Guinn, M. L., & Dohm, D. J. (2002). Potential vectors of West Nile virus in North America. In *Japanese Encephalitis and West Nile Viruses* (Vol. 267, pp. 241–252). Berlin: Springer-Verlag Berlin. doi:10.1007/978-3-642-59403-8_12

Vignolles, C., Lacaux, J.-P., Tourre, Y. M., Bigeard, G., Ndione, J.-A., & Lafaye, M. (2009). Rift Valley fever in a zone potentially occupied by Aedes vexans in Senegal: Dynamics and risk mapping. *Geospatial Health*, *3*(2), 211–220. doi:10.4081/gh.2009.221 PMID:19440963

Vignolles, C., Tourre, Y. M., Mora, O., Imanache, L., & Lafaye, M. (2010). TerraSAR-X high-resolution radar remote sensing: An operational warning system for Rift Valley fever risk. *Geospatial Health*, *5*(1), 23–31. doi:10.4081/gh.2010.184 PMID:21080318

Vilaly, A. E. E., Arora, M., Butterworth, M. K., Vilaly, M. A. M. E., Jarnagin, W., & Comrie, A. C. (2013). Climate, environment and disease: The case of Rift Valley fever. *Progress in Physical Geography*, *37*(2), 259–269. doi:10.1177/0309133313478315

Walter, C. T., & Barr, J. N. (2011). Recent advances in the molecular and cellular biology of bunyaviruses. *The Journal of General Virology*, *92*(Pt 11), 2467–2484. doi:10.1099/vir.0.035105-0 PMID:21865443

Wensman, J. J., Lindahl, J., Wachtmeister, N., Torsson, E., Gwakisa, P., Kasanga, C., & Misinzo, G. (2015). A study of Rift Valley fever virus in Morogoro and Arusha regions of Tanzania - serology and farmers' perceptions. *Infection Ecology & Epidemiology*, *5*(0), 30025. doi:10.3402/iee.v5.30025 PMID:26584830

Wong, G., Liu, W., Liu, Y., Zhou, B., Bi, Y., & Gao, G. F. (2015). MERS, SARS, and Ebola: The Role of Super-Spreaders in Infectious Disease. *Cell Host & Microbe*, *18*(4), 398–401. PMID:26468744

Woods, C. W., Karpati, A. M., Grein, T., McCarthy, N., Gaturuku, P., Muchiri, E., & Swanepoel, R. et al. (2002). An outbreak of Rift Valley fever in northeastern Kenya, 1997-98. *Emerging Infectious Diseases*, 8(12). PMID:11897064

World Bank. (2012). *People, pathogens, and our planet. Volume 2: the economics of one health.* Washington, DC. Retrieved from http://documents.worldbank.org/curated/en/2012/06/16360943/people-pathogens-planet-economics-one-health

Xie, H., You, L., Wielgosz, B., & Ringler, C. (2014). Estimating the potential for expanding smallholder irrigation in Sub-Saharan Africa. *Agricultural Water Management*, *131*, 183–193. doi:10.1016/j.agwat.2013.08.011

You, L., Xie, H., Wood-Sichra, U., Guo, Z., & Wang, L. (2014). Irrigation potential and investment return in Kenya. *Food Policy*, *47*, 34–45. doi:10.1016/j.foodpol.2014.04.006

You, L. Z. (2008). Africa: Irrigation investment Needs in Sub-Saharan Africa. Retrieved from https://openknowledge.worldbank.org/handle/10986/7870

Yuen, K. Y. (2015). From SARS to MERS and Ebola. *Journal of Microbiology, Immunology, and Infection*, *48*(2), S3. doi:10.1016/j.jmii.2015.02.002

Chapter 9
Climate Change Effects on Human Health with a Particular Focus on Vector-Borne Diseases and Malaria in Africa
A Case Study from Kano State, Nigeria Investigating Perceptions about Links between Malaria Epidemics, Weather Variables, and Climate Change

Salisu Lawal Halliru
Federal College of Education Kano, Nigeria

DOI: 10.4018/978-1-5225-0553-2.ch009

ABSTRACT

Malaria is currently affecting more people in the world than any other disease. On average, two members of each household suffered from malaria fever monthly, with females and children being most vulnerable to malaria attacks. This chapter assessed communities' perception about malaria epidemic, weather variable and climate change in metropolitan Kano. Information was extracted related to communities' perception about malaria epidemic and climate change. Socio demographic characteristics of respondents in the study areas were extracted and analyzed. 75% of the participants were males, while 25% were females, malaria disease affected 79.66% and 59.66% respondent perceived that heavy rainfall, floods and high temperature are better conditions to the breeding and spread of malaria vectors. Hospital records revealed that Month of March and April (2677 and 2464, respectively) has highest number of malaria cases recorded between December 2010 to June 2011. Further research is recommended for in-depth information from health officials related to raising awareness.

INTRODUCTION

This chapter covers the effect of climate change on human health with a particular emphasis on vector-borne diseases and malaria with focus in Kano metropolitan Nigeria with a view to investigate community perceptions about links between malaria epidemics, weather variables and climate change.

BACKGROUND

Malaria is currently affecting more people in the world than any other disease. It is currently endemic in over 100 countries and one of the 10 most prevalent and deadly diseases in the world (WHO, 2002). Apart from being one of the world's dreaded killer diseases, its prevalence has become a local and an international public problem. Malaria causes about 273 million clinical cases and 1.12 million deaths annually. More than 40% of the global population (> 2.1 billion people) are exposed to the malaria (Toure and Oduola, 2004). Malaria is caused by five distinct species of plasmodium parasite (*Plasmodium falciparum, Plasmodium vivax, Plasmodium malariae, Plasmodium ovale, Plasmodium knowlesi*) and is transmitted by Anopheline mosquitoes. (IPCC WGII AR5, 2013)

The current distribution of malaria is currently confined to tropical areas and poorer countries. The burden of mortality is unevenly distributed, with approximately 85% of all deaths and disease occurring in Africa. The distribution of malaria in the 1870s was at the peak of its global spread. This was when global trade routes had allowed the infections to be spread to the New World, and before the development of active control measures allowed disease reduction (McMichael et al., 2000). In Africa, malaria is responsible for up to 50% outpatient department (OPD) cases and 20% admission (WHO, 2006). Many families spend a significant portion of their income on malaria treatment. For example, estimates indicate that the direct cost of a single episode of malaria to a household was US$ 6.87 in Ghana, US$ 4.8 in Uganda and US$ 4.5 in Mali (WHO, 2006).

The relationship between economic development and malaria is two-way. Poor economic development is an effect of malaria as well as a cause. The direct costs of treating and preventing malaria morbidity and lost productivity are considerable, in relation to available funds in a developing country. Further, malaria has been shown to slow economic growth in low income African countries creating an ever-widening gap in prosperity between malaria-endemic and malaria-free countries. The reduced growth in countries with endemic malaria was estimated to be over 1% of GDP per year. The cumulative effect of this "growth penalty" is severe and restrains the economic growth of the entire region (Sachs, 2001).

The health impacts of climate variability and change are increasing. Studies in South Africa have focused on climate sensitive health outcomes including diarrheal, respiratory, cardiovascular health, and vector-borne infectious diseases such as malaria (Myers et al, 2013, Myers et al, 2011). Long-Term Adaptation Scenarios (LTAS) indicate climate change's potential impact on vector borne disease (DEA, 2013). Although the incidence of vector-borne infectious diseases, especially malaria, has declined in the South African context over recent years. Modeling predicts no overall increase in malaria incidence for sub-Saharan Africa but a shift from west to south and east driven by climate change impacts (Peterson, 2009; Byass, 2009).

Malaria was associated with rainfall and minimum temperature (with the strength of the association varying with altitude). In Ethiopia, temperature determines the timing and abundance of mosquitoes following adequate rainfall, which contributes to the availability of breeding site. For rainfall to have a positive effect on malaria cases, the temperature must be warm enough to support mosquito and parasite development (Bodker et al 2003). Worrall and colleagues (Worrall et al, 2007) used rainfall and maximum temperature at a lag of four months to fit a biological transmission model for malaria in a district in Zimbabwe. Craig and colleagues (Craig et al, 2004) linked inter-annual differences in malaria to rainfall and temperature in South Africa. In 2014, Evans and his colleague used poisson regression model and negative binomial regression model to study trends in malaria prevalence in Minna Niger state, Nigeria. The results revealed that malaria prevalence increased by 6% on a monthly basis (Evans & Adenomon, 2014).

In Nigeria, malaria is highly prevalent. 50% of the population has at least one episode of malaria each year and children under five years have up to 2-4 attacks annually (FMOH, 2005). The disease is responsible for 60% outpatient visit to health facilities, 30% of childhood mortality and 11% of maternal deaths (FMOH, 2005).

In Kano metropolis, a detailed study of 278 household made up of 3071 individuals (living around non – water outlet ponds) revealed that malaria is the most common disease (Yusuf, 2008). On average, two members of each household suffered from malaria fever monthly, with females and children being most vulnerable to malaria attacks (Maigari, 2005). A descriptive study of temporal variation of malaria occurrence in Kano municipal showed that high temperature, high congestion and improper waste disposal were contributors to disease transmission. In addition, the presence of stagnant water in ponds and open gutters contributed to mosquito breeding (Tukur, 2010)

Assessing the potential impact of climate change on human health requires an understanding of both the vulnerability of a population and its capacity to respond to new conditions. The vulnerability of a population to a health risk is dependent on various factors including:

- The local environment and social conditions.
- The level of resources.
- The effectiveness of governance and civil institutions.
- The quality of public health infrastructure.
- Access to relevant local information on extreme weather threats.
- Adaptive policies.

There is growing evidence that climate change will have profound effects on the health and wellbeing of citizens in countries throughout the world. The 2009 Lancet Commission report states that climate change will affect the key determinants of human health. Changing climate will inevitably affect the basic requirements for maintaining health: clean air and water, sufficient food and adequate shelter.

CLIMATE CHANGE AND ITS IMPACTS

Climate change is considered the most pervasive and global of all issues affecting humanity as it poses a serious threat to the environment as well as to economies and societies. Changes in land use (e.g. deforestation) and the burning of coal, oil and natural gases are some of the activities that increase the amounts of greenhouse gases, especially carbon dioxide, emitted into the atmosphere. The accumulation of greenhouse gases in the atmosphere due to human activities is changing the climate by enhancing the natural greenhouse effect, leading to an increase in the Earth's average surface temperature. Through economic and social development, industrial and agricultural emissions of carbon dioxide, methane and other greenhouse gases are increasing. Reducing these gases will require a fundamental shift in industrial and manufacturing processes, agriculture and energy production.

The Fourth Assessment Report of the Intergovernmental Panel on Climate Change (IPCC, 2007) forecasts an increase in average world temperature by 2100 within the range 1.4° – 5.8° C. The increase will be greater at higher altitudes and over land. In addition, IPCC projects that average annual rainfall will increase globally, with precipitation events and flooding becoming more severe, whereas, many middle and lower latitude land regions will become drier.

Climate change is expected to impact:

- Sea level – causing sea-level rise.
- Ecosystems.
- Human health.
- Agriculture.
- Tourism.
- Water resources.
- The poor.

Whilst it is clear that the impacts of climate change are varied, scientists have agreed that its effects will not be evenly distributed and that developing countries and Small Island developing States (SIDS) will be the first and hardest hit. Small island developing States, many of whom have fewer resources to adapt socially, technologically and financially to climate change, are considered to be most vulnerable to the potential impacts of climate change (IPCC, 2007).

Additionally, it is expected that climate change will pose a significant threat and drain public resources in many Caribbean States. Today, countries in the Caribbean continue to grapple with issues related to macroeconomic stability, reducing poverty levels and trade liberalization, to name a few (IPCC, 2007).

Climate change is expected to affect ecosystem services in ways that increase vulnerabilities with regard to food security, water supply, natural disasters, and human health and will, in turn, affect socioeconomic development. (IPCC, 2007b; Stern, 2006; WHO, 2003; WHO, 2008). Furthermore, it was reported in the IPCC Third Assessment that, while developing countries are expected to experience larger percentage losses in GDP, global mean losses could be between 1-5% GDP for 4° C of warming.

Climate change is also expected to cause a further rise in sea level of about 20cm by the year 2030 (IPCC, 2007). Forecasts of rising sea levels are based on climate model results which indicate that the earth's average surface temperatures may increase by 1.5° – 4.5° C over the next 100 years. The warming can cause the sea to rise in two ways: through thermal expansion of ocean water, and through the melting and sinking of ice caps and mountain glaciers.

Furthermore, extreme climate events are expected to become more frequent with climate change. These events are expected to have their greatest impact in poor and/or developing countries and in Small Island developing States. Climate change is also expected to affect the health of populations. In fact, the World Health Organization (WHO) in *Protecting Health from Climate Change* (2008) states that the continuation of current patterns of fossil fuel use, development and population growth will lead to ongoing climate change, with serious effects on the environment and, consequently, on human lives and health.

Climate change is considered by the Lancet Commission to be "the biggest global health threat of the twenty-first century". Their 2009 Report, entitled *Managing the Health Effects of Climate Change* states that the effects of climate change on health will affect most populations in the upcoming decades, and will put the lives and well-being of billions of people at increased risk.

Therefore, the development paths that the world chooses will have a strong influence on this increase in greenhouse gas emissions. According to the Stern Review (2006), the scientific evidence points to increasing risks of serious irreversible impacts from climate change associated with business-as usual (BAU) paths for emissions. If, for example, high priority is placed on sustainable energy use, temperatures are expected to rise by 1.8° C (likely range: 1.1°–2.9° C). If societies place a lower emphasis on sustainability measures, temperatures are expected to rise by about 4.0° C (2.4°–6.4° C), with a greater probability of abrupt or irreversible impacts (WHO, 2008).

Climate change is likely to be associated with multitude of effects:

1. Climate change has been associated with a shift in the composition and geographical distribution of many ecosystems (e.g. forests, deserts, coastal systems) as individual species respond to changing climatic conditions, with likely reduction in species diversity and agricultural yield.
2. Climate change will lead to an intensification of the global hydrological cycle and may affect regional water resources.
3. Additionally, climate change and the resulting sea-level rise can have a number of negative effects on energy, industry and transportation infrastructure, human settlements and tourism (IPCC, 2007).

Generally speaking, "the various potential health effects of global climate change upon human health can be divided into direct and indirect effects, according to whether they occur predominantly via the impacts of climate variable upon human biology, or are mediated by climate-induced changes in other biological and biogeochemical systems".(Martens et al, 1999).

Temperature and precipitation changes might influence the behaviour and geographical distribution of vectors, and thus change the incidence of vector borne diseases, which are major causes of mortality and morbidity in many tropical countries. Increase in non-vector-borne infectious diseases such as cholera, salmonellosis, and other food-and water-related infectious diseases could occur, particularly in tropical and sub-tropical regions, due to climatic impact on water distribution, temperature and proliferation of microorganisms (UNEP, 2003).

Measuring the health effects from climate change can only be very approximate. Nevertheless, a WHO assessment, taking into account only a subset of the possible health impacts, and assuming continued economic growth and health progress, concluded that climate change is expected to cause approximately 250 000 additional deaths per year from malnutrition, malaria, diarrhoea and heat stress (Okogun et al, 2005). The direct damage costs to health (that is excluding costs in health-determining sectors such as agriculture and water and sanitation), is estimated to be between US$ 2-4 billion/year by 2030 (Okongun et al, 2005).

Areas with weak health infrastructure – mostly in developing countries – will be the least able to cope without assistance to prepare and respond. Climatic conditions strongly affect water-borne diseases and diseases transmitted through insects, snails or other cold blooded animals. Changes in climate are likely to lengthen the transmission seasons of important vector-borne diseases and to alter their geographic range.

MALARIA AND WEATHER CONDITIONS

Malaria and its prevalence in Africa and Asia continue to be a public health concern. Climatic factors such as temperature and rainfall have great influence on the geographical spread and seasonal incidence (Epstein, 1999). Disease vectors reproduce within certain optimal climate conditions such as a combination of rainfall and temperature. Changes in these conditions can greatly alter disease transmission. Minimum as well as maximum temperatures are often critical to the survival of disease causing pathogens and their vectors. Temperature controls the rate of larval and parasite development. Higher temperatures shorten the development time of the larvae and parasites in the mosquitoes. Malaria epidemic risk is related to both warm temperatures and the following rainy season months. Climatic events that create these conditions could result in disease epidemics. Variability in precipitation may have direct consequences on disease outbreaks as increased precipitation may increase the size of the mosquito population by creating new breeding grounds and expanding the size of existent larval habitat (Caroline 2012, Wandiga et al., 2010).

The El Niño-Southern Oscillation (ENSO) cycle is a key factor affecting the emergence and prevalence of vector-borne infections (Pherez, 2007). Rainfall in particular is important, not only through its quantity and timing, but also through its effects on relative humidity and on human activity patterns, as well as its relevance to control activities. For example, the rainy season is the period of active farming in most rural communities in Nigeria (Opara et al, 2008) and is associated with higher mosquito net use compliance in Burkina Faso (Okrah et al,2002). Larviciding of breeding sites a month before onset of rainfall is recommended for vector control in Nigeria (Okogun et al 2005). While El Niño events have been suggested as a potential predictor for malaria epidemics in Sri Lanka, on the long-term, these associations break down (Zubair et al, 2008). ENSO is one of the numbers of climate indices, each reflecting temperature or pressure changes at different locations on the earth's surface. In western sub-Saharan Africa, rainfall is related to seasonal shifts of the Inter-Tropical Convergence Zone (ITCZ), which moves northward early in the year, retreating in the second half of the year. Onset and retreat dates for four locations in Nigeria— Ibadan, Ilorin, Kaduna and Kano have been reported in the literature (Odekunle, 2004) and were linked to variation in malaria transmission mediated by the vital role of rainfall on the distribution of mosquito breeding sites (Onwuemele, 2013).

CLIMATE CHANGE AND MALARIA

While climate change will alter the patterns and spread of malaria transmission, other important drivers of the disease include socioeconomic status, type of vectors available, population immigration and vector dispersal. Africa will continue to carry the greatest burden of the disease and in particular Eastern and Southern Africa (Andrew, 2009). Mountainous regions in South Asia are likely to experience an increase in transmission. The same applies to highland regions of South East Asia and Pacific countries. The United Kingdom, Europe, Canada and New Zealand are unlikely to be affected in the near future (Andrew, 2009).

Climate Change and Malaria in Africa

All the drivers of malaria transmission are found in abundance in Africa and climate change is a new addition. Since the late 1980s, there were reports of malaria epidemics particularly in the East African Highlands (Andrew, 2009). In many of these regions there was widespread parasite resistance to chloroquine, (Abeku et al 2003, Zhou et al 2004, Pascual et al 2006) and virtually no vector control. At that time, it was not clear whether the epidemics were a result of environmental change or due to the effects of drug resistance and lack of vector control. A detailed study of climate and malaria epidemics indicated that epidemics were associated with incidents of El Nino (Kovats et al 2003, Anyamba et al 2006), characterized by abnormal warming and wetness. Weak and late interventions failed to contain the outbreaks leading to severe health outcomes. Kenya, Uganda, Burundi, Tanzania, Eritrea, Ethiopia and Rwanda reported severe malaria epidemics in the late 1980s to 2003 (Andrew, 2009). This period had two very strong El Nino events and a number of weak events (1982, 1988, 1990-1994, 1997-8, 2003) (Andrew, 2009).

Climate change will contribute to increased malaria epidemics throughout Nigeria, with the highest impact in the Mangrove and Rainforest zones (Abiodun et al 2011). It should be noted that these zones have the highest population density in Nigeria, indicating population-based vulnerability in these regions. This implies that climate change would further strain the already poor public health resources in Nigeria, especially in the cities located in the Mangrove and Rainforest zones. We suggest that this might heighten the conflict in the Niger Delta region. Hence, adaptation options that would improve public health facilities and services are crucial in these areas (Abiodun et al 2011).

Statement of the Problem

Malaria parasite is growing at a rapid pace. The roll back malaria (RBM) campaign launched in 1998, with the stated goal to halve malaria deaths worldwide by 2010 was confronted by problems of drug resistance (Raghunath, 2004). In Kano state, the programme encountered the following challenges: poor symptom recognition, outdated malaria treatment, improper treatment practice and poor acceptance of insecticide treated Nets (ITNs). In addition to cross cutting issues such as ineffective advocacy and communication skills among malaria control programmes, low status and poor appreciation for communication, inadequate time and resources for strategic communication and lack of procedure evaluation. Currently, malaria prevalence in Kano state is increasing and a number of epidemics have been reported. In 2001-2005, 16,601 people were infected by malaria in Kano municipal (Tukur, 2010). A prevalence study in Wudil and Gaya hospitals in Kano state showed that the highest prevalence rate was detected in children aged 1-3 years (85.4% and 46.6%, respectively) (Yahaya et al,2014). A study in the community showed that 30.59% of children (1-5 years old) were infected with malaria (Gobir & Tukur, 2014).

- **General Objective of Study:** This study objective was to assess communities' perception about malaria epidemic, weather variable and climate change in metropolitan Kano.
- **Specific Objective:**
  - To assess communities' perception about malaria and climate change.
  - To find out the level of awareness and integration of climate change in public health.
  - To make recommendation based on the findings.
- **Significance of the Study:** It is hoped that the study could contribute to further research on malaria and climate change in Nigeria. Local governments' public health officials should consult research findings and include these in decision making processes, especially for effective malaria control and the health sector adaptation to climate change.
- **Study Area:** The study area is the Kano Metropolis comprising of the following local government areas of Dala, Tarauni, Kano Municipal, Gwale, Fagge, Kumbotso, Nasarawa and Ungogo. It is located at the central western part of the state between latitude 11°59'N and 12°2'North of the equator and between longitude 8°33'E and 8°31'East of the Greenwich meridian (see Figure 1). The population was 2826307 as at 2006 population census (NPC, 2007). The area has a tropical dry and wet climate. Almost 40% of the annual rainfall occurs in August, which is the peak of the wet season. The temperature regime is warm to hot throughout the year, even though there is a slightly cool period between November and February. The mean annual temperature is about 22°C in the coolest months (December or January) and 31°C in the hottest months (April or May). The mean annual potential evaporation, sunshine, and relative humidity are about 1,772mm, 8.5 hours per day, and 50% respectively (Olofin, 1987). The natural vegetation of the area and indeed most of the Kano region is Sudan savanna.

Methodology

In this study, qualitative and quantitative methodological approaches were used in order to assess the opinions of the local people. It involved interviews, discussions, and narratives to get more details about events, tendencies and situations regarding malaria outbreaks in relation to climate variability and other risk factors such as large influx of non-immune people into an area of transmission, high mortality rate etc., in addition to communities' awareness and perceptions and coping strategies. Quantitative research was used to look at relationships between age, education level, gender and place of residence and outcomes.

Figure 1. Map of Kano metropolitan showing the study area
Source: Geography Department, Bayero University Kano (2015).

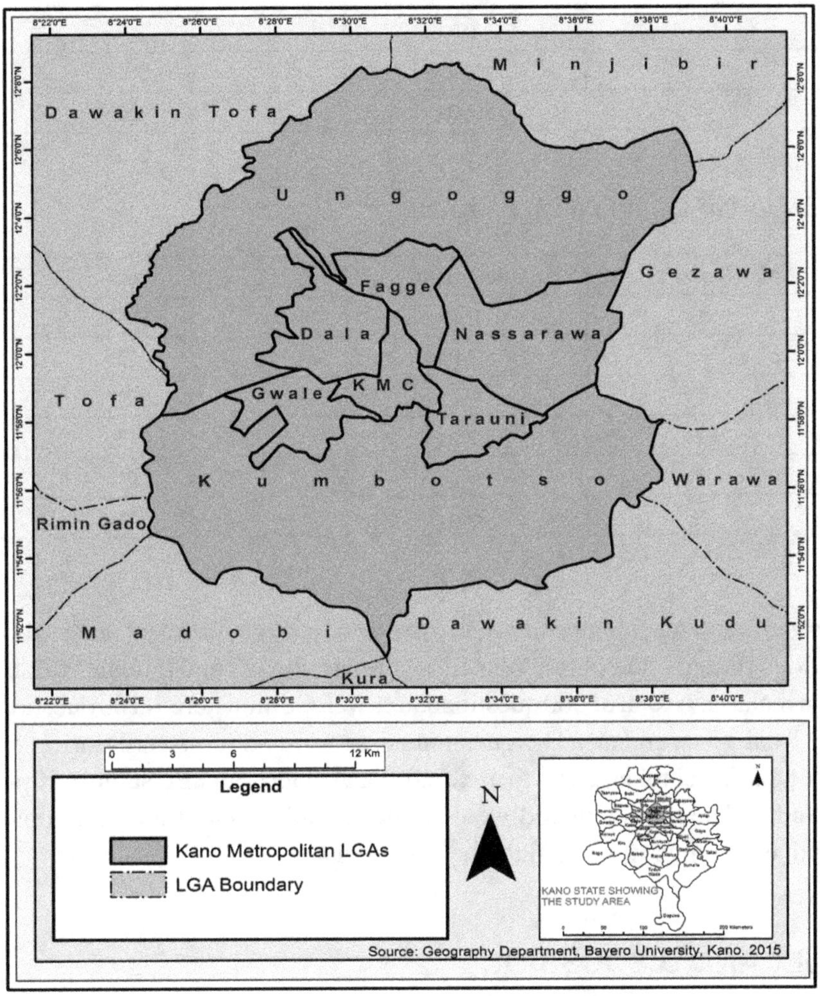

A survey using a questionnaire was used in six out of the eight local government areas that make up the Kano metropolitan (see figure 1). The six local governments were selected randomly through the lottery method (names of the local government were written on a paper, thoroughly mixed; put in a bowl and without looking six papers were selected). The selected areas for study were Nassarawa, Kano municipal, Gwale, Fagge, Tarauni and Dala. In each locality respondents were selected purposively because of their history of previous malaria infection. In addition, 6 health officials were interviewed 1 from each hospital namely; Sunusi hospital, Abdullahi wase hospital, Jidda hospital, Murtala Mohammed hospital, Arewa Surgery hospital

Table 1. Gender distribution of respondents

Gender	Respondents	%
Male	450	75
Female	150	25
Total	600	100

Table 2. Marital status of the respondents

Marital Status	Respondents	%
Single	105	17.5
Engaged	44	7.33
Married	410	68.33
Divorce	25	4.17
Widowed	16	2.67
Total	600	100

and Aminu Kano teaching hospital, across the 6 local government areas, about vulnerabilities strategy and policy related to climate change and human health, with a focus on malaria. A total of 600 questionnaires were administered (100 questionnaires at each local government). The questionnaires were administered during a period of 7 months (December 2010 to June 2011). A questionnaire sample is provided in appendix. The data gathered were collected and encoded into a computer and descriptive statistics were used to analyze the data.

RESULT AND DISCUSSION

Demographic Characteristics of the Respondents

75% of the participants were males, while 25% were females (Table 1). As shown on Table 2-5, the respondents are representative of the general population. However, the majority were married, aged 25-44 years old, with good educational level and professional.

The result of the investigations showed that 98.17% of people living in the metropolitan area consider that climate change poses a significant threat that result in the deterioration of human health, mostly due to the increased presence of vector-borne tropical diseases, such as malaria and dengue fever and prevalence of respiratory illness in urban Kano.

Table 3. Age structure of the respondents

Age	Respondents	%
16-24	47	7.83
25-34	235	39.17
35-44	205	34.17
45-54	76	12.67
55 or above	37	6.16
Total	600	100

Table 4. Educational status of respondents

Education	Respondents	%
Primary	85	14.17
Secondary	253	42.17
Tertiary	215	35.83
Informal	47	7.83
Total	600	100

Table 5. Occupational status of respondents from metropolitan

Occupation	Respondents	%
House-wife	71	11.83
Civil servant	207	33.5
Trading	223	37.17
Faming	80	13.33
student	25	4.17
Total	600	100

The question investigating the relevance of malaria in the study area showed that 79.66% respondents indicated that malaria disease affected them very much, 15.5% were affected a little and 4.84% not affected at all by malaria but they were aware that malaria cases are rampant in their community. The majority of respondents (78.66%) knew the effect of environmental factors on malaria prevalence, which is judged to be common knowledge.

Table 6. Perceptions about the effect of climate change on health

	Nassarawa	Kano Municipal	Gwale	Fagge	Tarauni	Dala	Number of Respondents	%
Strongly agree	80	81	75	90	85	80	491	81.84
Agree	18	17	21	10	13	19	98	16.33
Neither agree or disagreed	2	2	2	2	2	1	11	1.83
Disagree	0	0	0	0	0	0	00	-
Strongly disagreed	0	0	0	0	0	0	00	-
Total	100	100	100	100	100	100	600	100

When investigating the impact of high rainfall and hot temperature on increasing mosquito population, 41.84% of people living in metropolitan Kano strongly agree, 24.33% agree, 20.16% neither agree nor disagreed, 6.67% disagree and 7% strongly disagree (see Table 6).

Table 7 showed that 59.66% respondents agree that heavy rainfall, floods and high temperature are conducive conditions to the breeding and spread of malaria vectors. The respondents believe that malaria epidemics have partly coincided with periods of high rainfall and flood (El Nino occurrence).

Table 8 reveals that 431 (71.82%) respondents believe there is a link between malaria incidence and climate change in Kano metropolitan.

Table 7. Association of climate change with heavy rains and increase in temperature as conducive conditions for the breeding and spread of malaria vector

	Nassarawa	Kano Municipal	Gwale	Fagge	Tarauni	Dala	Number of Respondents	%
Strongly agree	29	35	32	34	71	40	241	40.16
Agree	12	24	20	22	12	27	117	19.5
Neither agree or disagreed	28	21	35	21	11	16	132	22
Disagree	13	10	8	9	1	8	49	8.17
Strongly disagreed	18	10	5	14	5	9	61	10.17
Total	100	100	100	100	100	100	600	100

Table 8. The relationship between malaria incidence and climate change in Kano metropolitan

	Nassarawa	Kano Municipal	Gwale	Fagge	Tarauni	Dala	Number of Respondents	%
Strongly agree	64	58	34	25	47	46	274	45.66
Agree	20	21	28	25	33	30	157	26.16
Neither agree nor disagreed	13	17	28	25	15	19	117	19.5
Disagree	2	3	6	14	4	3	32	5.33
Strongly disagreed	1	4	4	8	1	2	20	3.33
Total	100	100	100	100	100	100	600	100

Table 9. climate change integration in public health in Kano metropolitan and level of community awareness about climate change and malaria

	Nassarawa	Kano Municipal	Gwale	Fagge	Tarauni	Dala	Number of Respondents	%
Strongly agree	10	8	10	9	6	7	50	8.33
Agree	5	2	6	12	2	1	28	4.67
Neither agree or disagreed	9	14	10	13	8	8	62	10.33
Disagree	15	8	20	15	11	14	83	13.83
Strongly disagreed	61	68	54	51	73	70	377	62.83
Total	100	100	100	100	100	100	600	100

Table 9 showed that there are low levels of climate change integration and community awareness on issues of climate change and health due to lack of awareness. The implication of low level of awareness could put the communities involved at more risk of contracting malaria and delayed treatment seeking, which will contribute to increased disease burden.

To supplement the primary data, secondary data were collected for January-December 2011 to show trends of rainfall and temperatures as well as monthly records of malaria incidence from hospitals in Kano metropolitan (Tables 10 and 11).

Table 10. Climate data for Kano metropolitan (2011)

	Jan	Feb	Mar	Apr	May	Jun	Jul	Aug	Sep	Oct	Nov	Dec	Year
Average Temp.high °C	26	27	32	33	31	29	27	27	28	29	27	26	28.5
Average Temp. low °C	22	24	29	29	28	26	24	24	24	26	23	22	25.1
Average precipitation days	0	0	1	1	2	2	7	6	3	2	1	1	26

Source: Metrological station Aminu Kano International Airport (2012).

Table 11. Monthly malaria incidence in Kano metropolitan

	Nassarawa	Kano Municipal	Gwale	Fagge	Tarauni	Dala	Total
Dec. 2010	168	183	153	200	124	101	933
Jan. 2011	124	216	208	103	114	156	921
Feb. 2011	111	208	234	117	135	202	1007
Mar. 2011	328	406	397	412	581	616	2677
Apr. 2011	201	336	460	442	541	485	2465
May 2011	197	263	211	187	217	264	1339
Jun. 2011	174	210	283	209	235	214	1325

Source: Hospitals Records, 2011.

The hospital records of malaria and corresponding meteorological conditions showed a clear association. The highest number of cases were recorded during March and April (2677 and 2464, respectively) corresponding to the hottest months of the year in terms of both maximum and minimum temperature associated with low rainfall. It was observed that the study areas were very congested with an unplanned pattern of buildings and indiscriminate waste disposal which provide breeding sites for mosquitoes.

Health Officials' Perspective

In total six health officials were interviewed. Health officials were invited to participate in the interview voluntarily depending on their availability. The interviewees include one medical officer (medical doctor) and five nurses, all have worked in the hospital for more than ten years. The medical officer holds a bachelor degree in medicine and the other officials have either diploma or certificate in nursing. The age ranges from 28 to 61 years.

Most of the health officials interviewed at the local governments acknowledged the relationship between climatic factors and increasing malaria epidemic in the study area. Some reported that "malaria outbreaks occur during periods of dry seasons". A senior official from the malaria control unit reported that "high rainfall periods do not usually have mosquitoes because breeding grounds are disturbed" he further stated that mosquitoes increases significantly as temperature increases. A hospital worker with 20 years working experience observed that "Malaria today is common and increasing unlike in the past. I believe this has to do with changing climate in Kano, people are coming to hospital every day reporting sickness, after a test the result showed presence of malaria".

Some respondents explained that "there is a low level of awareness between climate variability and malaria". They attributed the low level of awareness in the communities to the "I don't care attitude of people" about awareness programs and funding challenges that these programs experience.

This case study showed that, overall, the respondents were well aware of the burden of malaria in their localities and its intimate association with weather variables. The impact of climate change on disease transmission and malaria incidence were well understood in a high proportion of respondents. However, better community knowledge through awareness and communication campaigns as well as integration of climate change in public health strategies are urgently needed.

CONCLUSION

Malaria occurrence will be influenced by climate change. Other factors such as personal hygiene, poor sanitation, and inadequate knowledge on the disease are likely to contribute significantly to the disease burden. At present, there is no clear answer to the problem of malaria because climate will keep on changing in the foreseeable future and influencing transmission. One solution is to integrate climate change in health policy and other relevant sectors in order to enable effective adaptation strategies.

Recommendation for a way forward are:

- NGOs, ministry of health and other organizations such as Society for Family Health, should continue to promote prompt and effective malaria case management at the health facility, community and household levels. In addition, vector control including Long Lasting Insecticidal Nets (LLINS), indoor residual spraying, use of intermittent preventive treatment (IPT) of malaria in pregnancy, insecticide treated mosquito nets (ITNs) and environmental management are also recommended.

- Climate change awareness and adaptation strategies in Kano metropolis are seriously needed. It is important that climate change and health policies are developed to protect communities. There should also be an effort to plan and establish a communication strategy detailing the risks, which should be targeted at vulnerable communities, NGOs, and all stakeholders.
- The author recommends further research in the study area by conducting more elaborate interviews with health officials in order to get more details and in depth information related to raising awareness and implementation strategies.

REFERENCES

Abeku, T. A., van Oortmarssen, G. J., Borsboom, G., de Vlas, S. J., & Habbema, J. D. F. (2003). Spatial and temporal variations of malaria epidemic risk in Ethiopia: Factors involved and implications. *Acta Tropica*, *87*(3), 331–340. doi:10.1016/S0001-706X(03)00123-2 PMID:12875926

Abiodun, B. J., Salami, A. T., & Tadross, M. (2011). Climate Change Scenarios for Nigeria: Understanding Biophysical Impacts. Climate Systems Analysis Group, Cape Town, for Building Nigeria's Response to Climate Change Project. Ibadan, Nigeria: Nigerian Environmental Study/Action Team (NEST).

Ahmad, E.I. (2007). The prevalence of malaria in Okigwe and Owerri. *Journal of community health*, 1.

Githeko, A.K. (2009). Malaria and climate change (Special feature). Commonwealth Health ministers' update.

Anyamba, A., Chretien, J. P., Small, J., Tucker, C. J., & Kelley, P. W. (2006). Developing global climate anomalies suggest potential disease risks for 2006-2007. *International Journal of Health Geographics*, *5*(1), 60. doi:10.1186/1476-072X-5-60 PMID:17194307

Bodker, R., Akida, J., Shayo, D., Kisinza, W., Msangeni, H. A., Pedersen, E. M., & Lindsay, S. W. (2003). Relationship between altitude and intensity of malaria transmission in the Usambara Mountains, Tanzania. *Journal of Medical Entomology*, *40*(5), 706–717. doi:10.1603/0022-2585-40.5.706 PMID:14596287

Byass, P. (2009). Climate change and population health in Africa: Where are the scientists? *Global Health Action*, *2009*, 2. PMID:20052421

Aboda, C. (2012): An assessment of the impact of climate variability on malaria in Uganda. A case study of Mount Elgon region of Bududa district [Master of Philosophy Thesis]. Norwegian University of Science and Technology.

Craig, M. H., Kleinschmidt, I., Nawn, J. B., Le Sueur, D., & Sharp, B. L. (2004). Exploring 30 years of malaria case data in KwaZulu-Natal, South Africa: Part I. The impact of climatic factors. *Tropical Medicine & International Health*, *9*(12), 1247–12. doi:10.1111/j.1365-3156.2004.01340.x PMID:15598256

Department of Environmental Affair DEA. (2013). *Long-Term Adaptation Scenarios Flagship Research Programme (LTAS) for South Africa*. Pretoria, South Africa: Department of Environmental Affairs.

Economic Commission for Latin America and the Caribbean (ECLAC) (2011), *An Assessment of the Economic impact of Climate Change on the health sector in Guyana*.

Environment Programme (UNEP). (2003). *Climate Change and Human Health - Risks and Responses: Summary. ISBN 92 4 159081 5*. Published by WHO.

Evans, O.P., & Adenomon, M.O. (2014). Modeling the prevalence of malaria in Niger State: An application of Poisson regression and negative binomial regression models International Journal of Physical Sciences, 2(4), 61-68. Retrieved from http://academeresearchjournals.org/journal/ijps

Federal Ministry of Health, National Malaria Control Programme/FMOH/2005 Annual Report. (2005, January).

Intergovernmental Panel on Climate Change (IPCC). (2007b). Human health. In M. L. Parry, O. F. Canziani, J. P. Palutikof, P. J. van der Linden, & C. E. Hanson (Eds.), *Climate Change 2007: Impacts, Adaptation and Vulnerability. Contribution of Working Group II to the Fourth Assessment Report of the Intergovernmental Panel on Climate Change* (pp. 391–431). Cambridge, UK: Cambridge University Press.

Intergovernmental panel on climate change fifth assessment report working group II final draft (2013). Human Health: Impacts, Adaptation, and Co-Benefits (Ch. 11). Retrieved from https://www.ipcc.ch/pdf/assessment-report/ar5/wg2/drafts/fd/WGIIAR5-Chap11_FGDall.pdf

Kabore, A. (2004). Overview of Malaria in West Africa. *WHO Newsletter*, 5(2).

Kovats, R. S., Bouma, M. J., Hajat, S., Worrall, E., & Haines, A. (2003). El Niño and health. *Lancet*, *362*(9394), 1481–1489. doi:10.1016/S0140-6736(03)14695-8 PMID:14602445

Maigari, A. I. (2005). *Surface water Resources management strategies in the metropolitan Kano, Wa'adallah Environmental consults (WADEC)*. Kano-Nigeria.

McMichael, A. J., Kovats, R. S., Martens, P., Nijhof, S., Livermore, M. T., Cawthorne, A., & deVries, P. (2000). Climate change and human health: final report to the department of environment, transport and the regions. London School of Hygiene and Tropical Medicine/ICIS, London/Maastricht.

Metrological station Aminu Kano International Airport. (2012). Kano metropolitan climatic Data

Myers, J., Young, T., Galloway, M., Manyike, P., & Tucker, T. (2011). A public health approach to the impact of climate change on health in southern Africa: Identifying priority modifiable risks. *South African Medical Journal, 101*, 817–819. PMID:22272963

Myers, J. E., & Rother, H. (2013). Public health impact of and response to climate change in South Africa. In A. Padarath & R. English (Eds.), *South African Health Review 2012/13*. Durban, South Africa: Health Systems Trust.

National Population Commission. (2007). *Federal Republic of Nigeria Official Gadget, 23*(94), 175–198.

Odekunle, T. O. (2004). Determining rainfall onset and retreat dates in Nigeria. *Journal of Human Ecology (Delhi, India), 16*, 239–247.

Okogun, G. R. A., Anosike, J. C., Okere, A. N., & Nwoke, B. E. B. (2005). Ecology of mosquitoes of Midwestern Nigeria. *Journal of Vector Borne Diseases, 42*, 1–8. PMID:15999454

Okrah, J., Traore, C., Pale, A., Somerfeld, J., & Muller, O. (2002). Community factors associated with malaria prevention by mosquito nets: An exploratory study in a rural Burkina Faso. *Tropical Medicine & International Health, 7*(3), 240–248. doi:10.1046/j.1365-3156.2002.00856.x PMID:11903986

Olofin, E.A. (1989) Some Aspect of Physical Geography of Kano and related human responses. Departmental lecture series N01, Department of Geography, Bayero University Kano-Nigeria.

Onwuemele, A. (2014). An assessment of the spatial pattern of malaria infection in Nigeria. *International Journal of Medicine and Medical Sciences*, 6(2), 80-86. Retrieved from http://www.academicjournals.org/IJMMS

Opara, K. N., Usip, L. P., & Akpabio, E. E. (2008). Transmission dynamics of *Simulium damnosum* in rural communities of Akwa Ibom State, Nigeria. *Journal of Vector Borne Diseases*, *45*, 225–230. PMID:18807379

Martens, P., & Rotmans, J. (Eds.). (1999). *Climate change: an integrated perspective Springer publishers Netherlands* (pp. 1–10). doi:10.1007/0-306-47982-6_1

Pascual, M., Ahumada, J. A., Chaves, L. F., Rodó, X., & Bouma, M. (2006). Malaria resurgence in the East African highlands: temperature trends revisited. *Proceedings of the National Academy of Sciences of the United States of America*, *103*(15), 5829–5834. doi:10.1073/pnas.0508929103 PMID:16571662

Peterson, A. T. (2009). Shifting suitability for malaria vectors across Africa with warming climates. *BMC Infectious Diseases*, *9*(1), 59–64. doi:10.1186/1471-2334-9-59 PMID:19426558

Pherez, F. M. (2007). Factors affecting the emergence and prevalence of vector borne infections (VBI) and the role of vertical transmission (VT). *Journal of Vector Borne Diseases*, *44*, 157–163. PMID:17896618

Raghunath, D. (2004) Malaria vaccine: Are we anywhere close? *Journal of postgraduate medicine*, 50(1).

Sachs, J. (2001). *Macroeconomics and health: Investing in health for economic development. Report of the Commission on Macro Economics and Health*. Geneva: World Health Organization.

Stern, N. (2006). The Stern Review on the Economics of Climate Change. HM Treasury. UK: Cambridge University press. Retrieved from http://www.hmtreasury.gov.uk/independent_reviews/stern_review_economics_climate_change/sternreview_index.cfm

Toure, Y. T., & Oduola, A. (2004). Focus: Malaria. *Nature Reviews. Microbiology*, *2*(4), 276–277. doi:10.1038/nrmicro870 PMID:15077609

Tukur, A. I. (2010): Temporal variation of malaria occurrence in Kano municipal local government area. *Bayero Journal of pure and applied science*, 3(1). Retrieved from www.ajol.info/index.php/bajopas/article/viewFile/58745/47071

Wandiga, S. O., Opondo, M., Olago, D., Githeko, A., Githui, F., Marshall, M., & Ouma, G. O. (2010). Vulnerability to epidemic malaria in the highlands of Lake Victoria basin: The role of climate change/variability, hydrology and socio-economic factors. *Climatic Change*, *99*(3), 473–497. doi:10.1007/s10584-009-9670-7

World Health Organization. (2006) The African malaria Report, Regional office for Africa and East Mediterranean.

World Health Organization (WHO). (2008). *Protecting Health from Climate Change - World Health Day 2008.* Geneva, Switzerland: WHO Press.

World health organization (WHO) year book 2002. World Health Organization (WHO)/ World Meteorological Organization (WMO)/ United Nations.

World Health Organization. (2013). Impact of Dengue World Health Organization, Geneva. IPCC WGII AR5, 2013.

Worrall, E., Connor, S. J., & Thomson, M. C. (2007). A model to simulate the impact of timing, coverage and transmission intensity on the effectiveness of indoor residual spraying (IRS) for malaria control. *Tropical Medicine & International Health, 12*(1), 75–88. doi:10.1111/j.1365-3156.2006.01772.x PMID:17207151

Yahaya, A., Tyav, Y. B., & Nura, M. S. (2014). Seroprevalence of Paediatric Malaria Infections in Two Hospitals in Kano State, Nigeria. *International Journal of Sciences: Basic and Applied Research, 14*(1). Retrieved from http://gssrr.org/index.php?journal=JournalOfBasicAndApplied&page=issue&op=view&path%5B%5D=82

Yusuf, D. M. (2008, August 25-28). An assessment of malaria prevalence in Bichi local government area Kano state. *Proceedings of the 50th Annual national conference The Association Nigerian Geographers (ANG)*, University of Calabar Cross River State.

Gobir, Z., & Tukur, Z. (2014). Prevalence of malaria parasitemia using rapid diagnostic test among apparently healthy children in Kano, Nigeria. *Journal of Medicine in the Tropic, 16*(1), 1–4. Retrieved from http://www.jmedtropics.org/article.asp?issn=22767096;year=2014;volume=16;issue=1;spage=1;epage=4;aulast=Gobir doi:10.4103/2276-7096.132566

Zhou, G., Minakawa, N., Githeko, A. K., & Yan, G. (2004). Association between climate variability and malaria epidemics in the East African highlands. *Proceedings of the National Academy of Sciences of the United States of America, 101*(8), 2375–2380. doi:10.1073/pnas.0308714100 PMID:14983017

Zubair, L., Galappaththy, G. N., Yang, H., Chandimala, J., Yahiya, Z., Amerasinghe, P., & Connor, S. J. et al. (2008). Epochal changes in the association between malaria epidemics and El Niño in Sri Lanka. *Malaria Journal, 7*(1), 140. doi:10.1186/1475-2875-7-140 PMID:18652697

APPENDIX: QUESTIONNAIRE ON PERCEPTIONS OF PEOPLE LIVING IN METROPOLITAN KANO ABOUT LINKS BETWEEN MALARIA EPIDEMICS, WEATHER VARIABLES AND CLIMATE CHANGE

Please tick appropriate on the options in Sections 1 and 2.

Section 1: Demographic Information

Table 12.

Serial Number	Questions	Coding Categories
01	Are you …………..	1. Male 2. Female
02	What is your current marital status?	1. Single 2. Engaged 3. Married 4. Divorced 5. Widowed
03	Please indicate the age bracket you are in:	1. 16-24 2. 25-34 3. 35-44 4. 45-54 5. 55 or above
04	What is your highest qualification (level of Education)	1. Primary 2. Secondary 3. Tertiary 4. Informal
05	Do you have an occupation which one is applicable to you	1. House-wife 2. Civil-servant 3. Trading 4. Farming 5. Student

Section 2

1. Changes in the weather (e.g. rainfall and temperature) have affected human health in your community.

Strongly Agreed (). Agree (). Neither agree nor disagree (). Disagree (). Strongly disagree ()

2. How does malaria disease affected you or any member your family?

Very much (). Little (). Not affected ().

3. Increase in rainfall and temperature brings an increase in mosquitoes in your area.

Strongly Agreed (). Agree (). Neither agree nor disagree (). Disagree (). Strongly disagree ()

4. Changing climate is associated with the warmer condition (increase in temperature) and it gives mosquito's better breeding ground and spread of malaria disease.

Strongly Agreed (). Agree (). Neither agree nor disagree (). Disagree (). Strongly disagree ()

5. There is a relationship between malarial incidence and climate variability in Kano.

Strongly Agreed (). Agree (). Neither agree nor disagree (). Disagree (). Strongly disagree ()

6. Climate change issue is fully integrated in public health sector and the communities in metropolitan Kano are informed about the links between malaria epidemics, weather variables and climate change.

Strongly Agreed (). Agree (). Neither agree nor disagree (). Disagree (). Strongly disagree ()

General comments:

....
....
....
....
....
....

Interview Guide for Health Officials

1. What can you say about the malaria disease occurrences in Kano?
2. What is your perception on the relationship between climatic factors and increase in malaria epidemic and when do you experience reports on malaria case?
3. Are people in the community aware about the relationship climate variability and malaria?
4. What period you usually have mosquitoes most in Kano?

Chapter 10
Are Climate Change Adaptation Policies a Game Changer?
A Case Study of Perspectives from Public Health Officials in Ontario, Canada

Chris G. Buse
University of Northern British Columbia, Canada

ABSTRACT

The health impacts of climate change have received significant attention in the international scholarly literature. Despite this, there is an absence of research evaluating existing policies aimed at promoting and protecting population health. This chapter provides an implementation analysis of the Ontario Public Health Standards (OPHS), 2008/2014--the provincial policy statement that governs mandatory public health activities in the province which includes taking action on climate change. This chapter responds to two specific questions: First, how are Ontario's 36 regional health units interpreting and implementing this policy statement; and second, how are those interpretations translated into practice. Using a web-scan and in-depth interviews with practitioners from twenty Ontario health units, this paper presents four interpretations of the OPHS, a typology of best practices related to regional adaptation, and policy recommendations to bolster domestic and international adaptive capacity to emerging infectious diseases associated with climate change, and a variety of other health-related climate impacts.

DOI: 10.4018/978-1-5225-0553-2.ch010

INTRODUCTION

Climate change is rapidly affecting human health around the globe and posing new challenges for public health infrastructure and systems (McMichael, 2013). This paper analyzes the degree to which existing public health policies are successful in bolstering adaptation to climate change in the public health sector, and whether the implementations of those policies create activities capable of responding to the emergence of new infectious diseases and a host of other climate-related health impacts. Specifically, this paper uses Ontario, Canada as a case study for examining how public health policies (i.e. the *Ontario Public Health Standards, 2008/2014*) are interpreted and made actionable by front-line practitioners. The paper begins with an overview of climate change and its health impacts across Ontario before describing the methods and presenting a typology of policy interpretations that speak to the strengths and weaknesses of a relatively broad policy mandate. More specifically, the paper seeks to understand how practitioners made climate change actionable in their work before and after the enactment of the *Ontario Public Health Standards,* and to what degree policy elicits adaptive programming and interventions. In doing so, the paper provides guidance for government officials and public health practitioners in other nations attempting to understand whether their policy infrastructure is robust enough to respond to climate-related health impacts.

BACKGROUND: CLIMATE CHANGE IN ONTARIO, CANADA

Canada's climate is rapidly changing, and as a result, so too are the health risks posed to Canadians (Séguin, 2008). While global average temperatures are projected to increase by between two and four degrees Celsius by the end of the twenty-first century, Canada will experience more rapid temperature increases (IPCC, 2013). Between 1948 and 2006, southern Ontario has already experienced a 1.3 degrees Celsius increase, and projections for the province indicate that it will experience a two to four degree Celsius increase by 2050 (Feltmate & Thistlewaite, 2012).

Climate change has been called the greatest public health threat of the twenty-first century (Costello et al., 2009). Accordingly, the Canadian experience of climate change will mirror that occurring in other democratic countries with a well-established public health sector. Primary risks include the direct biological consequences of extreme weather events (i.e., heat, cold, violent weather), temperature-enhanced air pollutants in urban areas, and increased exposure to UV radiation; secondary risks include risks mediated by biophysical or ecological processes including food security/foodborne disease, water scarcity/waterborne disease, and changes in disease vectors as previously inhospitable climates become warmer. For Ontario, Canada, there are significant concerns about emerging vectors contaminating drinking water and recreational water sources and food contamination, with increased risk of *E. coli, Cryptosporidium, Giardia, and S. typhi* all being linked to warmer temperatures (Health Canada, 2008). Changes in the ecology of various disease-carrying insects also pose significant concerns for the spread of mosquito borne vectors including the West Nile Virus (Paz, 2015) and Eastern Equine Encephalitis (Parham et al., 2015). There is also cause for concern surrounding the spread of Lyme disease spread by tick populations (Ostfeld et al., 2015) and the possible emergence of new diseases and those previously eradicated across the country (Cambell-Lendrum et al., 2015). Tertiary risks include a host of mental health issues, displacement and migration, and the exacerbation of existing health inequalities (Friel et al., 2011; McMichael, 2013; Séguin, 2008).

Climate change will also pose significant economic implications, particularly for major urban areas. For example, the cumulative cost of premature mortality risk attributable to heat and air quality impacts from climate change in Toronto between 2010 and 2100 is estimated at between $65 and $96 billion, and the healthcare costs attributable to air quality impacts from climate change alone in the same time period are estimated to be between $72 and $285 million (NRTEE, 2011). These estimates do not factor in the rising cost of extreme weather. Indeed, the July 2013 storm that struck much of southwestern Ontario has proven to be the province's costliest "natural" disaster with estimated damages totalling upwards of $850 million.

Some research has begun to explore how institutional policies and programs at the national level are fostering the country's capacity to adapt to climate change's health impacts (Lesnikowski et al., 2014; Panic and Ford, 2013). In the United States, empirical work demonstrates that public health staff may lack the appropriate training and skills to engage with climate change programmatically or substantively, and that adaptation and prevention have yet to be made a priority among American health departments as a result (Maibach et al., 2008). To remedy this, the Center for Disease Control developed the Building Resilience Against Climate Effects Framework which has been taken up across levels of government.

In the Canadian context, specific public health adaptation options have yet to be implemented in many regions across the country (Canadian Public Health Association, 2006), with the exception of perhaps Quebec which has been leading the charge in developing advanced climate change and health modelling scenarios with OURANOS. Following this, an investigation into emerging activities and best practices from other parts of Canada is warranted, but particularly those applying evidence informed public health approaches (Hess et al., 2014) which may have transferable lessons for other developed nation contexts. Further, while numerous frameworks and strategies for public health adaptation to climate change exist in the literature, there are fewer empirical examinations of existing policies and their resulting implementation, and how well they are able to address the health implications of climate change.

In the sections that follow, the author uses the public health system in Ontario as a case study for evaluating the programmatic outputs of public health policy specifically related to climate change adaptation. The author uses a web scan and in-depth interviews with key informants from Ontario's environmental public health practitioner community to 1) illustrate the markedly different interpretations of provincial public health policy protocols related to climate change both in the absence and presence of formal policy; and 2) develop a typology of adaptation activities being undertaken at the level of Ontario health units.

Why Study Ontario? The Ontario Public Health Standards (2008) and Climate Change Adaptation

Ontario is Canada's most populated province, yet despite a high degree of urbanization, it still holds a vast rural and northern population. The public health system is regulated at the provincial level by the Ministry of Health and Long-Term Care (MOHLTC), but program and service delivery occurs at the regional level with a cost-sharing mechanism that splits operational costs between the province and regional municipalities (Deber, 2003). The regional nature of Ontario's public health system is arguably well designed to engage with climate change because health units can respond to regionally specific health needs, as well as the regional specificity of climate change. The province has thirty-six regional health units (see Figure 1) which are governed under the *Ontario Public Health Standards* (OPHS).

The OPHS are published:

As the guidelines for the provision of mandatory health programs and services by the Minister of Health and Long-term Care pursuant to Section 7 of the Health Protection and Promotion Act, R.S.O. 1990, C.H.7. (MOHLTC, 2008, 1)

The OPHS are also one of the first pieces of Canadian policy creating a mandate to address the emergent health implications of climate change. The OPHS classifies climate change as an environmental health hazard, and broadly organizes the protocols for addressing environmental health hazards around preventing or reducing the burden of illness from hazards in the physical environment. Specific protocols include a mandate to:

1. Conduct surveillance of the environmental health status of the community;
2. Conduct epidemiological analysis of surveillance data, including monitoring trends over time, emerging trends, and priority populations;

Figure 1. Ontario's 36 regional public health units grouped by census division*

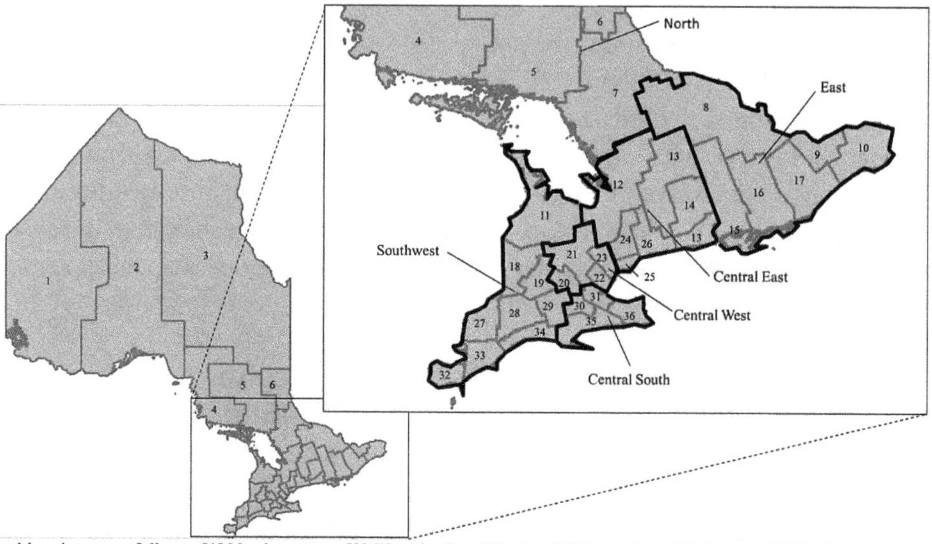

*Health units are as follows: [1] Northwestern; [2] Thunder Bay District; [3] Porcupine; [4] Algoma; [5] Sudbury and District; [6] Timiskaming; [7] North Bay-Parry Sound District; [8] Renfrew County and District; [9] City of Ottawa; [10] Eastern Ontario; [11] Grey-Bruce;; [12] Simcoe Muskoka District; [13] Haliburton, Kawartha, Pine Ridge District; [14] Peterborough County-City; [15] Hastings and Prince Edward Counties; [16] Kingston, Frontenac, Lennox and Addington; [17]Leeds, Grenville and Lanark; [18] Huron County; [19] Perth District; [20] Regional Municipality of Waterloo; [21] Wellington Dufferin-Guelph; [22] Halton Region; [23] Peel Region; [24] York Region; [25] City of Toronto; [26] Durham Region; [27] County of Lambton; [28] Middlesex-London; [29] Oxford County; [30] Brant County; [31] City of Hamilton ; [32] Windsor-Essex County; [33] Chatham-Kent; [34] Elgin-St.Thomas; [35] Haldimand-Norfolk; [36] Niagara Region

3. Increase public awareness of health risk factors associated with the following health hazards: indoor air quality, outdoor air quality, extreme weather, climate change, exposure to radiation, and other measures as emerging health issues arise. These efforts shall include adapting or supplementing national and provincial health communication strategies, and/or developing and implementing regional/local communications strategies; and
4. Assist community partners to develop healthy policies to reduce exposure to health hazards (MOHLTC, 2008, 46–47).

It merits attention that "climate change" appears in the OPHS document under only the third protocol where it is simultaneously identified as an environmental health hazard and sets a requirement to raise public awareness about the health risks associated with a changing climate.

An analysis of all the activities and protocols related to climate change in the OPHS is beyond the scope of this paper. A previous review by Clarke and Berry (2012) compared the OPHS with existing climate change and health adaptation frameworks.

Their findings indicate that the OPHS requires many traditional risk-management activities identified in pre-existing climate change adaptation frameworks related to emergency response and surveillance and monitoring.

However, these claims have not been empirically validated by environmental health managers mandated to work in the field of public health in Ontario, and some scholars are critical of whether "traditional" risk management activities are well suited to addressing the complexity of climate change's related health impacts (Morris, 2010). Clarke and Berry's results neither describe existing adaptation initiatives across the province nor how environmental health practitioners see existing protocols as linked with actions required on climate change adaptation. In moving somewhat closer to this end, Paterson and colleagues (2012) conducted interviews with ten health units in Ontario to assess public health adaptation activities. Their findings suggest that public health officials are particularly concerned with temperature, air quality, and extreme-weather-related impacts. Yet, their analysis is limited primarily to health units in the southwest which reduces the provincial generalizability of their results. Their work also does not describe existing adaptation activities that are either ongoing or in development throughout the province.

METHODS AND ANALYSIS

This study's methods are informed by qualitative methods and followed a two-phase data collection and analysis strategy. First, the author conducted a web scan of each of the 36 Ontario public health unit websites between January and April of 2013. For those health units with stand-alone websites, the Boolean search terms "climate change" or "greenhouse gas*" or "global warming" were used, making the documentation and collection of resulting hits straightforward. For health units' websites embedded in municipal or regional web domains, the author employed the same search terms with the additional search term "and health." The publishing dates of key documents were utilized to inform a temporal analysis of practices pre- and post-implementation of the *OPHS* in 2008, and if relevant 'hits' were undated or they were excluded from the temporal comparison, but included in the subsequent content analysis. Content analysis was used to document how health units describe the health impacts of climate change, and what programs each health unit tied directly to climate change.

Next, the author conducted twenty semi-structured interviews with public health professionals from twenty Ontario health units. While all thirty-six health units were contacted for an interview, twenty had a staff member who was willing to participate in the study, thirteen refused to participate on the grounds that they could not contribute anything to it (i.e., there was no climate-change orientation

Table 1. Interview sample by census division (N=20)

Health Units by Census Division (N)	Number of Health Units Represented in Interviews	% Health Units Reporting from Each Census Division
Central East (6)*	4	66.6%
Central West (4)	2	50.0%
Central South (4)	2	50.0%
East (6)	2	33.3%
Southwest (9)	6	66.6%
North (7)	4	57.1%
Total (36)	20	55.5%

*Includes Toronto for the purpose of analysis and anonymity.

to their work, nor were there plans to incorporate climate change into their work in the near future), and three health units did not respond. The author transcribed interviews following appropriate protocols (Poland, 1995).

The twenty completed interviews have strong regional representational by census division (see Table 1) and capture a wide-ranging spectrum of action on climate change. Interviews sought to uncover personal perspectives from active practitioners working in the field in an attempt to uncover how the then-five-year-old policy mandate was being translated into practice and what that meant for their work on a day-to-day basis (see interview questions in the Appendix). Accordingly, a limitation of this study is that it does not include the perspectives of provincial and federal level decision-makers because they are not necessarily front-line practitioners. Future work could seek to integrate perspectives across levels of government to improve the resolution of our understanding related to climate change policy and practice.

Due to the perceived political nature of climate change, the author protected research participants' anonymity through the use of pseudonyms, and their specific organizations are not identified by name. This study received ethical approval from the University of Toronto Research Ethics Board.

The analysis of interview transcripts followed an iterative strategy involving three distinct phases to organize and iteratively 'unpack' the data. As analytic strategies, each phase included identifying information and the 'uncleaned' data are therefore not presented here, but the three phases (described below) were utilized to enhance rigour and triangulate findings.

First, a cross-case display matrix was produced following Miles and Huberman (1994) that sought to produce a visual display of key variables and themes loosely organized around the interview questions. This method allows the researcher to draw comparisons across cases according to some of the distinctions present within core themes. Second, in an attempt to preserve the contextual nuance of each interview, case summaries (Poland et al., 2009) of each interview were produced that resemble a condensed transcript highlighting key programmatic areas of interests, comments on policy uptake, and other thematic issues that arose during the interview. Third and finally, the author employed "category zooming" — a form of analytic generalization that "zooms in" on a particular aspect of qualitative inquiry with the goal of going into greater interpretive depth about the complexities of a single point in a broader study (Halkier, 2011). This approach loosely resembles a targeted analysis of spoken text (in the form of interview transcripts) with the goal of understanding how discursive features of policy interpretation manifest in distinct and particular forms of policy implementation.

RESULTS

20 of 36 health units had a staff member volunteer to participate in this study, 13 refused to participate on the grounds that they had no climate change orientation to their existing work, and three did not respond (see Table 1). Completed interviews offer a geographically representative sample of Ontario health units when aggregated by census division: four of six health units from the central east (including Toronto for the purpose of analysis and anonymity); two of four health units from the central west; two of four from the central south; two of six from the east; six of nine from the southwest; and four of seven from the north.

Four Interpretations of OPHS Protocols Related to Climate Change Adaptation

The majority of interview participants (14 of 20) identified the existing guidelines as their entry point into engaging with climate change from a public health perspective. As one respondent put it: "The whole ministry put out the new ministry standards and I'm sure that you're probably familiar with those as well ... the standards and protocols, 2008, right? And in there they outlined climate control as one of our initiatives that needs to be addressed" (Syndra, Environmental Health Manager).

However, interpretations of what the OPHS required resulted in varied responses to program development across Ontario health units. The degree of engagement with climate change at the program level can be conceptualized in terms of four specific interpretations of the OPHS policy, including: inaction, communicating health risks, mainstreaming climate change into existing programs, and championing innovative activities.

Inaction

Inaction was the lowest level of engagement with climate change and it is notable that many of the public health unit staff interviewed for this project indicated that they did not have a comprehensive climate change program (N=8). Vladimir, an environmental program manager said "I'd have to say that if we were really to talk about climate change here as something we are sort of pursuing in a conscious way right now, I would have to say that we're really not" and that "climate change isn't up there as a label on anything that we do." This point is reiterated by other managers:

I have to say right off the bat that we don't directly call it that. We don't have a climate change management program [laughing]. (Riven, Associate Medical Officer of Health)

We haven't done a tonne on climate change really. (Lucian, Environmental Health Manager)

Well I don't know if we do it in a ... way that we conscientiously are doing this because of, you know, we're under the umbrella of climate change adaptation. But I think—but specifically—probably similar to other health units, where we've been involved in a number of ways ...I guess easily would be the extreme temperature ... types of programming and protocols that we have in place ... have had in place now for a number of years here. (Yorick, Environmental Health Manager)

Actually we have not really done much to address it. We are waiting for the ... province to come up with...I think, some province-wide...program or protocol that will apply. (Olaf, Environmental Health Manager)

Barriers to public health adaptation are well documented in the peer-reviewed literature and include perceived uncertainties about future climatic and socioeconomic conditions; the lack of financial, human, and technological resources; limited social capital to engage other sectors; the highly specialized nature of public health and other institutions that might limit or make invisible adaptations taking place in other sectors; and cognitive limitations around behaviour change (Huang et al., 2011). Wardekker et al. (2012) add that the multi-causality and complex cause-effect relationships between climate change and associated known and unknown impacts further complicate planning.

Table 2. Barriers to developing public health programs to adapt to climate change in Ontario

Barrier	Supporting Quote
Lack of Resources	"We don't really have the resources to start any kind of planning and anything in depth as the project goes to climate change" – Jayce, Environmental Health Manager
Problem Denial	"[CC] is hardly ever addressed as far as I hear. I'm not sure why and I think some of it might be personal. People aren't going to change, so maybe a bit of it is denial?" – Quinn, Health Promoter
Prioritization of Other Areas	"I don't think it's really seen as something that is important here." – Syndra, Environmental Health Manager
Blame Avoidance	"I think there's concern that it's just too much. It's just too far in the future. It's uncertain, it's scary, and is it really our business?" – Alistair, Medical Officer of Health

Interviews with public health practitioners in Ontario illuminated similar findings, whereby constraints in or lack of resources of time, energy, money, and political leadership to run new programs were identified as a primary reason for inaction (see Table 2). This is relative to health units with significantly more resources who may have greater ability to engage with climate change and its resulting health impacts. However, Ontario public health practitioners further identified that problem denial may be an issue within health units, and that other areas of research and practice are prioritized. A final barrier was the notion of "blame avoidance" (Howlett, 2014), whereby the uncertainty and time lag associated with climate change — while problematic from a programming perspective — were coupled with a question of what business or authority public health had engaging with climate change relative to other sectors.

One respondent claimed that public health's involvement might be conceived of as "the kindergarten kids trying to play with the big boys. I don't know, I suppose again, it's a question of what areas do you have a mandate over?" This respondent went on to clarify this point by adding:

Can we lead the charge on it? I'm not sure that would be the role of public health — to lead the charge on climate change — because the factors causing it are so, I mean they're individual choices, but they're also macro. They're systems, they're factories, they're regulations. You know, it's a whole different set of jurisdictions that deal with the major polluters. We can influence individual behaviour, to the extent that's even possible, but really only through information and education. We don't have a lot of levers to pull that would make a major difference systemically across the board. (Riven, Associate Medical Officer of Health)

Communication of Health Risks

A second interpretation of the OPHS highlights only the communication of health risks associated with climate change as the primary means of policy implementation, but none of the other OPHS-mandated protocols related to environmental health hazards. In this case, practitioners viewed public engagement and raising awareness of climate change's health impacts as their sole programmatic responsibility:

Under [the OPHS], it was mandated by the Board of Health outcomes ... the board of health shall increase public awareness of health risk factors associated with the following health hazards, and one of those health hazards was part of climate change — which was something that we were tasked to do. (Darius, Environmental Health Manager)

Namely [climate change programming] has been with our heat alerts and our cold alerts. Um, and through that mechanism, that's where most of the work has been associated with. (Zac, Environmental Health Manager)

That's what we do. So we make sure people are informed. So we have protocol here when those events happen to make sure that we communicate or issue immediate releases to ... the media we have around here. In addition to that, we do work with municipalities to make sure that the information is disseminated door to door. (Olaf, Environmental Health Manager)

So we're not at the stage where the public health standards say you have a role in prevention at the level of the health unit, because even if that word were mentioned, rather than just mitigating the appearance of risk, then there's mitigating risk, then there's preventing. So of the kind of three, we're only at the very beginning ... communicating risk. That's the first level. (Riven, Associate Medical Officer of Health)

The emphasis on risk communication was further highlighted through the web scan, where the majority of risk-communication activities were on an event-by-event basis (i.e., communicating the health risks of extreme heat during a heat alert). Some health units also linked to comprehensive climate change impact documents (see, for example, Séguin, 2008), or presented reviews of region-specific climate impacts through independent research or vulnerability assessments.

How climate change is tied to specific health effects is also highly dependent on the region in which a health unit is located. Figure 2 communicates which health impacts of climate change were described on health unit websites by census division. That is, this figure indicates what percentage of health units identify one or more

Figure 2. Percentage of climate change impact areas described on Ontario health unit websites according to census division

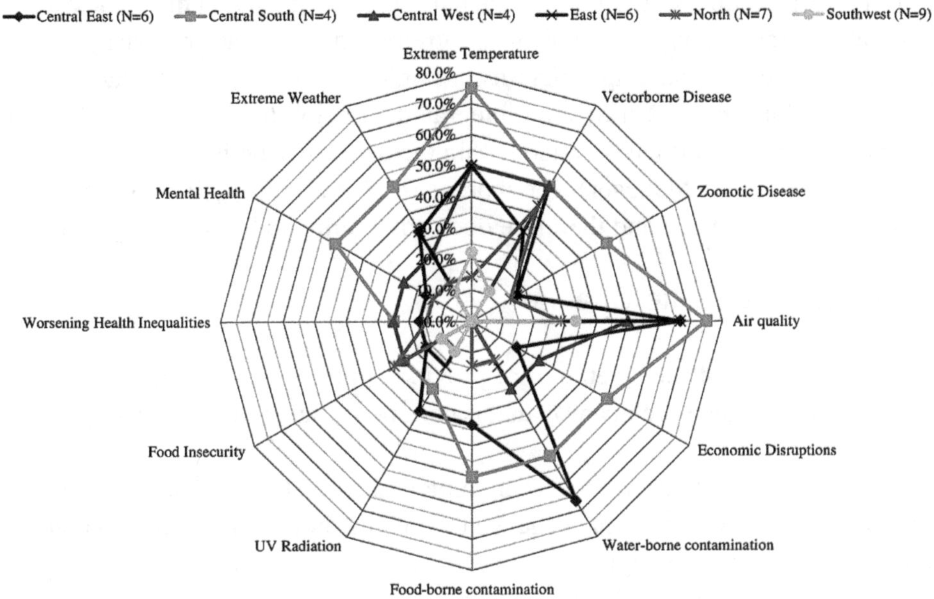

health impact in materials they have made publicly available via their websites. In other words, this figure merely serves to show the health impacts of greatest concern across census divisions, and across the province more generally.

Findings show that health units in the "Golden Horseshoe" area (i.e., Central West and Central South census divisions, which encompass the Greater Toronto Area [GTA] through to Niagara Falls) of the province tend to have more active outreach for identified health impacts. Perhaps because this is the most populous part of the province, these health units tend to be larger and well-resourced. Also, by virtue of the urban areas they occupy, there may be richer opportunities for collaboration with neighbouring health units and environmental not-for-profits. Finally, given the pre-existing concerns of extreme temperature and air quality issues in the Niagara-GTA corridor and the inseparability of these issues from the climate change discourse, the path dependence in program development before and after the development of the OPHS has led those health units to make natural connections to climate change.

Mainstreaming Climate Change into Existing Programs

A third interpretation of the OPHS mandate was to "mainstream," "backfill," or "integrate" climate change into existing programs. This approach primarily involves practitioners viewing existing programs through the lens of climate change, or "in other words, link[ing] them with an integrated strategy that says if there's an opportunity to promote climate change within the programs, do it at no cost" (Viktor, Environmental Health Manager). The approach of mainstreaming is well documented by Paterson et al. (2012) and therefore does not require additional analysis here.

Championing Climate Change

The final category of interpretation of provincial policy resembled a more active engagement with the issue in a way that is commensurate with the OPHS protocols and beyond.

The organizations that have so far "championed" climate change as a public health issue are actively linking climate change to existing and emerging health concerns in their regions, and are beginning to develop specific partnerships and programs to address climate change's health impacts.

A Typology of Adaptation Strategies among Climate Change "Champions": Pre- and Post-2008

Findings from the web scan and interviews indicate that prior to the implementation of the OPHS in 2008, climate change was rarely taken up by any health units in Ontario. However, climate change does factor into three unique streams of programmatic action prior to the 2008 policy mandate. First, instances where climate change, greenhouse gas emissions, or global warming were mentioned included activities around "Clean Air Days" and provincial "Idle free" legislation — particularly around schools, which demonstrates the inseparability of climate change with the broader air quality discourse. This was reiterated by one of the interview participants who said:

We still today support Clean Air Day and as an organization the municipality still offers free bus rides every first week of June. That happens on that Wednesday. So we still do the advertising. That never went away. We still continue to do that piece, but it's a very minor piece now in comparison to what we tried to make of it. (Syndra, Environmental Health Manager)

A second piece of programming prior to 2008 was the monitoring of health-related impacts from poor air quality and extreme heat, often in collaboration with the Ministry of Environment. In this instance, there is some evidence from the web scan that indicates the discourse of climate change was invoked to further communicate the health risks associated with exposure to poor air quality and high temperatures, although the active linking of these issues to climate change was primarily present in health units occupying the GTA-Niagara corridor.

Finally, although occurring to a lesser degree, some public health units began creating advocacy coalitions with research institutions and community organizations, such as the Clean Air Partnership and the GTA Clean Air Council, to begin to develop strategic plans for how best to approach climate change from a public health perspective. Numerous working groups were formed at the provincial level, including the participation by several health units in the development of the 2008 environmental health hazard protocols. Some regional municipalities entered into conversations around the development of regional climate change strategies and emissions inventories, with several health units playing a supporting role.

Following the enactment of the OPHS, however, climate-related practices began to proliferate across the province, largely driven by practitioners reading the protocols as an impetus to "champion" climate change at the regional level. Table 3 provides an overview of six specific areas of climate change adaptation from Ontario public health units broadly classified as: capacity building and knowledge exchange; management and planning; disease surveillance and analysis; communication and outreach; green infrastructure; and localized food production. It is notable to mention that health units have been required to monitor and report outbreaks of infectious disease prior to the implementation of the OPHS. However, it was not until

Table 3. Typology of adaptation strategies pursued by Ontario public health units

Adaptation Strategy	Examples
Capacity building and knowledge exchange	Training workshops; provincial meetings; roundtable exercises; information sharing
(Intersectoral) management and planning	Reviews of official plans; participation in regional climate change strategies; sourcing funding
Surveillance and analysis	Warning or observation systems (heat, air quality, vector-borne disease, water quality); epidemiological surveillance; vulnerability assessments
Communication and outreach	Dissemination of behaviour change messages (e.g., sun safety campaigns)
Green infrastructure	Organizational "greening"; built environment initiatives; urban forestry strategies
Localized food production	Development of local food charters

after 2008 that we begin to see infectious disease concerns tied directly to climate change, particularly those related to West Nile Virus and Lyme disease, although actions are primarily oriented towards the continued surveillance of these diseases, opting to primarily 'wait and see' what happens prior to designing new public health interventions to address possible increases in incidence and prevalence.

The majority of actions fall under the first four strategy types, with green infrastructure and localized food production emerging in but a few health units as upstream strategies to combat (i.e., mitigate) and adapt to climate change. In the interview data collected, it appears that the emerging "best practices" for climate change adaptation in Ontario currently fall under four programmatic areas. First, conducting vulnerability and adaptation assessments to assess baseline community vulnerability and the adaptive capacity of public health and community infrastructure seems to currently be in vogue among Ontario health units as an "aggressive" strategy to better understand the impacts of climate change across populations:

The health unit wanted to do something like other health units… [Health unit name omitted] has aggressively done the climate change work for their vulnerability assessment, so we wanted to be participants in that, so that we can better protect our community. (Darius, Environmental Health Manager)

At least two health units had completed comprehensive climate change and health vulnerability assessments that looked at health impacts at present and into the future using climate models to project future changes to health. An additional three health units conducted scoping reviews of the literature to broadly identify "vulnerable populations" that may or may not be present in their geographic area, but which are captured in the peer-review and grey literature.

Second, participating in regional climate change strategies is an adaptation option that was undertaken by at least eight health units. While such strategies are largely a capacity-building and collaborative exercise, public health units are active members of regional conversations on this subject and are aggressively considering how their existing competencies and expertise might contribute to an integrated regional strategy.

Third, "organizational greening" has been undertaken by at least two health units, where corporate practices have been analyzed to seek out energy efficiencies and lead by example before developing community interventions. This has proven to be a transformative step in fostering a broader culture of sustainability in the workplace:

It's certainly taking it internal, right, instead of...[absolutely] mitigation vs. adaptation in the community. It's sort of looking internally and saying what could we or should we be doing … to reduce our carbon footprint and other impacts on the environment. Leading by example. So, we've ... gradually implemented many of the recommendations about procurement, about reduced light usage, the temperature settings in our buildings, reflective coatings on windows. Sort of energy-conservation measures. We've really gone after transportation. So we've communicated repeatedly the importance of people carpooling and not travelling at all if they can webcast it instead or teleconference it instead or using public transportation or even active transportation if it fits. (Alistair, Medical Officer of Health)

A final emerging best practice relates to the built environment. Some health units highlighted their participation in urban forestry strategies, active transportation initiatives, and regional infrastructure development. However, several practitioners were critical of the emerging provincial emphasis on built environment strategies in relation to climate change (Dimoulas-Graham et al., 2012; OPHA, 2011; Perrotta, 2011), and questioned the degree to which the climate change discourse was actually present in built environment work:

And while, right now, the ministry and the Chief Medical Officer of Health for Ontario is promoting chronic disease prevention and obesity reduction strategies, and health units working with the municipalities on the built environment, like building walkable communities. I think that unfortunately they don't want to get message overload and they think they've got to target one message. That might be the strategy they are using, but I think there's such a unique opportunity right now if you are promoting walkabilty for physical activity, in just your second breath you could say you are reducing GHGs and addressing climate change and air quality. (Janna, Environmental Health Manager)

I think we at some point need to look at how you take it to the centre. There is an Ontario Public Health, public health sector strategic plan that was just launched by the Chief Medical Officer of Health in April, and it has a section on the built environment, and this is actually, environment, environmental health in general, and a specific area of focus on the built environment. It doesn't explicitly speak to climate change, and I think a future iteration of that plan should. It would be a worthy target that a future version of it actually speak to climate change. To actually become sort of a collective area of focus for all of public health. I would hate in a hundred years for people to look back and say "where were you?" (Alistair, Medical Officer of Health)

In both of these examples, practitioners signal that existing practices in public health may not go far enough with respect to the built environment, thereby actively questioning the degree to which co-benefits of climate change adaptation work are being stressed or emphasized. From their perspective, this raises an important question about how the work is focused and framed when climate change is left out of the conversation.

DISCUSSION: PUBLIC HEALTH POLICY AS CONSTRAINING AND ENABLING FOR FIELDS OF PUBLIC HEALTH PRACTICE

Results indicate that fourteen of twenty interview participants referred to the OPHS mandate to engage with climate change from a public health perspective, but participants had different interpretations of what the policy meant and how it should be implemented. The organizations that tended to be furthest ahead in terms of climate change work do not frequently invoke the OPHS, but describe particular programs they have developed and delivered. In this regard, the broad mandate present within the OPHS leaves room for a range of possible interpretations that can either limit or open up new opportunities to take action. For example, one environmental health manager indicated that the standards give practitioners new opportunities to explore climate change in relatively nuanced ways:

The public health standards now provide a little bit more latitude. So there's you know, more opportunities to do more health promotion around exposure. So we, you know, for example, we're responsible for increasing public awareness around health risks associated with a lot of health hazards, and one of them being climate change and extreme weather and outdoor air quality. So, that gives us a little more [laughs]. At least I feel like I have a little more latitude or opportunity to explore more than we have in the past and in — you know, there's more of a role on supporting public policy in the standards as well. (Evelyn, Environmental Health Manager)

However, the opposite can also be true, where a broad policy mandate is read in a way that stifles innovation and action is directed to meet only a minimum standard:

So the question's going to be, once the information is out there, how long is it going to take for the broader Ontario public health to start changing its attack on stuff? ...I've also said to people that the health standards talk about outreach and communication on climate change and extreme weather and leave it kind of broadly. (Udyr, Research and Policy Analyst)

What is clear from this discussion is that the existing policy mandate in Ontario has created space for practitioners to pursue a variety of strategic options related to climate change adaptation, and that a broad policy mandate can either enhance or inhibit innovation. However, it is probably fair to conclude — especially since thirteen health units declined to participate in this study given they had little to offer — that the OPHS have not translated into equal action on climate change across health units and that policy implementation is not necessarily commensurate with the intentions of the formal policy statement.

In this regard, the intensity of action on climate change adaptation can take the form of positive responses where health units use procedural tools to support existing actions and core functions of public health practice by integrating climate change into pre-existing programming. However, increased innovation can also take form through small-scale positive responses that use substantive programmatic "experiments" to carve out entirely new areas of practice. Such innovation recognizes the limitations of a health-protection focus that reactively responds to health impacts instead of engaging in a proactive agenda to mitigate and adapt to health impacts that are forecasted twenty-five or even fifty years into the future.

Successful pilot projects open the possibility for future policy change, while an alternative interpretation of existing policy protocols might reinforce the status quo of public health practice by re-emphasizing traditional responsibilities. Moreover, some health units may interpret the broad policy mandate as lacking clarity, or worse, as creating new responsibilities with the same amount of resources — which may effectively inhibit innovation and prevent health units from engaging with climate change entirely.

However, there are still significant challenges in advancing the climate change agenda in Ontario, particularly in relation to emerging infectious diseases. Findings above demonstrate that most of the work across the province is focused on disease surveillance and risk communication and more proactive actions in the face of climate change are limited. Indeed, the OPHS has a number of pre-established protocols to deal with infectious diseases that have a long history of affecting population health in Canada. The mandated activities are largely in line with what has been presented above: disease surveillance, health promotion (i.e. risk communication), and health protection (i.e. health inspections of high risk areas of potential transmission). This raises two particularly interesting questions. First, are existing public health activities in Canada adaptive enough to deal with the health impacts of climate change, and second, are the existing provincial policies robust enough (as suggested by some authors—see Clarke & Berry, 2012) to trigger action on emerging infectious diseases that do not have a long standing history of public health intervention in Ontario? These questions are perhaps, more relevant than ever as climate-sensitive infectious diseases (e.g. Lyme Disease, the West Nile Virus, Dengue, Malaria, endemic fungal diseases) pose new risks to the health of North American populations (Greer et al., 2008)

POLICY RECOMMENDATIONS TO BOLSTER ADAPTIVE CAPACITY

Several policy recommendations for local public health practitioners and regional/provincial decision-makers follow from these findings. These recommendations include:

1. Making the connections between climate change and health explicit for the practicing public health community;
2. Acquiring the resources and training necessary to conduct prospective disease modeling;
3. Conducting climate change and health vulnerability assessments to understand baseline exposure, sensitivity and adaptive capacity;
4. Utilizing adaptive management principles to help develop new programs, and evaluate their success; and making the political and economic case for programs that have co-benefits which improve community well-being and bolster adaptation to climate change at the community level.

While many existing actions on climate change are oriented toward building capacity to begin to address emergent health implications, there is still a lack of awareness of climate change as a public health issue, and a lack of clarity in the existing provincial protocols. It is therefore clear that policy should enhance the emphasis on the association between environmental issues and health, including the health-enhancing effects or co-benefits of pro-environmental behaviours (Plotkinoff et al., 2004).

The OPHS is a significant piece of legislation that outlines protocols for health surveillance and monitoring. However, there is a need to engage in the *prospective* surveillance of health impacts that are projected to increase under climate change (Ostry et al., 2010) which requires the provision of relevant high-resolution regional climate models to forecast a variety of scenarios into the future and engage in adaptive management based on those projections (e.g. the BRACE framework in the United States). This would require additional resources to be allocated and guidance provided to health units and increased awareness about how to use climate-specific information (Clarke and Berry, 2012). Indeed, Canadian federal support for climate change and health initiatives is currently limited relative to the United States.

Since climate change will strain health services unequally across the province, there is a need to understand existing vulnerabilities in the health sector, as well as broader community vulnerabilities. Some of the identified limitations of existing environmental health assessment methodologies might be overcome by conducting localized vulnerability assessments (Shin and Ha, 2012; WHO/PAHO, 2012). Several health units have embarked on this path. Undertaking a vulnerability and adaptation assessment requires practitioners to apply their in-depth local knowledge of human health in the context of climate change. Since the impacts of climate change will be regionally specific and health units already have regionally specific health needs, assessment methodologies should strive to identify adaptation options that are customizable for local conditions and populations (Ostry, et al., 2010) to better contextualize community vulnerabilities given that these are often intertwined in unique, place-based settings (Yardley, et al., 2010). Numerous frameworks exist to assess local vulnerabilities based on the conduct of a scoping review, the assessment of current health vulnerabilities under historical climate conditions, the assessment of future health vulnerabilities under future climate conditions, the prioritization of suitable adaptation options, and the evaluation of those options (Wardeklier, et al., 2012; WHO/PAHO, 2012).

However, undertaking a vulnerability and adaptation assessment requires that program managers are able to continually evaluate programs and policies in real time. Due to the uncertainty of climate change models and regional impacts, managers will be required to update baseline information from assessments and refine programs as patterns in health impacts continue to emerge. Adaptive management protocols have been forwarded as a potentially useful policy tool for engaging in climate change adaptation in the public health sector (Ebi, 2011; Hess, et al. 2012). Adaptive management relies on collaboration between multiple stakeholders that participate in co-defining particular problems and identifying particular solutions with the goal of projecting the possible consequences of pursuing different decisions over time and weighing stakeholder tolerance to outcomes over the longer term. This approach is broadly related to the developmental evaluation (Patton, 2011) of existing programs, and emphasizes the procedural "learning by doing" and evaluating program efficiency and effectiveness under changing climatic and social conditions. Adaptive management is well suited to addressing complex problems, and the OPHS policy statement could go to greater lengths to mandate the evaluation of environmental health hazard programs by the provincial governing body in order to reflect and assess the changing nature of best practices over time, while ensuring that information pertaining to promising practices is communicated back to health units in a timely fashion (Morris, 2010).

Finally, a significant focus in the international literature on public health responses to climate change emphasizes the health co-benefits of adaptations in other sectors through healthy public policy initiatives (Cheng and Berry, 2012; Haines, et al. 2009; Younger, et al. 2008). To that end, the use of health impact assessment methodologies in collaboration with other sectors (Patz, et al. 2008) is a useful mechanism to understand the influence of policy decisions on human health. This is particularly salient given that there is a relative paucity of attempts to incorporate climate-change-related health impacts as an input into existing environmental health impact assessment methodologies due to challenges with data collection, data quality, and persistent issues around stakeholder engagement (Turner, et al. 2013). While "built environment" initiatives are a growing area of focus for Ontario health units (OPHA, 2011; Dimoulas-Graham, 2012), climate change rarely enters into these discussions.

CONCLUSION

The relatively broad policy mandate from which to engage with climate change from a public health perspective leads to differing interpretations and programmatic outputs. First, the broad policy mandate has enabled public health officials to put off acting on the province's changing climate by waiting for more definitive protocols. Second, public health officials from health units that are relatively better resourced (i.e., have more staff and funding) have used the broad mandate to engage in and justify innovative adaptation activities.

Future research and provincial policy development should strive to be mindful of existing disparities between health units to ensure that programs can be adequately delivered across contexts. Governmental support for implementation and evaluation of public health activities both directly and indirectly related to climate change's health impacts should be considered, as modeled by the Building Resilience Against Climate Effects (BRACE) model funded by the Centers for Disease Control and Prevention (CDC) in the United States (Eidson et al., 2015). Moreover, it is clear that infectious diseases are a largely neglected topic in the Canadian context of public health adaptation to climate change, and the "wait and see" attitude expressed by some decision-makers is likely to limit the effectiveness of climate change adaptation policies. The belief that existing policies will be robust enough to treat emerging infectious diseases has therefore been called into question in this chapter. Instead, this chapter argues that in order to truly engage with the upstream determinants of health, public health decision makers will need to name climate change as a legitimate threat to public health and engage with the discourse in a way that shows how public health institutions can contribute to solving one of the most complex and challenging issues of the twenty-first century.

REFERENCES

Campbell-Lendrum, D., Manga, L., Bagayoko, M., & Sommerfeld, J. (2015). Climate change and vector-borne diseases: What are the implications for public health research and policy? *Philosophical Transactions B*, 370(1665).

Canadian Public Health Association. (2006). Snapshot of adaptation and response capacity to cliamte change in public health. Health Canada. Retrieved from http://www.cpha.ca/uploads/progs/env/ccph_final_e.pdf

Cheng, J. J., & Berry, P. (2012). Health co-benefits and risks of public health adaptation strategies to climate change: A review of current literature. *International Journal of Public Health*, *58*(2), 305–311. doi:10.1007/s00038-012-0422-5 PMID:23111371

Clarke, K.-L., & Berry, P. (2012). From theory to practice: A Canadian case study of the utility of climate change adaptation frameworks to address health impacts. *International Journal of Public Health*, *57*(1), 167–174. doi:10.1007/s00038-011-0292-2 PMID:21931977

Costello, A., Abbas, M., Allen, A., Ball, S., Bell, S., Bellamy, R., & Patterson, C. (2009). Managing the health effects of climate change. *Lancet*, *373*(9676), 1693–1733. doi:10.1016/S0140-6736(09)60935-1 PMID:19447250

Deber, R. B. (2003). Health Care Reform: Lessons From Canada. *American Journal of Public Health*, *93*(1), 20–24. doi:10.2105/AJPH.93.1.20 PMID:12511378

Dimoulas-Graham, P., Perotta, K., Behan, K., & Northcott, A. (Eds.), (2012). An Environmental Scan Of Built Environment Data Related to Walkability and Environmental Exposures in Urban Ontario. Kingston, Frontenac, Lennox and Addington Health Unit. Retrieved from http://www.kflapublichealth.ca/Files/Research/LCDPBuiltEnvironmentReport2012.pdf

Ebi, K. (2011). Climate Change And Health Risks: Assessing And Responding To Them Through "Adaptive Management". *Health Affairs*, *30*(5), 924–930. doi:10.1377/hlthaff.2011.0071 PMID:21555476

Eidson, M., Clancy, K.A., & Birkhead, G.S. (2015). Public health climate change adaptation planning using stakeholder feedback. *Journal of Public Health Management Practice.*

Feltmate, B., & Thistlethwaite, J. (2012). Climate Change Adaptation: A Priorities Plan for Canada. Report of the Climate Change Adaptation Project. University of Waterloo. Retrieved from https://uwaterloo.ca/environment/sites/ca.environment/files/uploads/files/CCAP-Report-30May-Final.pdf

Friel, S., Bowen, K., Campbell-Lendrum, D., Frumkin, H., McMichael, A. J., & Rasanathan, K. (2011). Climate Change, Noncommunicable Diseases, and Development: The Relationships and Common Policy Opportunities. *Annual Review of Public Health, 32*(1), 133–147. doi:10.1146/annurev-publhealth-071910-140612 PMID:21091194

Government of Canada. H. C. (2011, December 13). Adapting to Extreme Heat Events: Guidelines for Assessing Health Vulnerability [Health Canada, 2011]. Retrieved from http://www.hc-sc.gc.ca/ewh-semt/pubs/climat/adapt/index-eng.php

Greer, A., Ng, V., & Fisman, D. (2008). Climate change and infectious diseases in North America: The road ahead. *Canadian Medical Association Journal, 178*(6), 715–722.

Halkier, B. (2011). Methodological Practicalities in Analytical Generalization. *Qualitative Inquiry, 17*(9), 787–797. doi:10.1177/1077800411423194

Hess, J. J., Eidson, M., Tlumak, J. E., Raab, K. K., & Luber, G. (2014). An evidence-based public health approach to climate change adaptation. *Environmental Health Perspectives, 122*(11), 1177–1186. PMID:25003495

Hess, J. J., McDowell, J. Z., & Luber, G. (2012). Integrating climate change adaptation into public health practice: Using adaptive management to increase adaptive capacity and build resilience. *Environmental Health Perspectives, 120*(2), 171–179. doi:10.1289/ehp.1103515 PMID:21997387

Howlett, M. (2014). Why are policy innovations rare and so often negative? Blame avoidance and problem denial in climate change policy-making. *Global Environmental Change, 29*, 395–403. doi:10.1016/j.gloenvcha.2013.12.009

Huang, C., Vaneckova, P., Wang, X., FitzGerald, G., Guo, Y., & Tong, S. (2011). Constraints and Barriers to Public Health Adaptation to Climate Change. *American Journal of Preventive Medicine, 40*(2), 183–190. doi:10.1016/j.amepre.2010.10.025 PMID:21238867

Lesnikowski, A. C., Ford, J. D., Berrang-Ford, L., Barrera, M., Berry, P., Henderson, J., & Heymann, S. J. (2013). National-level factors affecting planned, public adaptation to health impacts of climate change. *Global Environmental Change, 23*(5), 1153–1163. doi:10.1016/j.gloenvcha.2013.04.008

Maibach, E. W., Chadwick, A., McBride, D., Chuk, M., Ebi, K. L., & Balbus, J. (2008). Climate Change and Local Public Health in the United States: Preparedness, Programs and Perceptions of Local Public Health Department Directors. *PLoS ONE, 3*(7), e2838. doi:10.1371/journal.pone.0002838 PMID:18665266

McMichael, A. J. (2013). Globalization, Climate Change, and Human Health. *The New England Journal of Medicine, 368*(14), 1335–1343. doi:10.1056/NEJMra1109341 PMID:23550671

Miles, M. B., & Huberman, A. M. (1994). Qualitative Data Analysis: An Expanded Sourcebook. *Sage (Atlanta, Ga.).*

Ministry of Health and Long-term Care. (2006). Ontario Public Health Unit and County Boundaries, 2006. MOHLTC. Retrieved from http://www.apheo.ca/resources/indicators/PHU_CD%20Map.pdf

Ministry of Health and Long-term Care. (2008, 2014). The Ontario Public Health Standards, 2008, Revised May 1, 2014. MOHLTC. Retrieved from http://www.health.gov.on.ca/en/pro/programs/publichealth/oph_standards/docs/ophs_2008.pdf

Morris, G. P. (2010). Ecological public health and climate change policy. *Perspectives in Public Health, 130*(1), 34–40. doi:10.1177/1757913909354149 PMID:20333948

National Roundtable for the Economy and Environment (NRTEE). (2011). Paying the price: The economic impacts of climate change for Canada. NRTEE. Retrieved from http://www.fcm.ca/Documents/reports/PCP/paying_the_price_EN.pdf

Ontario Public Health Association (OPHA). (2011). Public Health and Planning 101: An Education Module for Public Health and Planning Professionals in Ontario, Survey for Public Health Professionals – Highlights. OPHA. Retrieved from http://opha.on.ca/getmedia/ef89de0c-26ed-4617-b15c-28ba8360d3ac/Summary-of-survey-results-for-public-health-professionals.aspx

Ostfeld, R.S. and Brunner, J.L. (2015). Climate change and *Ixodes* tick-borne diseases of humans. *Philosophical Transactions of the Royal Society B*, 370(1665), online ahead of print.

Ostry, A., Ogborn, M., Bassil, K. L., Takaro, T. K., & Allen, D. M. (2010). Climate Change and Health in British Columbia: Projected Impacts and a Proposed Agenda for Adaptation Research and Policy. *International Journal of Environmental Research and Public Health, 7*(3), 1018–1035. doi:10.3390/ijerph7031018 PMID:20617016

Panic, M., & Ford, J. D. (2013). A Review of National-Level Adaptation Planning with Regards to the Risks Posed by Climate Change on Infectious Diseases in 14 OECD Nations. *International Journal of Environmental Research and Public Health, 10*(12), 7083–7109. doi:10.3390/ijerph10127083 PMID:24351735

Parham, P. E., Waldock, J., Christophides, G. K., Hemming, D., Agusto, F., ... (2015). Climate, environmental and socio-economic change: Weighing up the balance in vector-borne disease transmission. *Philosophical Transactions B, 370*(1665).

Paterson, J. A., Ford, J. D., Ford, L. B., Lesnikowski, A., Berry, P., Henderson, J., & Heymann, J. (2012). Adaptation to climate change in the Ontario public health sector. *BMC Public Health*, *12*(1), 452. doi:10.1186/1471-2458-12-452 PMID:22712716

Patton, M. Q. (2011). *Developmental Evaluation: Applying Complexity Concepts to Enhance Innovation and Use*. New York: Guilford Press.

Patz, J., Campbell-Lendrum, D., Gibbs, H., & Woodruff, R. (2008). Health impact assessment of global climate change: Expanding on comparative risk assessment approaches for policy making. *Annual Review of Public Health*, *29*(1), 27–39. doi:10.1146/annurev.publhealth.29.020907.090750 PMID:18173382

Paz, S. (2015). Climate change impacts on West Nile virus transmission in a global context. *Philosophical Transactions of the Royal Society B*, 370(1665).

Perotta, K. (2011). Public Health and Land Use Planning: How Ten Public Health Units Are Working to Create Healthy and Sustainable Communities. Clean Air Partnership. Retrieved from http://www.cleanairpartnership.org/files/CAP%20 PHLUP%20Background%20Report%20April%202011.pdf

Plotkinoff, R. C., Wright, M.-F., & Karunamuni, N. (2004). Knowledge, attitudes and behaviours related to climate change in Alberta, Canada: Implications for public health policy and practice. *International Journal of Environmental Health Research*, *14*(3), 223–229. doi:10.1080/09603120420002186 33 PMID:15203453

Poland, B. D. (1995). Transcription Quality as an Aspect of Rigor in Qualitative Research. *Qualitative Inquiry*, *1*(3), 290–310. doi:10.1177/107780049500100302

Poland, B., Gastaldo, D., Pancham, A., & Ferrence, R. (2009). The interpersonal management of environmental tobacco smoke in the home–a qualitative study. *Critical Public Health*, *19*(2), 203–221. doi:10.1080/09581590802395416

Seguin, J. (Ed.). (2008). Human health in a changing climate: A Canadian assessment of vulnerabilities and adaptive capacity. Minister of Health. Retrieved from http://publications.gc.ca/collections/collection_2008/hc-sc/H128-1-08-528E.pdf

Shin, Y. S., & Ha, J. (2012). Policy Directions Addressing the Public Health Impact of Climate Change in South Korea: The Climate-change Health Adaptation and Mitigation Program. *Environmental Health and Toxicology, 27.*

Smith, K. R., Jerrett, M., Anderson, H. R., Burnett, R. T., Stone, V., Derwent, R., & Thurston, G. et al. (2009). Public health benefits of strategies to reduce greenhouse-gas emissions: Health implications of short-lived greenhouse pollutants. *Lancet*, *374*(9707), 2091–2103. doi:10.1016/S0140-6736(09)61716-5 PMID:19942276

Stocker, T. F., Qin, D., Plattner, G.-K., Tignor, M., Allen, S. K., . . . (Eds.), (2013). Climate change 2013: The physical science basis. Contribution of working group I to the 5th assessment report of the IPCC. Cambridge University Press. Retrieved from https://www.ipcc.ch/pdf/assessment-report/ar5/wg1/WG1AR5_Frontmatter_FINAL.pdf

Tilley, S. A. (2003). "Challenging" Research Practices: Turning a Critical Lens on the Work of Transcription. *Qualitative Inquiry*, *9*(5), 750–773. doi:10.1177/1077800403255296

Turner, L. R., Alderman, K., Connell, D., & Tong, S. (2013). Motivators and Barriers to Incorporating Climate Change-Related Health Risks in Environmental Health Impact Assessment. *International Journal of Environmental Research and Public Health*, *10*(3), 1139–1151. doi:10.3390/ijerph10031139 PMID:23525029

Wardekker, J. A., de Jong, A., van Bree, L., Turkenburg, W. C., & van der Sluijs, J. P. (2012). Health risks of climate change: An assessment of uncertainties and its implications for adaptation policies. *Environmental Health*, *11*(1), 67. doi:10.1186/1476-069X-11-67 PMID:22992311

World Health Organization (WHO), & Pan-American Health Organization (PAHO). (2013). *Protecting health from climate change: vulnerability and adaptation assessment*. World Health Organization. Retrieved from http://apps.who.int//iris/handle/10665/104200

Yardley, J., Sigal, R. J., & Kenny, G. P. (2011). Heat health planning: The importance of social and community factors. *Global Environmental Change*, *21*(2), 670–679. doi:10.1016/j.gloenvcha.2010.11.010

Younger, M., Morrow-Almeida, H. R., Vindigni, S. M., & Dannenberg, A. L. (2008). The built environment, climate change, and health: Opportunities for co-benefits. *American Journal of Preventive Medicine*, *35*(5), 517–526. doi:10.1016/j.amepre.2008.08.017 PMID:18929978

APPENDIX: INTERVIEW QUESTIONS

1. Please tell me about the work that you do with your health unit?

[Probe: organizational structure; personal history with organization; training; education; work; how they became involved with climate change related work (OPHS as entry point?), if at all, and associated personal motivations and trajectory of that work]

2. What are you and your colleagues to address issues of adaptation to climate change?

[Probe: what does this work look like on a day-to-day basis (in terms of practice); focus and rationale of particular initiatives; who else are they working with]

3. What role does health equity play in the climate change work that you and your colleagues are conducting?

[Probe: priority populations; specific activities; specific foci]

4. How is the work you and your colleagues do on climate change positioned relative to other programs or work that your organization conducts on a day-to-day basis?

[Probe: how do other members of your organization think about and navigate climate change as a public health issue; does climate change adaptation require a different approach compared with other environmental health issues; what do you think a public health approach that engages with climate change adaptation should ideally look like?]

5. What has been challenging in climate change adaptation work from a public health perspective?

[Probe: push-back encountered (personal, organizational, external)]

6. What has worked well in your climate change adaptation work?

[Probe: practices, resources or assets that interviewee found useful/important; looking to the future, what resources or assets would be important to further this work?]

7. Based on our conversation today, are there any documents that it would be helpful for me to see and that would help me better understand the work you do?

Compilation of References

Abdo-Salem, S., Gerbier, G., Bonnet, P., Al-Qadasi, M., Tran, A., Thiry, E., & Roger, F. et al. (2006). Descriptive and spatial epidemiology of Rift valley fever outbreak in Yemen 2000-2001. *Annals of the New York Academy of Sciences*, *1081*(1), 240–242. doi:10.1196/annals.1373.028 PMID:17135517

Abeku, T. A., van Oortmarssen, G. J., Borsboom, G., de Vlas, S. J., & Habbema, J. D. F. (2003). Spatial and temporal variations of malaria epidemic risk in Ethiopia: Factors involved and implications. *Acta Tropica*, *87*(3), 331–340. doi:10.1016/S0001-706X(03)00123-2 PMID:12875926

Abiodun, B. J., Salami, A. T., & Tadross, M. (2011). Climate Change Scenarios for Nigeria: Understanding Biophysical Impacts. Climate Systems Analysis Group, Cape Town, for Building Nigeria's Response to Climate Change Project. Ibadan, Nigeria: Nigerian Environmental Study/ Action Team (NEST).

Aboda, C. (2012): An assessment of the impact of climate variability on malaria in Uganda. A case study of Mount Elgon region of Bududa district [Master of Philosophy Thesis]. Norwegian University of Science and Technology.

Acevedo-Whitehouse, K., & Duffus, A. L. (2009). Effects of environmental change on wildlife health. *Philosophical Transactions of the Royal Society of London. Series B, Biological Sciences*, *364*(1534), 3429–3438. doi:10.1098/rstb.2009.0128 PMID:19833653

Adger, W. N. (2001). Scales of governance and environmental justice for adaptation and mitigation of climate change. *Journal of International Development*, *13*(7), 921–931. doi:10.1002/jid.833

Adger, W. N., Barnett, J., Brown, K., Marshall, N., & O'Brien, K. (2013). Cultural dimensions of climate change impacts and adaptation. *Nature Climate Change*, *3*(2), 112–117. doi:10.1038/nclimate1666

Agence Nationale de l'Assurance Maladie (ANAM) (2013), Feuille de Route 2014-2018. Retrieved from http://www.anam.ma/upload/document/feuillederoute.pdf

Ahmad, E.I. (2007). The prevalence of malaria in Okigwe and Owerri. *Journal of community health*, 1.

Ahmad, A., Krumkamp, R., & Reintjes, R. (2009). Controlling SARS: A review on China's response compared with other SARS-affected countries. *Tropical Medicine & International Health, 14*(Suppl. 1), 36–45. doi:10.1111/j.1365-3156.2008.02146.x PMID:19508440

Ahmed, N. (2013). Peak oil, climate change and pipeline geopolitics driving Syria conflict. *The Guardian*. Retrieved from http://www.guardian.co.uk/environment/earth-insight/2013/may/13/1

Alexander, B., Agudelo, L. A., Navarro, F., Ruiz, F., Molina, J., Aguilera, G., & Quiñones, M. L. (2001). Phlebotomine sandflies and leishmaniasis risks in Colombian coffee plantations under two systems of cultivation. *Medical and Veterinary Entomology, 15*(4), 364–373. doi:10.1046/j.0269-283x.2001.00322.x PMID:11776455

Alexander, B., Agudelo, L. A., Navarro, J. F., Ruiz, J. F., Molina, J., Aguilera, G., & Quiñones, M. L. et al. (2009). Relationship between coffee cultivation practices in Colombia and exposure to infection with Leishmania. *Transactions of the Royal Society of Tropical Medicine and Hygiene, 103*(12), 1263–1268. doi:10.1016/j.trstmh.2009.04.018 PMID:19555985

Ali, M., & Lopez, A. L., AeYou, Y., EunKim, Y., Sah, B., Maskery, B., & Clemens, J. (2012). The global burden of cholera. *Bulletin of the World Health Organization, 90*(3), 209–218. doi:10.2471/BLT.11.093427 PMID:22461716

Allen, H. K., Donato, J., Wang, H. H., Cloud-Hansen, K. A., Davies, J., & Handelsman, J. (2010). Call of the wild: Antibiotic resistance genes in natural environments. *Nature Reviews. Microbiology, 8*(4), 251–259. doi:10.1038/nrmicro2312 PMID:20190823

Allen, M. R., & Ingram, W. J. (2002). Constraints on future changes in climate and the hydrologic cycle. *Nature, 419*(6903), 224–232. doi:10.1038/nature01092 PMID:12226677

Amal, L. (2012). *La prévention et la gestion des risques au Maroc, Ministère de l'Énergie, des Mines, de l'Eau et de l'Environnement*. Rabat: Département de l'Environnement.

Amman, B. R., Carroll, S. A., Reed, Z. D., Sealy, T. K., Balinandi, S., Swanepoel, R., & Towner, J. S. et al. (2012). Seasonal pulses of Marburg virus circulation in juvenile Rousettus aegyptiacus bats coincide with periods of increased risk of human infection. *PLoS Pathogens, 8*(10), e1002877. doi:10.1371/journal.ppat.1002877 PMID:23055920

Andersson, T., Bjelkmar, P., Hulth, A., Lindh, J., Stenmark, S., & Widerstrom, M. (2014). Syndromic surveillance for local outbreak detection and awareness: Evaluating outbreak signals of acute gastroenteritis in telephone triage, web-based queries and over-the-counter pharmacy sales. *Epidemiology and Infection, 142*(2), 303–313. doi:10.1017/S0950268813001088 PMID:23672877

Anonymous, . (1989). Health in the Greenhouse. *Lancet, 333*(8642), 819–820. doi:10.1016/S0140-6736(89)92275-7 PMID:2564905

Anthony, S. J., Epstein, J. H., Murray, K. A., Navarrete-Macias, I., Zambrana-Torrelio, C. M., Solovyov, A., & Lipkin, W. I. et al. (2013). A strategy to estimate unknown viral diversity in mammals. *mBio, 4*(5), e00598–e00513. doi:10.1128/mBio.00598-13 PMID:24003179

Antia, R., Regoes, R. R., Koella, J. C., & Bergstrom, C. T. (2003). The role of evolution in the emergence of infectious diseases[letters to nature]. *Nature*, *426*(6967), 658–661. doi:10.1038/nature02104 PMID:14668863

Anyamba, A., Chretien, J.-P., Small, J., Tucker, C. J., Formenty, P. B., & Richardson, J. H., … Linthicum, K. J. (2009). Prediction of a Rift Valley fever outbreak. *Proceedings of the National Academy of Sciences of the United States of America*, 106(3), 955–9. doi:doi:10.1073/pnas.0806490106 doi:10.1073/pnas.0806490106

Anyamba, A., Chretien, J. P., Small, J., Tucker, C. J., Formenty, P. B., Richardson, J. H., & Linthicum, K. J. et al. (2009). Prediction of a Rift Valley fever outbreak. *Proceedings of the National Academy of Sciences of the United States of America*, *106*(3), 955–959. doi:10.1073/pnas.0806490106 PMID:19144928

Anyamba, A., Chretien, J. P., Small, J., Tucker, C. J., & Kelley, P. W. (2006). Developing global climate anomalies suggest potential disease risks for 2006-2007. *International Journal of Health Geographics*, *5*(1), 60. doi:10.1186/1476-072X-5-60 PMID:17194307

Anyamba, A., Linthicum, K. J., Small, J., Britch, S. C., Pak, E., de La Rocque, S., & Swanepoel, R. et al. (2010). Prediction, assessment of the Rift Valley fever activity in East and Southern Africa 2006-2008 and possible vector control strategies. *The American Journal of Tropical Medicine and Hygiene*, *83*(2 Suppl.), 43–51. doi:10.4269/ajtmh.2010.09-0289 PMID:20682905

Anyamba, A., Linthicum, K. J., & Tucker, C. J. (2001). Climate-disease connections: Rift Valley fever in Kenya. *Cadernos de Saude Publica*, *17*, S133–S140. doi:10.1590/S0102-311X2001000700022 PMID:11426274

Anyamba, A., Small, J. L., Britch, S. C., Tucker, C. J., Pak, E. W., Reynolds, C. A., & Linthicum, K. J. et al. (2014). Recent weather extremes and impacts on agricultural production and vector-borne disease outbreak patterns. *PLoS ONE*, *9*(3), e92538. doi:10.1371/journal.pone.0092538 PMID:24658301

Ashbolt, N. J. (2004). Microbial contamination of drinking water and disease outcomes in developing regions. *Toxicology*, *198*(1-3), 229–238. doi:10.1016/j.tox.2004.01.030 PMID:15138046

Ashbolt, N. J., Grabow, W. O. K., & Snozzi, M. (2001). Indicators of microbial water quality. In L. Fewtrell & J. Bartram (Eds.), *Water Quality: Guidelines, Standards and Health* (pp. 289–316). London, UK: IWA Publishing.

Assessment, M. E. (2005). Ecosystems and Human Well-being. *Synthesis*.

Atherstone, C., Smith, E., Ochungo, P., Roesel, K., & Grace, D. (2015). Assessing the Potential Role of Pigs in the Epidemiology of Ebola Virus in Uganda. *Transboundary and Emerging Diseases*, n/a. doi:10.1111/tbed.12394 PMID:26310206

AU-IBAR. (2014). Standard Methods and Procedures (SMPs) for Control of Rift Valley Fever (RVF) in the Greater Horn of Africa. Nairobi.

Auld, H., MacIver, D., & Klaassen, J. (2004). Heavy rainfall and waterborne disease outbreaks: The Walkerton example. *Journal of Toxicology and Environmental Health. Part A.*, *67*(20-22), 1879–1887. doi:10.1080/15287390490493475 PMID:15371222

Aybar, C. A. V., Juri, M. J. D., Santana, M., de Grosso, M. S. L., & Spinelli, G. R. (2012). The spatio-temporal distribution patterns of biting midges of the genus Culicoides in Salta province, Argentina. *Journal of Insect Science*, *12*(1), 145. PMID:23461794

Azevedo, A. C., & Rangel, E. F. (1991). A study of sandfly species (Diptera: Psychodidae: Phlebotominae) in a focus of cutaneous leishmaniasis in the municipality of Baturité, Ceará, Brazil. *Memorias do Instituto Oswaldo Cruz*, *86*(4), 405–410. doi:10.1590/S0074-02761991000400005 PMID:1842431

Azevedo, A. C., Vilela, M. L., Souza, N. A., Andrade-Coelho, C. A., Barbosa, A. F., Firmo, A. L., & Rangel, E. F. (1996). The sand fly fauna (Diptera: Psychodidae: Phlebotominae) of a focus of cutaneous leishmaniasis in Ilhéus, state of Bahia, Brazil. *Memorias do Instituto Oswaldo Cruz*, *91*(1), 75–79. doi:10.1590/S0074-02761996000100012 PMID:8734952

Baize, S., Pannetier, D., Oestereich, L., Rieger, T., Koivogui, L., Magassouba, N., & Gunther, S. et al. (2014). Emergence of Zaire Ebola virus disease in Guinea. *The New England Journal of Medicine*, *371*(15), 1418–1425. doi:10.1056/NEJMoa1404505 PMID:24738640

Baker-Austin, C., Trinanes, J. A., Taylor, N. G. H., Hartnell, R., Siitonen, A., & Martinez-Urtaza, J. (2013). Emerging Vibrio risk at high latitudes in response to ocean warming. *Nature Climate Change*, *3*(1), 73–77. doi:10.1038/nclimate1628

Baker, M. L., Schountz, T., & Wang, L. F. (2013). Antiviral immune responses of bats: A review. *Zoonoses and Public Health*, *60*(1), 104–116. doi:10.1111/j.1863-2378.2012.01528.x PMID:23302292

Baldursson, S., & Karanis, P. (2011). Waterborne transmission of protozoan parasites: Review of worldwide outbreaks - an update 2004-2010. *Water Research*, *45*(20), 6603–6614. doi:10.1016/j.watres.2011.10.013 PMID:22048017

Baltimore, D. (2014). *Renato Dulbecco 1914 – 2012. Biographical Memoirs.* National Academy of Sciences. Retrieved from http://www.nasonline.org/publications/biographical-memoirs/memoir-pdfs/dulbecco-renato.pdf

Barkia, H., Barkia, A., Nhammi, H., & Belghyti, D. (2011). La schistosomiase au Maroc: de sa découverte à l'après-élimination. *La Revue de Santé de la Méditerranée orientale*, 17(3).

Barrette, R. W., Metwally, S. A., Rowland, J. M., Xu, L., Zaki, S. R., Nichol, S. T., & McIntosh, M. T. et al. (2009). Discovery of swine as a host for the Reston ebolavirus. *Science*, *325*(5937), 204–206. doi:10.1126/science.1172705 PMID:19590002

Barry, S. (2008). The plague of the third pandemic and its current reemergence. *Vesalius*, *14*(2), 69–73. PMID:19579344

Bartlett, J. M. S., & Stirling, D. (2003). A short history of the polymerase chain reaction. PCR protocols. *Methods in Molecular Biology (Clifton, N.J.), 226*, 3–6. PMID:12958470

Bayliss, H. R., Schindler, S., Essl, F., Rabitsch, W., & Pullin, A. S. (2015). What evidence exists for changes in the occurrence, frequency or severity of human health impacts resulting from exposure to alien invasive species in Europe? A systematic map protocol. *Environmental Evidence, 4*(1), 10. doi:10.1186/s13750-015-0037-4

Bayol, F., & Chevalier, J. F. (2004). Current state of the forest cover in Liberia.

Beer, K. D., Gargano, J. W., Roberts, V. A., Hill, V. R., Garrison, L. E., Kutty, P. K., & Yoder, J. S. et al. (2015). Surveillance for Waterborne Disease Outbreaks Associated with Drinking Water - United States, 2011-2012. *MMWR. Morbidity and Mortality Weekly Report, 64*(31), 842–848. doi:10.15585/mmwr.mm6431a2 PMID:26270059

Bell, D. M. (2004). Public health interventions and SARS spread, 2003. *Emerging Infectious Diseases, 10*(11), 1900–1906. doi:10.3201/eid1011.040729 PMID:15550198

Beniston, M. (2010). Climate change and its impacts: Growing stress factors for human societies. *International Review of the Red Cross, 92*(879), 557–568. doi:10.1017/S1816383110000342

Berbés-Blázquez, M., Oestreicher, J. S., Mertens, F., & Saint-Charles, J. (2014). Ecohealth and resilience thinking: A dialog from experiences in research and practice. *Ecology and Society, 19*(2), 24. doi:10.5751/ES-06264-190224

Bernard, S. M., & Anderson, S. A. (2006). Qualitative assessment of risk for monkeypox associated with domestic trade in certain animal species, United States. *Emerging Infectious Diseases, 12*(12), 1827–1833. doi:10.3201/eid1212.060454 PMID:17326932

Bevins, S. N., Pedersen, K., Lutman, M. W., Baroch, J. A., Schmit, B. S., Kohler, D., & DeLiberto, T. J. et al. (2014). Large-scale avian influenza surveillance in wild birds throughout the United States. *PLoS ONE, 9*(8), e104360. doi:10.1371/journal.pone.0104360 PMID:25116079

Bhattarai, A., Abdullah, S., Kachur, P., Martensson, A., Abbas, A., Khatib, R., & Bjorkman, A. et al. (2007). Impact of Artemisinin-Based Combination Therapy and Insecticide-Treated Nets on Malaria Burden in Zanzibar. *PLoS Medicine, 4*, e309. doi:10.1371/journal.pmed.0040309 PMID:17988171

Bird, B. H., Ksiazek, T. G., Nichol, S. T., & MacLachlan, N. J. (2009). Rift Valley fever virus. *Journal of the American Veterinary Medical Association, 234*(7), 883–893. doi:10.2460/javma.234.7.883 PMID:19335238

Black, R., Bennett, S. R. G., Thomas, S. M., & Beddington, J. R. (2011). Climate change: Migration as adaptation. *Nature, 478*(7370), 447-449. Doi:10.1038/478477a

Black, W. C. I. V., & Moore, C. G. (2005). Population biology as a tool to study vector-borne diseases. In W. C. Marquardt, B. C. Kondratieff, C. G. Moore, J. E. Freier, H. H. Hagedorn, W. C. I. V Black, … S. Higgs (Eds.), Biology of disease vectors (2nd ed., pp. 187–206). San Diego: Elsevier Academic Press.

Blanco, G., Gerlagh, R., Suh, S., Barrett, J., de Coninck, H. C., Diaz Morejon, C. F., . . . Zhou, P. (2014). Drivers, Trends and Mitigation. In O. Edenhofer, R. Pichs-Madruga, Y. Sokona, E. Farahani, S. Kadner, … J.C.M. (Eds.), Climate Change 2014: Mitigation of Climate Change. Contribution of Working Group III to the Fifth Assessment Report of the Intergovernmental Panel on Climate Change. New York, NY, USA: Cambridge University Press.

Bodker, R., Akida, J., Shayo, D., Kisinza, W., Msangeni, H. A., Pedersen, E. M., & Lindsay, S. W. (2003). Relationship between altitude and intensity of malaria transmission in the Usambara Mountains, Tanzania. *Journal of Medical Entomology*, *40*(5), 706–717. doi:10.1603/0022-2585-40.5.706 PMID:14596287

Bogoch, I. I., Brady, O., Kraemer, M. U. G., German, M., Creatore, M. I., Kulkarni, M. A., & Khan, K. et al. (2016). Anticipating the international spread of Zika virus from Brazil. *Lancet*, *387*(10016), 335–336. doi:10.1016/S0140-6736(16)00080-5 PMID:26777915

Borgdorff, M. W., & van Soolingen, D. (2013). The re-emergence of tuberculosis: What have we learnt from molecular epidemiology? *Clinical Microbiology and Infection*, *19*(10), 889–901. doi:10.1111/1469-0691.12253 PMID:23731470

Bounoua, L., Kahime, K., Houti, L., Blakey, T., Ebi, K. L., Zhang, P., & Boumezzough, A. et al. (2013). Linking climate to incidence of zoonotic cutaneous leishmaniasis (*L. major*) in Pre-Saharan North Africa. *International Journal of Environmental Research and Public Health*, *10*(8), 3172–3191. doi:10.3390/ijerph10083172 PMID:23912199

Bouzid, M. (2014). Detection of Food- and Waterborne Parasites: Conventional Methods and Recent Developments. In C. A. Batt & M. L. Tortorello (Eds.), *Encyclopedia of Food Microbiology*. Elsevier Ltd. doi:10.1016/B978-0-12-384730-0.00355-4

Bouzid, M., Colon-Gonzalez, F. J., Lung, T., Lake, I. R., & Hunter, P. R. (2014). Climate change and the emergence of vector-borne diseases in Europe: Case study of dengue fever. *BMC Public Health*, *14*(1), 781. doi:10.1186/1471-2458-14-781 PMID:25149418

Bouzid, M., Hunter, P. R., Chalmers, R. M., & Tyler, K. M. (2013). Cryptosporidium pathogenicity and virulence. *Clinical Microbiology Reviews*, *26*(1), 115–134. doi:10.1128/CMR.00076-12 PMID:23297262

Bowles, D. C., Butler, C. D., & Friel, S. (2014). Climate change and health in Earth's future. *Earth's Future*(2), 60-67.

Bowles, D. C., Braidwood, M., & Butler, C. D. (2014). Unholy Trinity: Climate change, conflict and ill health. In C. D. Butler (Ed.), *Climate change and global health*. Wallingford: Cabi. doi:10.1079/9781780642659.0144

Bowles, D. C., & Butler, C. D. (2014). Socially, politically and economically mediated health effects of climate change: Possible consequences for Africa. *South African Medical Journal*, *104*(98). PMID:26307807

Bowles, D. C., Butler, C. D., & Morisetti, N. (2015). Climate change, conflict and health. *Journal of the Royal Society of Medicine*, *108*(10), 390–395. doi:10.1177/0141076815603234 PMID:26432813

Bowles, D. C., Reuveny, R., & Butler, C. D. (2014). Moving to a better life? Climate, migration and population health. In C. D. Butler (Ed.), *Climate change and global health*. Wallingford: CABI. doi:10.1079/9781780642659.0135

Boxall Alistair B.A. et al. (2009). Impacts of Climate Change on Indirect Human Exposure to Pathogens and Chemicals from Agriculture. *Environmental Health Perspectives*, 117(4), 508-514.

Boyles, J. G., Cryan, P. M., McCracken, G. F., & Kunz, T. H. (2011). Conservation. Economic importance of bats in agriculture. *Science*, *332*(6025), 41–42. doi:10.1126/science.1201366 PMID:21454775

Brachman, P. (2003). Infectious diseases – past, present, and future[Editorial]. *International Journal of Epidemiology*, *32*(5), 684–686. doi:10.1093/ije/dyg282 PMID:14559728

Bradbury, R. (2012, July 13). A world without coral reefs. *New York Times*.

Bradley, D., Evison, L. M., James, A., Pescod, M. B., & White, A. U. et al. (1977). Health aspects of water supplies in tropical countries. In *Water, wastes and health in hot climates*. London: John Wiley.

Braga, I. A., & Valle, D. (2007). Aedes aegypti: Histórico do controle no Brasil. *Epidemiologia e Serviços de Saúde*, *16*(2), 113–118.

Braks, M. A. H., & Husman, A. M. R. (2013). Dimensions of effects of climate change on water-transmitted infectious diseases. *Air & Water Borne Diseases*, 2, 109. doi:10.4172/2167-7719.10001092: 1000109.

Britch, S. C., Binepal, Y. S., Ruder, M. G., Kariithi, H. M., Linthicum, K. J., Anyamba, A., & Wilson, W. C. et al. (2013). Rift Valley fever risk map model and seroprevalence in selected wild ungulates and camels from Kenya. *PLoS ONE*, *8*(6), e66626. doi:10.1371/journal.pone.0066626 PMID:23840512

Brock, T. (Ed.). (1961). *Milestones in Microbiology: 1546 - 1940*. Washington, DC: American Society for Microbiology.

Brownstein, J. S., Holford, T. R., & Fish, D. (2005). Effect of Climate Change on Lyme Disease Risk in North America. *EcoHealth*, *2*(1), 38–46. doi:10.1007/s10393-004-0139-x PMID:19008966

Bruinsma, J. (2009). The resource outlook to 2050: By how much do land, water use and crop yields need to increase by 2050? *Paper presented at the Expert Meeting on How to Feed the World in 2050*, Rome.

Brumpt, E., & Pedroso, A. (1913). Pesquisas epidemiológicas sobre a leishmaniose americana das florestas no estado de São Paulo. *Anais Paulistas de Medicina e Cirurgia*, *1*, 97–136.

Bryant, J., Wang, H., Cabezas, C., Ramirez, G., Watts, D., Russell, K., & Barrett, A. (2003). Enzootic transmission of yellow fever virus in Peru. *Emerging Infectious Diseases*, *9*(8), 926–933. doi:10.3201/eid0908.030075 PMID:12967489

Buhaug, H., Nordkvelle, J., Bernauer, T., Böhmelt, T., Brzoska, M., & ... J.S.A. Link, (2014). One effect to rule them all? A comment on climate and conflict. *Climatic Change*, *127*, 391–397. doi:10.1007/s10584-014-1266-1

Burge, C. A., Mark Eakin, C., Friedman, C. S., Froelich, B., Hershberger, P. K., Hofmann, E. E., & Harvell, C. D. et al. (2014). Climate change influences on marine infectious diseases: Implications for management and society. *Annual Review of Marine Science*, *6*(1), 249–277. doi:10.1146/annurev-marine-010213-135029 PMID:23808894

Butler, C. D. (2010). Climate change, crop yields, and the future. *SCN News*, *38*, 18–25.

Butler, C. D. (2014a). Famine, hunger, society and climate change. In C. D. Butler (Ed.), *Climate Change and Global Health*. Wallingford: Cabi. doi:10.1079/9781780642659.0124

Butler, C. D. (Ed.). (2014b). *Climate Change and Global Health*. Wallingford: CABI. doi:10.1079/9781780642659.0000

Butler, C. D., & Harley, D. (2010). Primary, secondary and tertiary effects of eco-climatic change: The medical response. *Postgraduate Medical Journal*, *86*(1014), 230–234. doi:10.1136/pgmj.2009.082727 PMID:20354046

Byass, P. (2009). Climate change and population health in Africa: Where are the scientists? *Global Health Action*, *2009*, 2. PMID:20052421

Cabral, A. C., Fe, N. F., Suarez-Mutis, M. C., Boia, M. N., & Carvalho-Costa, F. A. (2010). Increasing incidence of malaria in the Negro River basin, Brazilian Amazon. *Transactions of the Royal Society of Tropical Medicine and Hygiene*, *104*(8), 556–562. doi:10.1016/j.trstmh.2010.03.008 PMID:20462621

Cabral, J. P. (2010). Water microbiology. Bacterial pathogens and water. *International Journal of Environmental Research and Public Health*, *7*(10), 3657–3703. doi:10.3390/ijerph7103657 PMID:21139855

Campbell, L. P., Luther, C., Moo-Llanes, D., Ramsey, J. M., Danis-Lozano, R., & Peterson, A. T. (2015). Climate change influences on global distributions of dengue and chikungunya virus vectors. *Philosophical Transactions of the Royal Society of London B: Biological Sciences*, 370(1665), 20140135.

Campbell-Lendrum, D., Manga, L., Bagayoko, M., & Sommerfeld, J. (2015). Climate change and vector-borne diseases: What are the implications for public health research and policy? *Philosophical Transactions B*, 370(1665).

Campbell-Lendrum, D. H. (2005). How much does the health community care about global environmental change? *Global Environmental Change*, *15*(4), 296–298. doi:10.1016/j.gloenvcha.2005.09.001

Canadian Coalition for Green Health Care and the Nova Scotia Department of Environment and Health Canada. (2013), *Health Care Facility Climate Change Resiliency Toolkit.* Retrieved from http://www.greenhealthcare.ca/climateresilienthealthcare/CCGHC-HealthCareFacilityResiliencyChecklist.pdf

Canadian Public Health Association. (2006). Snapshot of adaptation and response capacity to cliamte change in public health. Health Canada. Retrieved from http://www.cpha.ca/uploads/progs/env/ccph_final_e.pdf

Cann, K. F., Thomas, D. R., Salmon, R. L., Wyn-Jones, A. P., & Kay, D. (2013). Extreme water-related weather events and waterborne disease. *Epidemiology and Infection, 141*(4), 671–686. doi:10.1017/S0950268812001653 PMID:22877498

Canziani, O. F., & Diaz, S. (1998). Latin America. In R. T. Watson, M. C. Zinyowera, & R. H. Moss (Eds.), *The Regional Impacts of Climate Change: An Assessment of Vulnerability* (pp. 187–230). Cambridge, UK: Cambridge University Press.

Caqui, K. B., Quispe, C. L. H. A., & Zegarra, R. O. (2013). Disaster risk management in the framework of the UNASUR health. In E. H. Carmo, A. Gemal, & S. Oliveira (Eds.), Health Surveillance in South America: epidemiological, sanitary and environmental (pp. 41-54). Rio de Janeiro: ISAGS.

Carbajo, A. E., Cardo, M. V., & Vezzani, D. (2012). Is temperature the main cause of dengue rise in non-endemic countries? The case of Argentina. *International Journal of Health Geographics, 11*(1), 26. doi:10.1186/1476-072X-11-26 PMID:22768874

Carbajo, A. E., Vera, C., & González, P. L. (2009). Hantavirus reservoir *Oligoryzomys longicaudatus* spatial distribution sensitivity to climate change scenarios in Argentine Patagonia. *International Journal of Health Geographics, 8*(1), 44. doi:10.1186/1476-072X-8-44 PMID:19607707

Carbajo, A. E., & Vezzani, D. (2015). Waiting for chikungunya fever in Argentina: Spatio-temporal risk maps. *Memorias do Instituto Oswaldo Cruz, 110*(2), 259–262. doi:10.1590/0074-02760150005 PMID:25946252

Cárdenas, R., Sandoval, C. M., Rodriguez-Morales, A. J., & Vivas, P. (2008). Zoonoses and climate variability: The example of Leishmaniasis in southern departments of Colombia. *Animal Biodiversity and Emerging Diseases, 1149*, 326–330.

Carpenter, S., Groschup, M. H., Garros, C., Felippe-Bauer, M. L., & Purse, B. V. (2013). Culicoides biting midges, arboviruses and public health in Europe. *Antiviral Research, 100*(1), 102–113. doi:10.1016/j.antiviral.2013.07.020 PMID:23933421

CDC - Center for Disease Control and Prevention. (2010). *The 2009 H1N1 Pandemic: Summary Highlights, April 2009 - April 2010.* Retrieved from http://www.cdc.gov/h1n1flu/cdcresponse.htm

CDC - Center for Disease Control and Prevention. (2014, November18). *Plague*. Retrieved from www.cdc.gov/plague/histoty

Centers for Disease Control and Prevention. (2014). Outbreaks Chronology: Ebola Virus Disease. Retrieved from http://www.cdc.gov/vhf/ebola/outbreaks/history/chronology.html

Centers for Disease Control and Prevention. (2015, June 26). 2014 Ebola Outbreak in West Africa - Outbreak Distribution Map. Retrieved from http://www.cdc.gov/vhf/ebola/outbreaks/2014-west-africa/distribution-map.html

Cesario, M., & Cesario, R. R. (2005). Infecção bacteriana rumo ao Brasil: Endêmica nos Andes, Bartonelose se alastra com abertura de estradas e degradação ambiental. *Scientific American Brasil, 3*(34), 10–11.

Chagas, C. (1907). Adenda: Lassance. In J. R. Cançado & M. Chuster (Eds.), *Cardiopatia Chagásica* (pp. 391–413). Belo Horizonte: Fundação Carlos Chagas.

Chalmers, R. M. (2012). Waterborne outbreaks of cryptosporidiosis. *Annali dell'Istituto Superiore di Sanita, 48*(4), 429–446. doi:10.4415/ANN_12_04_10 PMID:23247139

Chan, E. H., Brewer, T. F., Madoff, L. C., Pollack, M. P., Sonricker, A. L., & Keller, M., ... Brownstein, J. S. (2010). Global capacity for emerging infectious disease detection. *Proceedings of the National Academy of Sciences* (Vol. 107, pp. 21701–21706). doi:doi:10.1073/pnas.1006219107 doi:10.1073/pnas.1006219107

Chaves, L. F., & Pascual, M. (2006). Climate cycles and forecasts of cutaneous leishmaniasis, a nonstationary vector-borne disease. *PLoS Medicine, 3*(8), e295. PMID:16903778

Cheng, J. J., & Berry, P. (2012). Health co-benefits and risks of public health adaptation strategies to climate change: A review of current literature. *International Journal of Public Health, 58*(2), 305–311. doi:10.1007/s00038-012-0422-5 PMID:23111371

Chevalier, V., Pépin, M., Plée, L., & Lancelot, R. (2010). Rift Valley fever--a threat for Europe? *Euro Surveillance: Bulletin Européen Sur Les Maladies Transmissibles. European Communicable Disease Bulletin, 15.*

Chinga-Alayo, E., Huarcaya, E., Nasarre, C., Del Aguila, R., & Llanos-Cuentas, A. (2004). The influence of climate on the epidemiology of Bartonellosis in Ancash, Peru. *Transactions of the Royal Society of Tropical Medicine and Hygiene, 98*(2), 116–124. doi:10.1016/S0035-9203(03)00017-8 PMID:14964812

Chivian, E., & Bernstein, A. (2008). *Sustaining life: How human health depends on biodiversity.* New York: Oxford University Press.

Choi, B. C. K. (2012). The past, present, and future of public health surveillance. *Scientific, 2012*, 26 p.,10.6064/2012/875253

Choi, B. C. K., & Pak, A. W. P. (2001). Lessons for surveillance in the 21st century: A historical perspective from the past five millennia. *Sozial- und Praventivmedizin, 46*(6), 361–368. doi:10.1007/BF01321662 PMID:11851070

Chowell, G., Cazelles, B., Broutin, H., & Munayco, C. V. (2011). The influence of geographic and climate factors on the timing of dengue epidemics in Perú, 1994-2008. *BMC Infectious Diseases*, *11*(1), 1. doi:10.1186/1471-2334-11-164 PMID:21651779

Chua, K. B. (2003). Nipah virus outbreak in Malaysia. *Journal of Clinical Virology*, *26*(3), 265–275. http://www.ncbi.nlm.nih.gov/pubmed/12637075 doi:10.1016/S1386-6532(02)00268-8 PMID:12637075

Chughtai, A. A., Barnes, M., & Macintyre, C. R. (2016). Persistence of Ebola virus in various body fluids during convalescence: Evidence and implications for disease transmission and control. *Epidemiology and Infection*, 144(8), 1652-1660. doi:10.1017/S0950268816000054 PMID:26808232

Clarke, K.-L., & Berry, P. (2012). From theory to practice: A Canadian case study of the utility of climate change adaptation frameworks to address health impacts. *International Journal of Public Health*, *57*(1), 167–174. doi:10.1007/s00038-011-0292-2 PMID:21931977

Cohn, S. K. (2005). The Black Death 1346-1353: The Complete History (Book Reviews). *England Journal of Medicine*, *352*(10), 1054–1055. doi:10.1056/NEJM200503103521026

Cohuet, A., Harris, C., Robert, V., & Fontenille, D. (2010). Evolutionary forces on Anopheles: What makes a malaria vector? *Trends in Parasitology*, *26*(3), 130–136. doi:10.1016/j.pt.2009.12.001 PMID:20056485

Collaborative Partnership on Sustainable Wildlife Management (CPW). (March 18, 2015). *Bushmeat Sourcebook.*

Colwell, R. R., & Patz, J. A. (1998). *Climate, infectious disease and human health: an interdisciplinary perspective.* Washington, USA: American Academy of Microbiology.

Comisión Económica para América Latina y el Caribe. (2014). Anuario estadístico de América Latina y el Caribe. Santiago de Chile, Chile. Retrieved from http://www.cepal.org/en/publications/37647-anuario-estadistico-de-america-latina-y-el-caribe-2014-statistical-yearbook-latin

Confalonieri, U. E., & Costa Neto, C. (2012). Diversity of mosquito vectors (Diptera: Culicidae) in Caxiuanã, Pará, Brazil. *Interdisciplinary Perspectives on Infectious Diseases*, 2012, id 741273.

Confalonieri, U. E. C., Mascarenhas, B. M., Pinto, C. S., Santa Brigida, M., Chagas, E., & Guimarães, D. G. (2013). Distribuição Vertical e Temporal de Culicídeos (Diptera: Culicidae) em sítio na Floresta Nacional de Caxiuanã, Melgaço, Pará. In *P.L. Lisboa (Org.), Caxiuanã: paraíso ainda preservado* (pp. 117–122). Belém: Editora do Museu Paraense Emilio Goeldi.

Confalonieri, U. E. C., Menezes, J. A., & de Souza, C. M. (2015). Climate change and adaptation of the health sector: The case of infectious diseases. *Virulence*, *6*(6), 550–553. doi:10.1080/21505594.2015.1023985 PMID:26177788

Confalonieri, U. E., Margonari, C., & Quintão, A. F. (2014). Environmental change and the dynamics of parasitic diseases in the Amazon. *Acta Tropica*, *129*, 33–41. doi:10.1016/j.actatropica.2013.09.013 PMID:24056199

Conn, J. E., Wilkerson, R. C., Segura, M. N. O., de Souza, R. T., Schlichting, C. D., Wirtz, R. A., & Póvoa, M. M. (2002). Emergence of a new neotropical malaria vector facilitated by human migration and changes in land use. *The American Journal of Tropical Medicine and Hygiene*, *66*(1), 18–22. PMID:12135261

Connolly, M. A., Gayer, M., Ryan, M. J., Salama, P., Spiegel, P., & Heymann, D. L. (2004). Communicable diseases in complex emergencies: Impact and challenges. *Lancet*, *364*(9449), 1974–1983. doi:10.1016/S0140-6736(04)17481-3 PMID:15567014

Consultative Group for RVF Decision Support. (2010). Decision-support tool for prevention and control of Rift Valley fever epizootics in the Greater Horn of Africa. *The American Journal of Tropical Medicine and Hygiene*, *83*(2 Suppl.), 75–85. PMID:20682910

Convention on Biological Diversity. (2014). *Global Biodiversity Outlook 4.*

Cooley, S. R., & Doney, S. C. (2009). Anticipating ocean acidification's economic consequences for commercial fisheries. *Environmental Research Letters*, *4*(2), 024007. doi:10.1088/1748-9326/4/2/024007

Cordel, H., Quatresous, I., Paquet, C., & Couturier, E. (2006). Imported cases of chikungunya in metropolitan France, April 2005–February 2006. *Eurosurveillance*, *11*(4), E060420. PMID:16809828

Cordovez, J. M., Rendon, L. M., Gonzalez, C., & Guhl, F. (2014). Using the basic reproduction number to assess the effects of climate change in the risk of Chagas disease transmission in Colombia. *Acta Tropica*, *129*, 74–82. doi:10.1016/j.actatropica.2013.10.003 PMID:24416781

Cornell, S., Berkhout, F., Tuinstra, W., Tàbara, J. D., Jäger, J., Chabay, I., & van Kerkhoff, L. et al. (2013). Opening up knowledge systems for better responses to global environmental change. *Environmental Science & Policy*, *28*, 60–70. doi:10.1016/j.envsci.2012.11.008

Costa, J., Dornak, L. L., Almeida, C. E., & Peterson, A. T. (2014). Distributional potential of the Triatoma brasiliensis species complex at present and under scenarios of future climate conditions. *Parasites and Vectors*, *7*(1), 238. doi:10.1186/1756-3305-7-238 PMID:24886587

Costello, A., Abbas, M., Allen, A., Ball, S., Bell, S., Bellamy, R., & Patterson, C. et al. (2009). Managing the health effects of climate change: Lancet and University College London Institute for Global Health Commission. *Lancet*, *373*(9676), 1693–1733. doi:10.1016/S0140-6736(09)60935-1 PMID:19447250

Coura, J. R. (2015). The main scenarios of Chagas disease transmission. The vectors, blood and oral transmissions - A comprehensive review. *Memorias do Instituto Oswaldo Cruz*, *110*(3), 277–282. doi:10.1590/0074-0276140362 PMID:25466622

Coura, J. R., & Junqueira, A. C. (2012). Risks of endemicity, morbidity and perspectives regarding the control of Chagas disease in the Amazon Region. *Memorias do Instituto Oswaldo Cruz*, *107*(2), 145–154. doi:10.1590/S0074-02762012000200001 PMID:22415251

Cowling, B. J., Park, M., Fang, V. J., Wu, P., Leung, G. M., & Wu, J. T. (2015). Preliminary Epidemiological Assessment Of Mers-Cov Outbreak In South Korea, May To June 2015. *Eurosurveillance*, *20*(25), 21163. doi:10.2807/1560-7917.ES2015.20.25.21163 PMID:26132767

Craig, M. H., Kleinschmidt, I., Le Sueur, D., & Sharp, B. L. (2004). Exploring 30 Years of Malaria Case Data in KwaZulu-Natal, South Africa: Part II. The Impact of Non-Climatic Factors. *Tropical Medicine & International Health*, *9*(12), 1258–1266. doi:10.1111/j.1365-3156.2004.01341.x PMID:15598257

Craig, M. H., Kleinschmidt, I., Nawn, J. B., Le Sueur, D., & Sharp, B. L. (2004). Exploring 30 years of malaria case data in KwaZulu-Natal, South Africa: Part I. The impact of climatic factors. *Tropical Medicine & International Health*, *9*(12), 1247–12. doi:10.1111/j.1365-3156.2004.01340.x PMID:15598256

Cribb, J. (2010). *The coming famine: The global food crisis and what we can do to avoid it.* Berkeley: University of California Press.

Cunha, R. V., & Nogueira, R. M. R. (2005). Dengue e Dengue Hemorrágico. In J. R. Coura (Ed.), *Dinâmica das doenças infecciosas e parasitárias* (pp. 1755–1766). Rio de Janeiro: Guanabara Koogan.

Cupples, A. M., Rose, J. B., & Xagoraraki, I. (2010). New Molecular Methods for Detection of waterborne Pathogens. In R. Mitchell & J.-D. Gu (Eds.), Environmental Microbiology (pp. 57-94). Hoboken, New Jersey: Wiley-Blackwell.

Curriero, F. C., Patz, J. A., Rose, J. B., & Lele, S. (2001). The association between extreme precipitation and waterborne disease outbreaks in the United States, 1948-1994. *American Journal of Public Health*, *91*(8), 1194–1199. doi:10.2105/AJPH.91.8.1194 PMID:11499103

Cyril, C., Sari, K., Joacim, R., Adrian, M., Tompkins, A. M., Andrew, P., & Simon, J. L. et al. (2014). Impact of climate change on global malaria distribution. 3286–3291. *Proceedings of the National Academy of Sciences of the United States of America*, *11*(9), 3286–3291. doi:10.1073/pnas.1302089111 PMID:24596427

da Silva-Nunes, M., Codeço, C. T., Malafronte, R. S., da Silva, N. S., Juncansen, C., Muniz, P. T., & Ferreira, M. U. (2008). Malaria on the Amazonian frontier: Transmission dynamics, risk factors, spatial distribution, and prospects for control. *The American Journal of Tropical Medicine and Hygiene*, *79*(4), 624–635. PMID:18840755

da Silva, O., de Sousa, M. E., dos Santos, F. A., da Silava, P., Gazin, P., & da Silva, P. (1999). American cutaneous leishmaniasis in the sugar cane-producing area of Pernambuco, North-East Brazil. *Sante (Montrouge, France)*, *10*(2), 123–126. PMID:10960810

Daszak, P., Cunningham, A. A., & Hyatt, A. D. (2000). Emerging infectious diseases of wildlife--threats to biodiversity and human health. *Science*, *287*(5452), 443–449. http://www.ncbi.nlm.nih.gov/pubmed/10642539 doi:10.1126/science.287.5452.443 PMID:10642539

Daszak, P., Zambrana-Torrelio, C., Bogich, T. L., Fernandez, M., Epstein, J. H., Murray, K. A., & Hamilton, H. (2013). Interdisciplinary approaches to understanding disease emergence: The past, present, and future drivers of Nipah virus emergence. *Proceedings of the National Academy of Sciences of the United States of America*, *110*(Suppl. 1), 3681–3688. doi:10.1073/pnas.1201243109 PMID:22936052

Daubney, R., Hudson, J. R., & Garnham, P. C. (1931). Enzootic hepatitis or rift valley fever. An undescribed virus disease of sheep cattle and man from east africa. *The Journal of Pathology and Bacteriology*, *34*(4), 545–579. doi:10.1002/path.1700340418

Davies, F. G. (2010). The historical and recent impact of Rift Valley fever in Africa. *The American Journal of Tropical Medicine and Hygiene*, *83*(2 Suppl.), 73–74. doi:10.4269/ajtmh.2010.83s2a02 PMID:20682909

Davies, F. G., & Martin, V. (2006). Recognizing Rift Valley Fever. *Veterinaria Italiana*, *42*(1), 31–53. PMID:20429078

de Araújo Pedrosa, F., & de Alencar Ximenes, R. A. (2009). Sociodemographic and environmental risk factors for American Cutaneous Leishmaniasis (ACL) in the State of Alagoas, Brazil. *The American Journal of Tropical Medicine and Hygiene*, *81*(2), 195–201. PMID:19635869

De Vos, A., Cumming, G., Cumming, D., Ament, J., Baum, J., Clements, H., & Moore, C. et al. (2016). Pathogens, disease, and the social-ecological resilience of protected areas. *Ecology and Society*, *21*(1), art20. doi:10.5751/ES-07984-210120

Deber, R. B. (2003). Health Care Reform: Lessons From Canada. *American Journal of Public Health*, *93*(1), 20–24. doi:10.2105/AJPH.93.1.20 PMID:12511378

Degallier, N., Rosa, A. T. D., Hervé, J. P., Vasconcelos, P. F., Rosa, E. S., Rodrigues, S. G., & Dias, L. B. et al. (1992). A comparative study of yellow fever in Africa and South America. *Ciencia e Cultura*, *44*(2/3), 143–151.

Delgado-Petrocelli, L., Córdova, K., Camardiel, A., Aguilar, V. H., Hernández, D., & Ramos, S. (2012). Analysis of the El Niño/La Niña-southern oscillation variability and Malaria in the Estado Sucre, Venezuela. *Geospatial Health*, *6*(3), 51–57. doi:10.4081/gh.2012.122 PMID:23032283

Delpla, I., Jung, A. V., Baures, E., Clement, M., & Thomas, O. (2009). Impacts of climate change on surface water quality in relation to drinking water production. *Environment International*, *35*(8), 1225–1233. doi:10.1016/j.envint.2009.07.001 PMID:19640587

Department of Defense. (2014). *Quadrennial Defense Review 2014*. Washington, D.C.: Department of Defense.

Department of Environmental Affair DEA. (2013). *Long-Term Adaptation Scenarios Flagship Research Programme (LTAS) for South Africa*. Pretoria, South Africa: Department of Environmental Affairs.

Despommier, D. D., Gwadz, R. W., Hotez, P. J., & Knirsch, C. A. (2006). Parasitic Diseases (5th ed.). New York: Apple Tree Productions, LLC.

Dimoulas-Graham, P., Perotta, K., Behan, K., & Northcott, A. (Eds.), (2012). An Environmental Scan Of Built Environment Data Related to Walkability and Environmental Exposures in Urban Ontario. Kingston, Frontenac, Lennox and Addington Health Unit. Retrieved from http://www.kflapublichealth.ca/Files/Research/LCDPBuiltEnvironmentReport2012.pdf

Dobson, A. P. (2005). Virology. What links bats to emerging infectious diseases? *Science*, *310*(5748), 628–629. doi:10.1126/science.1120872 PMID:16254175

Dobson, A. P., & Carper, E. R. (1996). Infectious diseases and human population history. *Bioscience*, *46*(2), 115–126. doi:10.2307/1312814

Dora, C., Haines, A., Balbus, J., Fletcher, E., Adair-Rohani, H., Alabaster, G., & Neira, M. et al. (2015). Indicators linking health and sustainability in the post-2015 development agenda. *Lancet*, *385*(9965), 380–391. doi:10.1016/S0140-6736(14)60605-X PMID:24923529

Downs, W. G. (1981). The known and the unknown in yellow fever ecology and epidemiology. *Ecology of Disease*, *1*(2-3), 103–110. PMID:6148230

Drexler, J. F., Corman, V. M., Muller, M. A., Maganga, G. D., Vallo, P., Binger, T., & Drosten, C. et al. (2012). Bats host major mammalian paramyxoviruses. *Nature Communications*, *3*, 796. doi:10.1038/ncomms1796 PMID:22531181

Dutra, F. R. L. S., Valadão, R. C., Confalonieri, U. E., Muller, G. V., & da Quadro, M. F. L. (2015). A influência da variabilidade da precipitação no padrão de distribuição dos casos de leptospirose em Minas Gerais, no período de 1998-2012. *Hygeia; Revista Brasileira de Geografia Médica e da Saúde*, *11*(20), 106.

Dye, C. (2014). After 2015: Infectious diseases in a new era of health and development. *Philosophical Transactions of the Royal Society of London. Series B, Biological Sciences*, *369*(1645), 20130426. doi:10.1098/rstb.2013.0426 PMID:24821913

Earth, F. (2014). *Strategic Research Agenda 2014*. Retrieved from http://www.futureearth.org/media/strategic-research-agenda-2014

Ebi, K. (2011). Climate Change And Health Risks: Assessing And Responding To Them Through "Adaptive Management". *Health Affairs*, *30*(5), 924–930. doi:10.1377/hlthaff.2011.0071 PMID:21555476

Ebi, K. L., Hartman, J., Chan, N., McConnell, K. J., Schlesinger, M., & Weyant, J. (2005). Climate Suitability for Sta ble Malaria Transmission in Zimbabwe under Different Climate Change Scenarios. *Climatic Change*, *73*(3), 375–393. doi:10.1007/s10584-005-6875-2

Ebi, K. L., Lindgren, E., Suk, J. E., & Semenza, J. C. (2013). Adaptation to the infectious disease impacts of climate change. *Climatic Change*, *118*(2), 355–365. doi:10.1007/s10584-012-0648-5

Ebi, K. L., Woodruff, R., von Hildebrand, A., & Corvalan, C. (2007). Climate change-related health impacts in the Hindu Kush–Himalayas. *EcoHealth*, *4*(3), 264–270. doi:10.1007/s10393-007-0119-z

Echegaray, F., & Afonso, M. H. F. (2014). Respostas às mudanças climáticas: Inovação tecnológica ou mudança de comportamento individual? *Estudos Avançados*, *28*(82), 155–174. doi:10.1590/S0103-40142014000300010

Echeverría, V. I. (2010). Girolamo Fracastoro and the invention of syphilis.[PubMed]. *Historia, Ciencias, Saude--Manguinhos*, *17*(4), 877–884. doi:10.1590/S0104-59702010000400002

Economic Commission for Latin America and the Caribbean (ECLAC) (2011), *An Assessment of the Economic impact of Climate Change on the health sector in Guyana.*

Edberg, S. C., Rice, E. W., Karlin, R. J., & Allen, M. J. (2000). *Escherichia coli*: The best biological drinking water indicator for public health protection. *Journal of Applied Microbiology*, *88*(Suppl. 1), 1068–1168. doi:10.1111/j.1365-2672.2000.tb05338.x PMID:10880185

Edge, T. A., Khan, I. U. H., Bouchard, R., Guo, J., Hill, S., Locas, A., & Watson, S. et al. (2013). Occurrence of waterborne pathogens and *Escherichia coli* at offshore drinking water intakes in Lake Ontario. *Applied and Environmental Microbiology*, *79*(19), 5799–5813. doi:10.1128/AEM.00870-13 PMID:23835181

EHC - Research Triangle Environmental Health Collaborative. (2010). *Environmental Health Summit: Recommendations from the Research Triangle Environmental Health Collaborative.* Retrieved from http://www.pewtrusts.org/~/media/Assets/0001/EnvironmentalHealthResearch-TriangleCollaborative2010EnvironmentalHealthSummit2010Recommendations.pdf

Eidson, M., Clancy, K.A., & Birkhead, G.S. (2015). Public health climate change adaptation planning using stakeholder feedback. *Journal of Public Health Management Practice.*

Elliott, R. M. (1997). Emerging viruses: The Bunyaviridae. *Molecular Medicine (Cambridge, Mass.)*, 3. PMID:9323708

Engelkirk, P. G., & Duben-Engelkirk, J. (2011). *Burton's microbiology for the health sciences* (9th ed.). Philadelphia, PA: Wolters Kluwer / Lippincott Williams & Wilkins.

Enright, M. C., Robinson, D. A., Randle, G., Feil, E. J., Grundmann, H., & Spratt, B. G. (2002). The evolutionary history of methicillin-resistant Staphylococcus aureus (MRSA). *Proceedings of the National Academy of Sciences of the United States of America*, *99*(11), 7687–7692. doi:10.1073/pnas.122108599 PMID:12032344

Environment Programme (UNEP). (2003). *Climate Change and Human Health - Risks and Responses: Summary. ISBN 92 4 159081 5.* Published by WHO.

Enzler, S. M. (n. d.). *History of water treatment*. Lennteck. Retrieved from http://www.lenntech.com/history-water-treatment.htm

Epstein, P. R. (2001). West Nile virus and the climate. *Journal of Urban Health*, *78*(2), 367–371. doi:10.1093/jurban/78.2.367 PMID:11419587

Erickson, M. C., & Doyle, M. P. (2012). *Plant Food Safety Issues: Linking Production Agriculture with One Health Improving food safety through a One Health approach*. Washington, DC: The National Academies Press.

Erkoreka, A. (2009). Origins of the Spanish Influenza pandemic (1918-1920) and its relation to the First World War. *Journal of Molecular and Genetic Medicine*, *3*(2), 190–194. PMID:20076789

European Parliament. (2006). Directive 2006/7/EC of the European Parliament and of the Council of 15 February 2006 concerning the management of bathing water quality and repealing Directive 76/160/EEC. *Official Journal of the European Union, 2006*, 64.

Evans, O.P., & Adenomon, M.O. (2014). Modeling the prevalence of malaria in Niger State: An application of Poisson regression and negative binomial regression models International Journal of Physical Sciences, 2(4), 61-68. Retrieved from http://academeresearchjournals.org/journal/ijps

Evans, A., Gakuya, F., Paweska, J. T., Rostal, M., Akoolo, L., Van Vuren, P. J., & Feikin, D. R. et al. (2008). Prevalence of antibodies against Rift Valley fever virus in Kenyan wildlife. *Epidemiology and Infection*, *136*(09), 1261–1269. doi:10.1017/S0950268807009806 PMID:17988425

Fa, J. E., Juste, J., Del Val, J. P., & Castroviejo, J. (1995). Impact of Market Hunting on Mammal Species in Equatorial Guinea. *Conservation Biology*, *9*(5), 1107–1115. doi:10.1046/j.1523-1739.1995.951107.x

Fa, J. E., Peres, C. A., & Meeuwig, J. (2001). Bushmeat exploitation in tropical forest: An intercontinental comparison. *Conservation Biology*, *16*(1), 232–237. doi:10.1046/j.1523-1739.2002.00275.x

Faraj, C., Ouahabi, S., Adlaoui, E., Boccolini, D., Romi, R., & El Aouad, R. (2008). Risque de reemergence du paludisme au Maroc étude de la capacité vectorielle d'*Anopheles labranchiae* dans une zone rizicole au nord du pays. *Parasite (Paris, France)*, *15*(4), 605–610. doi:10.1051/parasite/2008154605 PMID:19202769

Faria, N. R., Rambaut, A., Suchard, M. A., Baele, G., Bedford, T., Ward, M. J., & Lemey, P. et al. (2014). HIV epidemiology. The early spread and epidemic ignition of HIV-1 in human populations. *Science*, *346*(6205), 56–61. doi:10.1126/science.1256739 PMID:25278604

Federal Ministry of Health, National Malaria Control Programme/FMOH/2005 Annual Report. (2005, January).

Feldmann, H. (2014). Ebola--a growing threat? *The New England Journal of Medicine*, *371*(15), 1375–1378. doi:10.1056/NEJMp1405314 PMID:24805988

Feldmann, H., Sanchez, A., & Geisbert, T. W. (2013). Filoviridae: Marburg and Ebola viruses. In D. M. Knipe & P. M. Howley (Eds.), *Fields Virology* (6th ed., pp. 923–956). Philadelphia: Lippincott Williams & Wilkins.

Feldmann, H., Slenczka, W., & Klenk, H. D. (1996). Emerging and reemerging of filoviruses. *Archives of Virology. Supplementum*, *11*, 77–100. Retrieved from http://www.ncbi.nlm.nih.gov/pubmed/8800808 PMID:8800808

Feliciangeli, M. D., & Rabinovich, J. E. (1998). Abundance of *Lutzomyia ovallesi* but not *Lu. gomezi* (Diptera: Psychodidae) correlated with cutaneous leishmaniasis incidence in north-central Venezuela. *Medical and Veterinary Entomology*, *12*(2), 121–131. doi:10.1046/j.1365-2915.1998.00072.x PMID:9622364

Feltmate, B., & Thistlethwaite, J. (2012). Climate Change Adaptation: A Priorities Plan for Canada. Report of the Climate Change Adaptation Project. University of Waterloo. Retrieved from https://uwaterloo.ca/environment/sites/ca.environment/files/uploads/files/CCAP-Report-30May-Final.pdf

Femia, F., & Werrell, C. (2013). Climate change before and after the Arab Awakening: The cases of Syria and Libya. In C. E. Werrell & F. Femia (Eds.), *The Arab Spring and Climate Change*. Washington, DC: Center for American Progress.

Ferreira, I. B., Pereira, L. E., Rocco, I. M., Marti, A. T., Souza, L., & Iversson, L. B. (1994). Surveillance of arbovirus infections in the Atlantic forest region, State of São Paulo, Brazil: I. detection of hemagglutination-inhibition antibodies in wild birds between 1978 and 1990. *Revista do Instituto de Medicina Tropical de Sao Paulo*, *36*(3), 265–274. doi:10.1590/S0036-46651994000300011 PMID:7855491

Ferreira, K. V., Rocha, K. C., Caputto, L. Z., Fonseca, A. L. A., & Fonseca, F. L. A. (2011). Histórico da febre amarela no Brasil e a importância da vacinação antiamarílica. *Arquivos Brasileiros de Ciências da Saúde*, *36*(1), 40–47.

Field, J. S., & Lape, P. V. (2010). Paleoclimates and the emergence of fortifications in the tropical Pacific islands. *Journal of Anthropological Archaeology*, *29*(1), 113-124. doi:10.1016/j.jaa.2009.11.001

Figueiredo, M. L. G. D., & Figueiredo, L. T. M. (2014). Emerging alphaviruses in the Americas: Chikungunya and Mayaro. *Revista da Sociedade Brasileira de Medicina Tropical*, *47*(6), 677–683. doi:10.1590/0037-8682-0246-2014 PMID:25626645

Findlay, G. M. (1932). Rift valley fever or enzootic hepatitis. *Transactions of the Royal Society of Tropical Medicine and Hygiene*, *25*(4), 229–IN11. doi:10.1016/S0035-9203(32)90042-X

Flint, H. S., Franson, P. L., Dusek, J. C., Derksen, R. J., Gill, D. V., Jr, R. E., & Rothe, T. C. et al. (2008). Prevalence of Influenza A viruses in wild migratory birds in Alaska: Patterns of variation in detection at a crossroads of intercontinental flyways. *Virology Journal*, *5*(1), 71. doi:10.1186/1743-422X-5-71 PMID:18533040

Foley, J. A., Ramankutty, N., Brauman, K. A., Cassidy, E. S., Gerber, J. S., Johnston, M., & Zaks, D. P. et al. (2011). Solutions for a cultivated planet. *Nature*, *478*(7369), 337–342. doi:10.1038/nature10452 PMID:21993620

Food and Agriculture Organization of the UN. (2009). *How to Feed the World in 2050*. Retrieved from http://www.fao.org/fileadmin/templates/wsfs/docs/expert_paper/How_to_Feed_the_World_in_2050.pdf

Formenty, P., Hatz, C., Le Guenno, B., Stoll, A., Rogenmoser, P., & Widmer, A. (1999). Human infection due to Ebola virus, subtype Cote d'Ivoire: Clinical and biologic presentation. *The Journal of Infectious Diseases*, *179*(Suppl. 1), S48–S53. doi:10.1086/514285 PMID:9988164

Fouque, F., Gaborit, P., Issaly, J., Carinci, R., Gantier, J. C., Ravel, C., & Dedet, J. P. (2007). Phlebotomine sand flies (Diptera: Psychodidae) associated with changing patterns in the transmission of the human cutaneous leishmaniasis in French Guiana. *Memorias do Instituto Oswaldo Cruz*, *102*(1), 35–40. doi:10.1590/S0074-02762007000100005 PMID:17293996

Friel, S., Trevor, H., Tord, K., Gordon, M., Monge, P., & Joyashree R. (2011). Urban Health Inequities and the Added Pressure of Climate Change: An Action-Oriented Research Agenda. *Journal of Urban Health*, 88(5), 886-895.

Friel, S., Bowen, K., Campbell-Lendrum, D., Frumkin, H., McMichael, A. J., & Rasanathan, K. (2011). Climate Change, Noncommunicable Diseases, and Development: The Relationships and Common Policy Opportunities. *Annual Review of Public Health*, *32*(1), 133–147. doi:10.1146/annurev-publhealth-071910-140612 PMID:21091194

Frith, J. (2012). The history of plague – Part 1. The three great pandemics. *Journal of Military and Veterans'Health*, *20*(2), 11–16.

Frommel, A. Y., Maneja, R., Lowe, D., Malzahn, A. M., Geffen, A. J., Folkvord, A., et al. (2012). Severe tissue damage in Atlantic cod larvae under increasing ocean acidification. *Nature Clim. Change*, *2*(1), 42-46. Doi:10.1038/nclimate1324

Frumkin, H., Hess, J., Luber, G., Malilay, J., & McGeehin, M. (2008). Climate change: The public health response. *American Journal of Public Health*, *98*(3), 435–445. doi:10.2105/AJPH.2007.119362 PMID:18235058

Funari, E., Manganelli, M., & Sinisi, L. (2012). Impact of climate change on waterborne diseases. *Annali dell'Istituto Superiore di Sanita*, *48*(4), 473–487. doi:10.4415/ANN_12_04_13 PMID:23247142

Gage, K. L., Burkot, T. R., Eisen, R. J., & Hayes, E. B. (2008). Climate and vectorborne diseases. *American Journal of Preventive Medicine*, *35*(5), 436–450. doi:10.1016/j.amepre.2008.08.030 PMID:18929970

Gall, A. M., Marinas, B. J., Lu, Y., & Shisler, J. L. (2015). Waterborne Viruses: A Barrier to Safe Drinking Water. *PLoS Pathogens*, *11*(6), e1004867. doi:10.1371/journal.ppat.1004867 PMID:26110535

GBD 2013 Mortality and Causes of Death Collaborators. (2015). Global, regional, and national age-sex specific all-cause and cause-specific mortality for 240 causes of death, 1990-2013: a systematic analysis for the Global Burden of Disease Study 2013. *Lancet, 385*(9963), 117-171. doi:10.1016/S0140-6736(14)61682-2

Gerber, P. J., Steinfeld, H., Henderson, B., Mottet, A., Opio, C., Dijkman, J., . . . Tempio, G. (2013). *Tackling climate change through livestock – A global assessment of emissions and mitigation opportunities.*

Gerland, P., Raftery, A. E., ev ikova, H., Li, N., Gu, D., Spoorenberg, T., & Wilmoth, J. et al. (2014). World population stabilization unlikely this century. *Science*, *346*(6206), 234–237. doi:10.1126/science.1257469 PMID:25301627

Gest, H. (2004). The discovery of microorganisms by Robert Hooke and Antoni van Leeuwenhoek, Fellows of The Royal Society.[PubMed]. *Notes and Records of the Royal Society of London*, *58*(2), 187–201. doi:10.1098/rsnr.2004.0055

Ge, X. Y., Li, J. L., Yang, X. L., Chmura, A. A., Zhu, G., Epstein, J. H., & Shi, Z. L. et al. (2013). Isolation and characterization of a bat SARS-like coronavirus that uses the ACE2 receptor. *Nature*, *503*(7477), 535–538. doi:10.1038/nature12711 PMID:24172901

Ghobarah, H. A., Huth, P., & Russett, B. (2004). The post-war public health effects of civil conflict. *Social Science & Medicine*, *59*(4), 869–884. doi:10.1016/j.socscimed.2003.11.043 PMID:15177842

Gire, S. K., Goba, A., Andersen, K. G., Sealfon, R. S., Park, D. J., Kanneh, L., & Sabeti, P. C. et al. (2014). Genomic surveillance elucidates Ebola virus origin and transmission during the 2014 outbreak. *Science*, *345*(6202), 1369–1372. doi:10.1126/science.1259657 PMID:25214632

Githeko, A.K. (2009). Malaria and climate change (Special feature). Commonwealth Health ministers' update.

Githeko, A. K., Lindsay, S. W., Confalonieri, U. E., & Patz, J. A. (2000). Climate change and vector-borne diseases: A regional analysis. *Bulletin of the World Health Organization*, *78*, 1136–1147. PMID:11019462

Gleditsch, N. P. (2012). Whither the weather? Climate change and conflict. *Journal of Peace Research*, *49*(1), 3–9. doi:10.1177/0022343311431288

Gleick, P. H. (2014). Water, drought, climate change, and conflict in Syria. *Weather, Climate, and Society*, *6*(3), 331–340. doi:10.1175/WCAS-D-13-00059.1

Gobir, Z., & Tukur, Z. (2014). Prevalence of malaria parasitemia using rapid diagnostic test among apparently healthy children in Kano, Nigeria. *Journal of Medicine in the Tropic*, *16*(1), 1–4. Retrieved from http://www.jmedtropics.org/article.asp?issn=22767096;year=2014;volume=16;issue=1;spage=1;epage=4;aulast=Gobir doi:10.4103/2276-7096.132566

Godden Lee et al. (2013), *Governance arrangements for climate change adaptation and natural disaster risk management - An initial analysis of legal and regulatory issues*, Victorian Centre for Climate Change Adaptation Research, available at: <http://www.vcccar.org.au/sites/default/files/publications/VCCCAR%20report%20Governance%20Arrangements%20for%20web%20060913.pdf

Golden, C. D., Fernald, L. C., Brashares, J. S., Rasolofoniaina, B. J., & Kremen, C. (2011). Benefits of wildlife consumption to child nutrition in a biodiversity hotspot. *Proceedings of the National Academy of Sciences of the United States of America*, *108*(49), 19653–19656. doi:10.1073/pnas.1112586108 PMID:22106297

Goldstein, S. T., Juranek, D. D., Ravenholt, O., Hightower, A. W., Martin, D. G., Mesnik, J. L., & Herwaldt, B. L. et al. (1996). Cryptosporidiosis: An outbreak associated with drinking water despite state-of-the-art water treatment. *Annals of Internal Medicine*, *124*(5), 459–468. doi:10.7326/0003-4819-124-5-199603010-00001 PMID:8602703

Gonçalves Neto, V. S., Barros Filho, A. K. D., Santos, A. M. D., Prazeres, M. P. C. D. S., Bezerril, A. C. R., Fonseca, A. V. D. L., & Rebelo, J. M. M. (2013). An analysis of the spatiotemporal distribution of American cutaneous leishmaniasis in counties located along road and railway corridors in the State of Maranhao, Brazil. *Revista da Sociedade Brasileira de Medicina Tropical*, *46*(3), 322–328. doi:10.1590/0037-8682-0056-2012 PMID:23856875

Goodin, D. G., Koch, D. E., Owen, R. D., Chu, Y. K., Hutchinson, J. M., & Jonsson, C. B. (2006). Land cover associated with Hantavirus presence in Paraguay. *Global Ecology and Biogeography*, *15*(5), 519–527. doi:10.1111/j.1466-822X.2006.00244.x

Gostin, L. O., & Friedman, E. A. (2015). A retrospective and prospective analysis of the west African Ebola virus disease epidemic: Robust national health systems at the foundation and an empowered WHO at the apex. *Lancet*, *385*(9980), 1902–1909. doi:10.1016/S0140-6736(15)60644-4 PMID:25987158

Gottdenker, N. L., Chaves, L. F., Calzada, J. E., Saldaña, A., & Carroll, C. R. (2012). Host life history strategy, species diversity, and habitat influence Trypanosoma cruzi vector infection in changing landscapes. *PLoS Neglected Tropical Diseases*, 6(11): e1884.

Gottdenker, N. L., Calzada, J. E., Saldaña, A., & Carroll, C. R. (2011). Association of anthropogenic land use change and increased abundance of the Chagas disease vector Rhodnius pallescens in a rural landscape of Panama. *The American Journal of Tropical Medicine and Hygiene*, *84*(1), 70–77. doi:10.4269/ajtmh.2011.10-0041 PMID:21212205

Gottdenker, N. L., Streicker, D. G., Faust, C. L., & Carroll, C. R. (2014). Anthropogenic land use change and infectious diseases: A review of the evidence. *EcoHealth*, *11*(4), 619–632. doi:10.1007/s10393-014-0941-z PMID:24854248

Government of Canada. H. C. (2011, December 13). Adapting to Extreme Heat Events: Guidelines for Assessing Health Vulnerability [Health Canada, 2011]. Retrieved from http://www.hc-sc.gc.ca/ewh-semt/pubs/climat/adapt/index-eng.php

Grace, D., & Bett, B. (2014). Zoonotic diseases and their drivers in Africa. In C. D. Butler (Ed.), Climate Change and Global Health (pp. 228–236). Boston: CABI. Retrieved from https://books.google.com/books?hl=en&lr=&id=qVDNBAAAQBAJ&pgis=1 doi:doi:10.1079/9781780642659.0228 doi:10.1079/9781780642659.0228

Grace, D., Gilbert, J., Randolph, T., & Kang'ethe, E. (2012). The multiple burdens of zoonotic disease and an Ecohealth approach to their assessment. *Tropical Animal Health and Production*, *44*(S1), S67–S73. doi:10.1007/s11250-012-0209-y PMID:22886445

Gray, N. F. (2014). The Implications of Global Warming and Climate Change on Waterborne Diseases. In S. Percival, M. V. Yates, D. Williams, R. Chalmers, & N. Gray (Eds.), *Microbiology of Waterborne Diseases* (2nd ed., pp. 653–666). London: Academic Press, Elsevier. doi:10.1016/B978-0-12-415846-7.00036-6

Greer, A., Ng, V., & Fisman, D. (2008). Climate change and infectious diseases in North America: The road ahead. *Canadian Medical Association Journal*, *178*(6), 715–722.

Gubler, D. J. (2002). The global emergence/resurgence of arboviral diseases as public health problems. *Archives of Medical Research*, *33*(4), 330–342. doi:10.1016/S0188-4409(02)00378-8 PMID:12234522

Gubler, D. J., Reiter, P., Ebi, K. L., Yap, W., Nasci, R., & Patz, J. A. (2001). Climate variability and change in the United States: Potential impacts on vector- and rodent-borne diseases. *Environmental Health Perspectives*, *109*, 223–233. doi:10.2307/3435012 PMID:11359689

Guilvard, E., Rioux, J. P., Gallego, M., Pratlong, F., Mahjour, J., Martinezorlega, E., & Martini, A. et al. (1991). *Leishmania tropica* au Maroc III- Rôle de *Phlebotomus sergenti*. A propos de 89 isolats1. *Annales de Parasitologie Humaine et Comparee*, *66*, 96–99. PMID:1776784

Haas, J. (2009). The origins of war and ethnic violence. In J. Carman & A. Harding (Eds.), *Ancient Warfare: archaeological perspectives*. Gloucestershire: The History Press.

Hack, W. H. (1955). Estudios sobre biología del *Triatoma infestans* (Klug, 1834) (Hemiptera, Reduviidae). *An Inst Med Regional*, *4*, 125–147.

Hadorn, D. C., & Stark, K. D. (2008). Evaluation and optimization of surveillance systems for rare and emerging infectious diseases. *Veterinary Research*, *39*(6), 57. doi:10.1051/vetres:2008033 PMID:18651991

Haensch, S., Bianucci, R., Signoli, M., Rajerison, M., Schultz, M., & Kacki, S. … Bramanti, B. (2010). Distinct Clones of Yersinia pestis Caused the Black Death. Plos Pathogens, 6(10). Retrieved from http://www.plospathogens.org

Hahn, M. B., Gangnon, R. E., Barcellos, C., Asner, G. P., & Patz, J. A. (2014). Influence of deforestation, logging, and fire on Malaria in the Brazilian Amazon. *PLoS ONE*, *9*(1), e85725. doi:10.1371/journal.pone.0085725 PMID:24404206

Haines-Young, R., & Potschin, M. (2010). The links between biodiversity, ecosystem services and human well-being. In D. C. F. Raffaelli (Ed.), *Ecosystem Ecology: a new synthesis* (pp. 110–139). Cambridge. doi:10.1017/CBO9780511750458.007

Halkier, B. (2011). Methodological Practicalities in Analytical Generalization. *Qualitative Inquiry*, *17*(9), 787–797. doi:10.1177/1077800411423194

Halpin, K., Young, P. L., Field, H., & Mackenzie, J. S. (1999). Newly discovered viruses of flying foxes. *Veterinary Microbiology*, *68*(1-2), 83–87. Retrieved from http://www.ncbi.nlm.nih.gov/pubmed/10501164 doi:10.1016/S0378-1135(99)00063-2 PMID:10501164

Hankins, C. A., Friedman, S. R., Zafar, T., & Strathdee, S. A. (2002). Transmission and prevention of HIV and sexually transmitted infections in war settings: Implications for current and future armed conflicts. *AIDS (London, England)*, *16*(17), 2245–2252. doi:10.1097/00002030-200211220-00003 PMID:12441795

Harvard University Library. (n. d.). *The Great Plague of London,1665. Contagium Historical Views of Disease and Epidemics.* Harvard University Library Open Collection Program. Retrieved from http://ocp.hul.harvard.edu/contagion/plague.html

Hassan, O. A., Ahlm, C., Sang, R., & Evander, M. (2011). The 2007 Rift Valley fever outbreak in Sudan. *PLoS Neglected Tropical Diseases*, *5*(9), e1229. doi:10.1371/journal.pntd.0001229 PMID:21980543

Haug, G. H., Günther, D., Peterson, L. C., Sigman, D. M., Hughen, K. A., & Aeschlimann, B. (2003). Climate and the collapse of Maya civilization. *Science*, *299*(5613), 1731–1735. doi:10.1126/science.1080444 PMID:12637744

Hays, J. N. (2005). *Epidemics and Pandemics: Their Impact in Human History*. Santa Barbara, California: ABC-CLIO.

Hennessey, M., Fischer, M., & Staples, J. E. (2016). Zika Virus Spreads to New Areas - Region of the Americas, May 2015-January 2016. *MMWR. Morbidity and Mortality Weekly Report*, *65*(3), 55–58. doi:10.15585/mmwr.mm6503e1 PMID:26820163

Herrador, B. R. G., de Blasio, B. F., MacDonald, E., Nichols, G., Sudre, B., Vold, L., & Nygard, K. et al. (2015). Analytical studies assessing the association between extreme precipitation or temperature and drinking water-related waterborne infections: A review. *Environmental Health*, 2015. doi:10.1186/s12940-015-0014-y

Herrera-Martinez, A. D., & Rodriguez-Morales, A. J. (2010). Potential influence of climate variability on dengue incidence registered in a western pediatric Hospital of Venezuela. *Tropical Biomedicine*, *27*(2), 280–286. PMID:20962726

Herrero, M., & Thornton, P. K. (2013). Livestock and global change: Emerging issues for sustainable food systems. *Proceedings of the National Academy of Sciences of the United States of America*, *110*(52), 20878–20881. doi:10.1073/pnas.1321844111 PMID:24344313

Hess, J. J., Eidson, M., Tlumak, J. E., Raab, K. K., & Luber, G. (2014). An evidence-based public health approach to climate change adaptation. *Environmental Health Perspectives*, *122*(11), 1177–1186. PMID:25003495

Hess, J. J., McDowell, J. Z., & Luber, G. (2012). Integrating climate change adaptation into public health practice: Using adaptive management to increase adaptive capacity and build resilience. *Environmental Health Perspectives*, *120*(2), 171–179. doi:10.1289/ehp.1103515 PMID:21997387

Heymann, D. L., & Dar, O. A. (2014). Prevention is better than cure for emerging infectious diseases. *BMJ (Clinical Research Ed.)*, *348*, g1499. PMID:24563451

Hoegh-Guldberg, O. (2011). Coral reef ecosystems and anthropogenic climate change. *Regional Environmental Change*, *11*(0), 215–227. doi:10.1007/s10113-010-0189-2

Hoegh-Guldberg, O., Mumby, P., Hooten, A., Steneck, R., Greenfield, P., Gomez, E., & Hatziolos, M. E. et al. (2007). Coral reefs under rapid climate change and ocean acidification. *Science*, *318*(5857), 1737–1742. doi:10.1126/science.1152509 PMID:18079392

Hofstra, N. (2011). Quantifying the impact of climate change on enteric waterborne pathogen concentrations in surface water. *Current Opinion in Environmental Sustainability*, *3*(6), 471–479. doi:10.1016/j.cosust.2011.10.006

Holdstock, D. (1995). Climate, health and war. *Medicine and War, 11*(4), 145-147. doi:10.1080/07488009508409234

Homer-Dixon, T. F. (1994). Environmental scarcities and violent conflict: Evidence from cases. *International Security*, *19*(1), 5–40. doi:10.2307/2539147

Hosseini, P. R., Fuller, T., Harrigan, R., Zhao, D., Arriola, C. S., Gonzalez, A., & Daszak, P. et al. (2013). Metapopulation dynamics enable persistence of influenza A, including A/H5N1, in poultry. *PLoS ONE*, *8*(12), e80091. doi:10.1371/journal.pone.0080091 PMID:24312455

Hotez, P. J., & Kamath, A. (2009). Neglected tropical diseases in sub-saharan Africa: Review of their prevalence, distribution, and disease burden. *PLoS Neglected Tropical Diseases*, *3*(8), e412. doi:10.1371/journal.pntd.0000412 PMID:19707588

Howlett, M. (2014). Why are policy innovations rare and so often negative? Blame avoidance and problem denial in climate change policy-making. *Global Environmental Change*, *29*, 395–403. doi:10.1016/j.gloenvcha.2013.12.009

Hsiang, S. M., Meng, K. C., & Cane, M. A. (2011). Civil conflicts are associated with the global climate. *Nature, 476*(7361), 438-441. Doi:10.1038/nature10311

Huang, C., Vaneckova, P., Wang, X., FitzGerald, G., Guo, Y., & Tong, S. (2011). Constraints and Barriers to Public Health Adaptation to Climate Change. *American Journal of Preventive Medicine*, *40*(2), 183–190. doi:10.1016/j.amepre.2010.10.025 PMID:21238867

Hulden, L., & Hulden, L. (2009). The decline of malaria in Finland--the impact of the vector and social variables. *Malaria Journal*, *8*(1), 94. doi:10.1186/1475-2875-8-94 PMID:19422702

Hunter, P. R. (1998). Cyanobacterial toxins and human health. *Symposium Series (Society for Applied Microbiology), 27*(S1), 35S–40S. doi:10.1046/j.1365-2672.1998.0840s135S.x PMID:9750360

Hunter, P. R., Waite, M., & Ronchi, E. (2003). *Drinking water and infectious disease establishing the links*. London: CRC Press IWA Publishing.

Huttenhower, C., Gevers, D., Knight, R., Abubucker, S., Badger, J. H., Chinwalla, A. T., & White, O. et al.Human Microbiome Project Consortium. (2012). Structure, function and diversity of the healthy human microbiome. *Nature, 486*(7402), 207–214. doi:10.1038/nature11234 PMID:22699609

Ibarra, A. M. S., Ryan, S. J., Beltrán, E., Mejía, R., Silva, M., & Muñoz, Á. (2013). Dengue vector dynamics (*Aedes aegypti*) influenced by climate and social factors in Ecuador: Implications for targeted control. *PLoS ONE, 8*(11), e78263. doi:10.1371/journal.pone.0078263 PMID:24324542

Impoinvil, D. E., Solomon, T., Schluter, W. W., Rayamajhi, A., Bichha, R. P., Shakya, G., & Baylis, M. et al. (2011). The Spatial Heterogeneity between Japanese Encephalitis Incidence Distribution and Environmental Variables in Nepal. *PLoS ONE*, 6(7). PMID:21811573

Institute of Medicine and National Research Council. (2008). *Current Surveillance Systems for Detecting Zoonoses in Animals*.

Intergovernmental Panel on Climate Change (IPCC). (2007b). Human health. In M. L. Parry, O. F. Canziani, J. P. Palutikof, P. J. van der Linden, & C. E. Hanson (Eds.), *Climate Change 2007: Impacts, Adaptation and Vulnerability. Contribution of Working Group II to the Fourth Assessment Report of the Intergovernmental Panel on Climate Change* (pp. 391–431). Cambridge, UK: Cambridge University Press.

Intergovernmental Panel on Climate Change (IPCC). (2014). *Summary for policymakers*.

Intergovernmental panel on climate change fifth assessment report working group II final draft (2013). Human Health: Impacts, Adaptation, and Co-Benefits (Ch. 11). Retrieved from https://www.ipcc.ch/pdf/assessment-report/ar5/wg2/drafts/fd/WGIIAR5-Chap11_FGDall.pdf

Intergovernmental Panel on Climate Change. (2013). *The physical science basis*. Geneva: Intergovernmental Panel on Climate Change.*Climatic Change*, 2013.

International Committee of the Red Cross. (2011). *Health Care in Danger: Making the case*. Geneva: International Committee of the Red Cross.

IOM (Institute of Medicine) European Centre for Disease Prevention and Control. (2016). *Rapid Risk Assessment. Zika virus disease epidemic: potential association with microcephaly and Guillain–Barré syndrome.*

IUCN. (2008). The IUCN Red List of Threatened Species (Version 2014.2). Retrieved from http://www.iucnredlist.org

Jacobi, P. R. (2014). Mudanças climáticas e ensino superior: A combinação entre pesquisa e educação. *Educational Review, 3*(spe3), 57–72. doi:10.1590/0104-4060.38107

Janes, C. R., Corbett, K. K., Jones, J. H., & Trostle, J. (2012). Emerging infectious diseases: The role of social sciences. *Lancet, 380*(9857), 1884–1886. doi:10.1016/S0140-6736(12)61725-5 PMID:23200487

Jarvis, L., Montgomery, H., Morisetti, N., & Gilmore, I. (2011). Climate change, ill health, and conflict. *BMJ (Clinical Research Ed.)*, April. PMID:21467102

Johnson, K. M. (1993). Emerging viruses in context: an overview of viral hemorrhagic fevers. In S. S. Morse (Ed.), *Emerging viruses* (pp. 46–47). New York: Oxford University Press.

Jones, K. E., Patel, N. G., Levy, M. A., Storeygard, A., Balk, D., Gittleman, J. L., & Daszak, P. (2008). Global trends in emerging infectious diseases. *Nature, 451*(7181), 990–993. doi:10.1038/nature06536 PMID:18288193

Jones, K. E., Purvis, A., MacLarnon, A., Bininda-Emonds, O. R., & Simmons, N. B. (2002). A phylogenetic supertree of the bats (Mammalia: Chiroptera). *Biological Reviews of the Cambridge Philosophical Society, 77*(2), 223–259. Retrieved from http://www.ncbi.nlm.nih.gov/pubmed/12056748 doi:10.1017/S1464793101005899 PMID:12056748

Jost, C. C., Nzietchueng, S., Kihu, S., Bett, B., Njogu, G., Swai, E. S., & Mariner, J. C. (2010). Epidemiological assessment of the Rift Valley fever outbreak in Kenya and Tanzania in 2006 and 2007. *The American Journal of Tropical Medicine and Hygiene, 83*(2 Suppl.), 65–72. doi:10.4269/ajtmh.2010.09-0290 PMID:20682908

Kabore, A. (2004). Overview of Malaria in West Africa. *WHO Newsletter*, 5(2).

Kahime, K., Boussaa, S., Laamrani-El Idrissi, A., Nhammi, H., and Boumezzough, A. (2015b). Epidemiological study on acute cutaneous leishmaniasis in Morocco. *Journal of Acute Diseases*. doi:10.1016/j.joad.2015.08.004.

Kahime, K., Bounoua, B., Messouli, M., Boussaa, S., & Boumezzough, A. (in press). Evaluation of Eco-Adaptation Strategies of Health to Climate change: Case of Zoonotic Cutaneous Leishmaniasis (ZCL) as Vulnerability Indicator in Pre-Saharan region of Morocco. In M. Behnassi & K. Mcglade (Eds.), *Environmental Change and Human Security in the Middle East and Africa*. Netherlands: Springer.

Kahime, K., Boussaa, S., Bounoua, L., Fouad, O., Messouli, M., & Boumezzough, A. (2014). Leishmaniasis in Morocco: Diseases and vectors. *Asian Pacific Journal of Tropical Disease, 4*, 530–534. doi:10.1016/S2222-1808(14)60671-X

Kahime, K., Boussaa, S., Ouanaimi, F., & Boumezzough, A. (2015a). Entomological Assessment of Human Visceral Leishmaniasis Risk in area with sporadic cases, Morocco. *Acta Tropica, 48*, 58–65. doi:10.1016/j.actatropica.2015.04.006 PMID:25930188

Kahl, C. H. (2006). *States, Scarcity and Civil Strife in the Developing World*. Princeton: Princeton University Press.

Kaldor, M. (2012). New and Old Wars: Organized violence in a global era (3rd ed.). Stanford: Stanford University Press.

Karesh, W. B., & Cook, R. A. (2005). The Human-Animal Link. *Foreign Affairs*, July/August.

Karesh, W. B., Cook, R. A., Bennett, E. L., & Newcomb, J. (2005). Wildlife trade and global disease emergence. *Emerging Infectious Diseases*, *11*(7), 1000–1002. doi:10.3201/eid1107.050194 PMID:16022772

Karesh, W. B., Dobson, A., Lloyd-Smith, J. O., Lubroth, J., Dixon, M. A., Bennett, M., & Heymann, D. L. et al. (2012). Ecology of zoonoses: Natural and unnatural histories. *Lancet*, *380*(9857), 1936–1945. doi:10.1016/S0140-6736(12)61678-X PMID:23200502

Keesing, F., Belden, L. K., Daszak, P., Dobson, A., Harvell, C. D., Holt, R. D., & Ostfeld, R. S. et al. (2010). Impacts of biodiversity on the emergence and transmission of infectious diseases. *Nature*, *468*(7324), 647–652. doi:10.1038/nature09575 PMID:21124449

Keesing, F., Holt, R. D., & Ostfeld, R. S. (2006). Effects of species diversity on disease risk. *Ecology Letters*, *9*(4), 485–498. doi:10.1111/j.1461-0248.2006.00885.x PMID:16623733

Keim, M. E. (2008). Building human resilience: The role of public health preparedness and response as an adaptation to climate change. *American Journal of Preventive Medicine*, *35*(5), 508–516. doi:10.1016/j.amepre.2008.08.022 PMID:18929977

Keiser, J., De Castro, M. C., Maltese, M. F., Bos, R., Tanner, M., Singer, B. H., & Utzinger, J. (2005). Effect of irrigation and large dams on the burden of malaria on a global and regional scale. *The American Journal of Tropical Medicine and Hygiene*, *72*(4), 392–406. PMID:15827275

Kelley, C. P., Mohtadi, S., Cane, M. A., Seager, R., & Kushnir, Y. (2015). Climate change in the Fertile Crescent and implications of the recent Syrian drought. *Proceedings of the National Academy of Sciences of the United States of America*, *112*(11), 3241–3246. doi:10.1073/pnas.1421533112 PMID:25733898

Kennett, D. J., Breitenbach, S. F., Aquino, V. V., Asmerom, Y., Awe, J., Baldini, J. U., & Haug, G. H. et al. (2012). Development and disintegration of Maya political systems in response to climate change. *Science*, *338*(6108), 788–791. doi:10.1126/science.1226299 PMID:23139330

Keune, H., Kretsch, C., De Blust, G., Gilbert, M., Flandroy, L., Van Den Berge, K., & Bauler, T. et al. (2013). Science–policy challenges for biodiversity, public health and urbanization: Examples from Belgium. *Environmental Research Letters*, *8*(2), 025015. doi:10.1088/1748-9326/8/2/025015

Khasnis, A. A., & Nettleman, M. D. (2005). Global warming and infectious disease. *Archives of Medical Research*, *36*(6), 689–696. doi:10.1016/j.arcmed.2005.03.041 PMID:16216650

Kilpatrick, A. M., & Altizer, S. (2010). Disease Ecology. *Nature Education Knowledge*, *3*(10), 55. PMID:19836101

Kilpatrick, A. M., & Randolph, S. E. (2012). Drivers, dynamics, and control of emerging vector-borne zoonotic diseases. *Lancet*, *380*(9857), 1946–1955. doi:10.1016/S0140-6736(12)61151-9 PMID:23200503

Kim, K. H., Kabir, E., & Ara Jahan, S. (2014). A review of the consequences of global climate change on human health. *J Environ Sci Health C Environ Carcinog Ecotoxicol Rev*, *32*(3), 299–318. doi:10.1080/10590501.2014.941279 PMID:25226222

Kortekaas, J., Kant, J., Vloet, R., Cêtre-Sossah, C., Marianneau, P., Lacote, S., & Brun, A. et al. (2013). European ring trial to evaluate ELISAs for the diagnosis of infection with Rift Valley fever virus. *Journal of Virological Methods*, *187*(1), 177–181. doi:10.1016/j.jviromet.2012.09.016 PMID:23022325

Kothavade, R. J. (2012). Potential molecular tools for assessing the public health risk associated with waterborne Cryptosporidium oocysts. *Journal of Medical Microbiology*, *61*(Pt 8), 1039–1051. doi:10.1099/jmm.0.043158-0 PMID:22628454

Kotloff, K. L., Nataro, J. P., Blackwelder, W. C., Nasrin, D., Farag, T. H., Panchalingam, S., & Levine, M. M. et al. (2013). Burden and aetiology of diarrhoeal disease in infants and young children in developing countries (the Global Enteric Multicenter Study, GEMS): A prospective, case-control study. *Lancet*, *382*(9888), 209–222. doi:10.1016/S0140-6736(13)60844-2 PMID:23680352

Kovats, R. S., Bouma, M. J., Hajat, S., Worrall, E., & Haines, A. (2003). El Niño and health. *Lancet*, *362*(9394), 1481–1489. doi:10.1016/S0140-6736(03)14695-8 PMID:14602445

Kramer, L. D., & Ebel, G. D. (2003). Dynamics of flavivirus infection in mosquitoes. In T. J. Chambers & T. P. Monath (Eds.), *Flaviviruses: Pathogenesis and Immunity* (Vol. 60, pp. 187–232). San Diego: Academic Press Inc.

Kumar, V., & Raul, D. (2014). History and evolution of surveillance in public health. *Global Journal of Medicine and Public Health*, 3(1). Retrieved from http://www.gjmedph.org/uploads/O6-Vo3No1.pdf

Lafferty, K. D. (2009). The ecology of climate change and infectious diseases. *Ecology*, *90*(4), 888–900. doi:10.1890/08-0079.1 PMID:19449681

Lam, S. K., & Chua, K. B. (2002). Nipah Virus Encephalitis Outbreak in Malaysia. *Clinical Infectious Diseases*, *34*(Suppl. 2), S48–S51. doi:10.1086/338818 PMID:11938496

Landin, R., & Giatti, L. L. (2014). Política de mudança do clima no município de São Paulo, Brasil: Reflexividade e permeabilidade do Setor Saúde. *Ciencia & Saude Coletiva*, *19*(10), 4149–4156. doi:10.1590/1413-812320141910.08972014 PMID:25272124

Lape, P. V., & Chin-Yung, C. (2008). Fortification as a human response to late Holocene climate change in East Timor. *Archaeology in Oceania, 42*(4), 11-21.

Laurance, W. F., Camargo, J. L., Luizão, R. C., Laurance, S. G., Pimm, S. L., Bruna, E. M., & Vasconcelos, H. L. et al. (2011). The fate of Amazonian forest fragments: A 32-year investigation. *Biological Conservation*, *144*(1), 56–67. doi:10.1016/j.biocon.2010.09.021

Laurance, W. F., Clements, G. R., Sloan, S., O'Connell, C. S., Mueller, N. D., Goosem, M., & Arrea, I. B. et al. (2014). A global strategy for road building. *Nature*, *513*(7517), 229–232. doi:10.1038/nature13717 PMID:25162528

Leclerc, H., Mossel, D. A. A., Edberg, S. C., & Struijk, C. B. (2001). Advances in the bacteriology of the Coliform group: Their suitability as Markers of microbial water safety. *Annual Review of Microbiology*, *55*(1), 201–234. doi:10.1146/annurev.micro.55.1.201 PMID:11544354

Leclerc, H., Schwartzbrod, L., & Dei-Cas, E. (2002). Microbial agents associated with waterborne diseases. *Critical Reviews in Microbiology*, *28*(4), 371–409. doi:10.1080/1040-840291046768 PMID:12546197

Lederberg, J. (2000). Infectious history. *Science*, *288*(5464), 287–293. doi:10.1126/science.288.5464.287 PMID:10777411

Lee, B., Preston, F., Kooroshy, J., & Bailey, R. (2012). *G., L.* Resources Futures.

Leroy, E. M., Epelboin, A., Mondonge, V., Pourrut, X., Gonzalez, J. P., Muyembe-Tamfum, J. J., & Formenty, P. (2009). Human Ebola outbreak resulting from direct exposure to fruit bats in Luebo, Democratic Republic of Congo, 2007. *Vector Borne and Zoonotic Diseases (Larchmont, N.Y.)*, *9*(6), 723–728. doi:10.1089/vbz.2008.0167 PMID:19323614

Leroy, E. M., Kumulungui, B., Pourrut, X., Rouquet, P., Hassanin, A., Yaba, P., & Swanepoel, R. et al. (2005). Fruit bats as reservoirs of Ebola virus. *Nature*, *438*(7068), 575–576. doi:10.1038/438575a PMID:16319873

Leroy, E. M., Rouquet, P., Formenty, P., Souquiere, S., Kilbourne, A., Froment, J. M., & Rollin, P. E. et al. (2004). Multiple Ebola virus transmission events and rapid decline of central African wildlife. *Science*, *303*(5656), 387–390. doi:10.1126/science.1092528 PMID:14726594

Lesnikowski, A. C., Ford, J. D., Berrang-Ford, L., Barrera, M., Berry, P., Henderson, J., & Heymann, S. J. (2013). National-level factors affecting planned, public adaptation to health impacts of climate change. *Global Environmental Change*, *23*(5), 1153–1163. doi:10.1016/j.gloenvcha.2013.04.008

Levinson, J., Bogich, T. L., Olival, K. J., Epstein, J. H., Johnson, C. K., Karesh, W., & Daszak, P. (2013). Targeting surveillance for zoonotic virus discovery. *Emerging Infectious Diseases*, *19*(5), 743–747. doi:10.3201/eid1905.121042 PMID:23647732

Lieshout, V. M., Kovats, R. S., Livermore, M. T. J., & Martens, P. (2004). Climate change and malaria: Analysis of the SRES climate and socio-economic scenarios. *Global Environmental Change*, *14*(1), 87–99. doi:10.1016/j.gloenvcha.2003.10.009

Lima, G. F. D. C., & Layrargues, P. P. (2014). Climate change, education and environment: beyond Dynamic Conservatism. *Educar em Revista*, (SPE3), 73-88.

Lindahl, J. F., & Grace, D. (2015). The consequences of human actions on risks for infectious diseases: a review. *Infection Ecology & Epidemiology, 5*.

Linthicum, J. J., Davies, F. G., Bailey, C. L., & Kairo, A. (1984). Mosquito species encountered in a flooded grassland dambo in Kenya. *Mosquito News*, *44*(2), 228–232.

Linthicum, K. J., Anyamba, A., Tucker, C. J., Kelley, P. W., Myers, M. F., & Peters, C. J. (1999). Climate and satellite indicators to forecast Rift Valley fever epidemics in Kenya. *Science*, *285*(5426), 397–400. http://www.ncbi.nlm.nih.gov/pubmed/10411500 doi:10.1126/science.285.5426.397 PMID:10411500

Linthicum, K. J., Davies, F. G., Kairo, A., & Bailey, C. L. (1985). Rift Valley fever virus (family Bunyaviridae, genus Phlebovirus). Isolations from Diptera collected during an inter-epizootic period in Kenya. *The Journal of Hygiene*, *95*(1), 197–209. doi:10.1017/S0022172400062434 PMID:2862206

Lipp, E. K., Huq, A., & Colwell, R. R. (2002). Effects of global climate on infectious disease: the cholera model. *Clin Microbiol Rev*, *15*(4), 757-+. doi: 10.1128/Cmr.15.4.757-770.2002

Liu, Q., Cao, L., & Zhu, X. Q. (2014). Major emerging and re-emerging zoonoses in China: A matter of global health and socioeconomic development for 1.3 billion. *International Journal of Infectious Diseases*, *25*, 65–72. doi:10.1016/j.ijid.2014.04.003 PMID:24858904

Logan, T. M., Linthicum, K. J., Thande, P. C., Wagateh, J. N., Nelson, G. O., & Roberts, C. R. (1991). Egg hatching of Aedes mosquitoes during successive floodings in a Rift Valley fever endemic area in Kenya. *Journal of the American Mosquito Control Association*, *7*(1), 109–112. PMID:2045800

Loh, E. H., Olival, K. J., Zambrana-Torellio, C., Bogich, T. L., Johnson, C. K., Mazet, J. A. K., & Daszak, P. et al. (2015). Targeting transmission pathways for emerging zoonotic disease surveillance and control. *Vector Borne and Zoonotic Diseases (Larchmont, N.Y.)*, *15*(7), 432–437. doi:10.1089/vbz.2013.1563 PMID:26186515

Lopes, O. D. S., Coimbra, T. L., Sacchetta, L. D. A., & Calisher, C. H. (1978). Emergence of a new arbovirus disease in Brazil. I. Isolation and characterization of the etiologic agent, Rocio virus. *American Journal of Epidemiology*, *107*(5), 444–449. PMID:665659

Lovell, S. A. (2011). Health governance and the impact of climate change on Pacific small island developing states. *IHDP Update Issue*, *1*, 50–55.

Luby, S. P., Gurley, E. S., & Hossain, M. J. (2009). Transmission of human infection with Nipah virus. *Clinical Infectious Diseases*, *49*(11), 1743–1748. doi:10.1086/647951 PMID:19886791

Machalaba, C. C., Daszak, P., Karesh, W. B., & Shrivastava, P. (2015b). Future Earth and Eco-Health: A New Paradigm Toward Global Sustainability and Health. *EcoHealth*, *12*(4), 553–554. doi:10.1007/s10393-015-1076-6 PMID:26631383

Machalaba, C. C., Elwood, S. E., Forcella, S., Smith, K. M., Hamilton, K., Jebara, K. B., & Karesh, W. B. et al. (2015c). Global avian influenza surveillance in wild birds: A strategy to capture viral diversity. *Emerging Infectious Diseases*, *21*(4), e1–e7. doi:10.3201/eid2104.141415 PMID:25811221

Machalaba, C. C., & Karesh, W. B. (2015). Envisioning a World without Emerging Disease Outbreaks. *Solutions*, *6*(2), 63–71.

Machalaba, C., Romanelli, C., Stoett, P., Baum, S. E., Bouley, T. A., Daszak, P., & Karesh, W. B. (2015a). Climate Change and Health: Transcending Silos to Find Solutions. *Ann Glob Health*, *81*(3), 445–458. doi:10.1016/j.aogh.2015.08.002 PMID:26615080

Mackay, I. M., & Arden, K. E. (2015). Ebola virus in the semen of convalescent men. *The Lancet Infectious Diseases*, *15*(2), 149–150. doi:10.1016/S1473-3099(14)71033-3 PMID:25467652

Mackenzie, W. R., Hoxie, N. J., Proctor, M. E., Gradus, M. S., Blair, K. A., Peterson, D. E., & Davis, J. P. et al. (1994). A Massive Outbreak in Milwaukee of Cryptosporidium Infection Transmitted through the Public Water-Supply. *The New England Journal of Medicine*, *331*(3), 161–167. doi:10.1056/NEJM199407213310304 PMID:7818640

Maco, V., Maguiña, C., Tirado, A., Maco, C., & Vidal, J. E. (2004). Carrion's disease (*Bartonellosis bacilliformis*) confirmed by histopathology in the High Forest of Peru. *Revista do Instituto de Medicina Tropical de Sao Paulo*, *46*(3), 171–174. doi:10.1590/S0036-46652004000300010 PMID:15286824

Magrin, G. O., Marengo, J. A., Boulanger, J. P., Buckeridge, M. S., Castellanos, E., Poveda, G., & Vicuña, S. (2014). Central and South America. In V. R. Barros, C. B. Field, D. J. Dokken, M. D. Mastrandrea, K. J. Mach, T. E. Bilir, & L. L. White et al. (Eds.), *Climate Change 2014: Impacts, Adaptation, and Vulnerability. Part B: Regional Aspects. Contribution of Working Group II to the Fifth Assessment Report of the Intergovernmental Panel on Climate Change* (pp. 1499–1566). Cambridge, UK: Cambridge University Press.

Maibach, E. W., Chadwick, A., McBride, D., Chuk, M., Ebi, K. L., & Balbus, J. (2008). Climate Change and Local Public Health in the United States: Preparedness, Programs and Perceptions of Local Public Health Department Directors. *PLoS ONE*, *3*(7), e2838. doi:10.1371/journal.pone.0002838 PMID:18665266

Maier, R. M., Pepper, I. L., & Gerba, C. P. (2009). Introduction to Environmental Microbiology. In R. M. Maier, I. L. Pepper, & C. P. Gerba (Eds.), *Environmental Microbiology* (2nd ed., pp. 3–7). Burlington, MA: Elsevier/Academic Press. [etc.] doi:10.1016/B978-0-12-370519-8.00001-8

Maigari, A. I. (2005). *Surface water Resources management strategies in the metropolitan Kano, Wa'adallah Environmental consults (WADEC)*. Kano-Nigeria.

Manganelli, M., Scardala, S., Stefanelli, M., Palazzo, F., Funari, E., Vichi, S., & Testai, E. et al. (2012). Emerging health issues of cyanobacterial blooms. *Annali dell'Istituto Superiore di Sanita*, *48*(4), 415–428. doi:10.4415/ANN_12_04_09 PMID:23247138

Mantilla, G., Oliveros, H., & Barnston, A. G. (2009). The role of ENSO in understanding changes in Colombia's annual malaria burden by region, 1960-2006. *Malaria Journal*, *8*(6), 1–11. PMID:19133152

Mari Saez, A., Weiss, S., Nowak, K., Lapeyre, V., Zimmermann, F., Dux, A., & Leendertz, F. H. et al. (2015). Investigating the zoonotic origin of the West African Ebola epidemic. *EMBO Molecular Medicine*, *7*(1), 17–23. doi:10.15252/emmm.201404792 PMID:25550396

Marineli, F., Tsoucalas, G., Karamanou, M., & Androutsos, G. (2013). Mary Mallon (1869-1938) and the history of typhoid fever. *Annals of Gastroenterology*, *26*(2), 132–134. PMID:24714738

Marlar, R. A., Leonard, B. L., Billman, B. R., Lambert, P. M., & Marlar, J. E. (2000). Biochemical evidence of cannibalism at a prehistoric Puebloan site in southwestern Colorado. *Nature*, *407*(6800), 74-78. Doi:10.1038/35024064

Marshall, B. M., & Levy, S. B. (2011). Food animals and antimicrobials: Impacts on human health. *Clinical Microbiology Reviews*, *24*(4), 718–733. doi:10.1128/CMR.00002-11 PMID:21976606

Martens, P., & Rotmans, J. (Eds.). (1999). *Climate change: an integrated perspective Springer publishers Netherlands* (pp. 1–10). doi:10.1007/0-306-47982-6_1

Martens, W. J. M., Jetten, T. H., & Focks, D. A. (1997). Sensitivity of malaria, schistosomiasis and dengue to global warming. *Climatic Change*, *35*(2), 145–156. doi:10.1023/A:1005365413932

Martens, W. J. M., Jetten, T. H., Rotmans, J., & Niessen, L. W. (1995). Climate change and vector-borne diseases: A global modeling per-spective. *Global Environmental Change*, *5*(3), 195–209. doi:10.1016/0959-3780(95)00051-O

Martin, A. (2005). Environmental conflict between refugee and host communities. *Journal of Peace Research*, *42*(3), 329–346. doi:10.1177/0022343305052015

Martin, V., Chevalier, V., Ceccato, P. N., Anyamba, A., Simone, L., De, , & Lubroth, J., … Domenech, J. (. (2008). The impact of climate change on the epidemiology and control of Rift Valley fever. *Revue Scientifique et Technique (International Office of Epizootics)*, *27*(2), 413–426. doi:10.20506/rst.27.2.1802 PMID:18819669

Mazo, J. (2010). *Climate Conflict: How global warming threatens security and what to do about it.* London: Routledge, the International Institute for Strategic Studies.

Mazza, G., Tricarico, E., Genovesi, P., & Gherardi, F. (2014). Biological Invaders are Threats to Human Health: An Overview. *Ethology Ecology and Evolution*, *26*(2-3), 112–129. doi:10.1080/03949370.2013.863225

McCreesh, N., & Booth, M. (2013). Challenges in predicting the effects of climate change on *Schistosoma mansoni* and *Schistosoma haematobium* transmission potential. *Trends in Parasitology*, *29*(11), 548–555. doi:10.1016/j.pt.2013.08.007 PMID:24064438

McFarlane, R. A., Sleigh, A. C., & McMichael, A. J. (2013). Land-Use Change and Emerging Infectious Disease on an Island Continent. *International Journal of Environmental Research and Public Health*, *10*(7), 2699–2719. doi:10.3390/ijerph10072699 PMID:23812027

McMichael, A. J., Kovats, R. S., Martens, P., Nijhof, S., Livermore, M. T., Cawthorne, A., & deVries, P. (2000). Climate change and human health: final report to the department of environment, transport and the regions. London School of Hygiene and Tropical Medicine/ICIS, London/Maastricht.

McMichael, A. J. (2013). Globalization, climate change, and human health. *The New England Journal of Medicine*, *368*(14), 1335–1343. doi:10.1056/NEJMra1109341 PMID:23550671

McMichael, A. J., & Beaglehole, R. (2000). The changing global context of public health. *Lancet*, *356*(9228), 495–499. doi:10.1016/S0140-6736(00)02564-2 PMID:10981904

McMichael, A. J., & Lindgren, E. (2011). Climate change: Present and future risks to health, and necessary responses. *Journal of Internal Medicine*, *270*(5), 401–413. doi:10.1111/j.1365-2796.2011.02415.x PMID:21682780

McMichael, A. J., Powles, J. W., Butler, C. D., & Uauy, R. (2007). Food, livestock production, energy, climate change, and health. *Lancet*, *370*(9594), 1253–1263. doi:10.1016/S0140-6736(07)61256-2 PMID:17868818

McMichael, A. J., Woodruff, R. E., & Hales, S. (2006). Climate change and human health: Present and future risks. *Lancet*, *367*(9513), 859–869. doi:10.1016/S0140-6736(06)68079-3 PMID:16530580

McMichael, C., Barnett, J., & McMichael, A. J. (2012). An ill wind? Climate change, migration, and health. *Environmental Health Perspectives*, *120*(5), 646–654. doi:10.1289/ehp.1104375 PMID:22266739

McMichael, T., Montgomery, H., & Costello, A. (2012). Health risks, present and future, from global climate change. *BMJ (Clinical Research Ed.)*, *344*, e1359. doi:10.1136/bmj.e1359 PMID:22431661

Medact. (2008). *Rehavilitation under fire: Health care in Iraq 2003-7*. London: Medact.

Medley, K. E. (1993). Primate Conservation along the Tana River, Kenya: An Examination of the Forest Habitat. *Conservation Biology*, *7*(1), 109–121. doi:10.1046/j.1523-1739.1993.07010109.x

Medley, K. E. (2009). Patterns of forest diversity along the Tana River, Kenya. *Journal of Tropical Ecology*, *8*(04), 353. doi:10.1017/S0266467400006684

Medone, P., Ceccarelli, S., Parham, P. E., Figuera, A., & Rabinovich, J. E. (2015). The impact of climate change on the geographical distribution of two vectors of Chagas disease: implications for the force of infection. *Philosophical Transactions of the Royal Society of London B: Biological Sciences*, 370(1665), 20130560.

Meentemeyer, R. K., Haas, S. E., & Vaclavik, T. (2012). Landscape epidemiology of emerging infectious diseases in natural and human-altered ecosystems. *Annual Review of Phytopathology*, *50*(1), 379–402. doi:10.1146/annurev-phyto-081211-172938 PMID:22681449

Mellor, P. S., & Leake, C. J. (2000). Climatic and geographic influences on arboviral infections and vectors. *Revue Scientifique Et Technique De L Office International Des Epizooties, 19*, 41–54.

Membrive, N. A., Rodrigues, G., Gualda, K. P., Bernal, M. V. Z., Oliveira, D. M., Lonardoni, M. V. C., & Silveira, T. G. V. et al. (2012). Environmental and animal characteristics as factors associated with American cutaneous Leishmaniasis in rural locations with presence of dogs, Brazil. *PLoS ONE, 7*(11), e47050. doi:10.1371/journal.pone.0047050 PMID:23144799

Métras, R., Collins, L. M., White, R. G., Alonso, S., Chevalier, V., Thuranira-McKeever, C., & Pfeiffer, D. U. (2011). Rift Valley fever epidemiology, surveillance, and control: What have models contributed? *Vector Borne and Zoonotic Diseases (Larchmont, N.Y.), 11*(6), 761–771. doi:10.1089/vbz.2010.0200 PMID:21548763

Metrological station Aminu Kano International Airport. (2012). Kano metropolitan climatic Data

Mikota, S. K., & Maslow, J. N. (2011). Tuberculosis at the human-animal interface: An emerging disease of elephants. *Tuberculosis (Edinburgh, Scotland), 91*(3), 208–211. doi:10.1016/j.tube.2011.02.007 PMID:21397564

Miles, M. B., & Huberman, A. M. (1994). Qualitative Data Analysis: An Expanded Sourcebook. *Sage (Atlanta, Ga.).*

Mills, J. N. (2006). Biodiversity loss and emerging infectious disease: An example from the rodent-borne hemorrhagic fevers. *Biodiversity, 7*(1), 9–17. doi:10.1080/14888386.2006.9712789

Ministry of Health and Long-term Care. (2006). Ontario Public Health Unit and County Boundaries, 2006. MOHLTC. Retrieved from http://www.apheo.ca/resources/indicators/PHU_CD%20Map.pdf

Ministry of Health and Long-term Care. (2008, 2014). The Ontario Public Health Standards, 2008, Revised May 1, 2014. MOHLTC. Retrieved from http://www.health.gov.on.ca/en/pro/programs/publichealth/oph_standards/docs/ophs_2008.pdf

Mogi, M. (1996). Overwintering strategies of mosquitoes (Diptera: Culicidae) on warmer islands may predict impact of global warming on Kyushu, Japan. *Journal of Medical Entomology, 33*(3), 438–444. doi:10.1093/jmedent/33.3.438 PMID:8667392

Mohamed, M., Mosha, F., Mghamba, J., Zaki, S. R., Shieh, W.-J., Paweska, J., & Njenga, M. K. et al. (2010). Epidemiologic and clinical aspects of a Rift Valley fever outbreak in humans in Tanzania, 2007. *The American Journal of Tropical Medicine and Hygiene, 83*(2 Suppl.), 22–27. doi:10.4269/ajtmh.2010.09-0318 PMID:20682902

Monath, T. P. (1993). Arthropod-borne viruses. In S. S. Morse (Ed.), *Emerging Viruses* (pp. 138–148). New York: Oxford University Press.

Monath, T. P., & Vasconcelos, P. F. (2015). Yellow fever. *Journal of Clinical Virology, 64*, 160–173. doi:10.1016/j.jcv.2014.08.030 PMID:25453327

Moors, E., Singh, T., Siderius, C., Balakrishnan, S., & Mishra, A. (2013). Climate change and waterborne diarrhoea in northern India: Impacts and adaptation strategies. *The Science of the Total Environment*, *468-469*(Suppl.), S139–S151. doi:10.1016/j.scitotenv.2013.07.021 PMID:23972324

Morales, M. A., Barrandeguy, M., Fabbri, C., Garcia, J. B., Vissani, A., Trono, K., & Enría, D. et al. (2006). West Nile virus isolation from equines in Argentina, 2006. *Emerging Infectious Diseases*, *12*(10), 1559–1561. doi:10.3201/eid1210.060852 PMID:17176571

Morgan, J. A. T., Dejong, R. J., Snyder, S. D., Mkoji, G. M., & Loker, E. S. (2001). *Schistosoma mansoni* and *Biomphalaria*: Past history and future trends. *Parasitology*, *123*(7 Suppl.), 211–228. doi:10.1017/S0031182001007703 PMID:11769285

Morisetti, N. (2012). Climate change and resource security. *BMJ (Clinical Research Ed.)*, March. PMID:22431658

Moroccan Ministry of Health (MMH). (1980). *Revue bibliographique des publications et travaux sur la bilharziose au Maroc* (Vol. 1 et 2).

Moroccan Ministry of Health (MMH). (2004). *Direction de l'épidémiologie et de lutte contre les maladies (1966–2004) Données épidémiologiques des maladies sous surveillance. Bulletin épidémiologique*. Maroc: Ministère de la Santé.

Moroccan Ministry of Health (MMH). (2012). La Stratégie Sectorielle "Santé" 2012-2016, http://srvweb.sante.gov.ma/Docs/Documents/secteur%20santé.pdf

Moroccan Ministry of Health (MMH). (2014). A report on progress of control programs against parasitic diseases. Directorate of Epidemiology and Disease Control, Ministry of Health, Rabat, Morocco. Retrieved from http://www.sante.gov.ma/departements/delm/index-delm.htm

Morper-Busch, L., Kienberger, S., & Hagenlocher, M. (2015). Healthy Futures Atlas: An Open-source WebGIS to Support Infectious Disease Intervention Planning in Eastern Africa. *Journal for Geographic Information Science*, 1, 460–463.

Morris, G. P. (2010). Ecological public health and climate change policy. *Perspectives in Public Health*, *130*(1), 34–40. doi:10.1177/1757913909354149 PMID:20333948

Morse, S. S., Mazet, J. A., Woolhouse, M., Parrish, C. R., Carroll, D., Karesh, W. B., & Daszak, P. et al. (2012). Prediction and prevention of the next pandemic zoonosis. *Lancet*, *380*(9857), 1956–1965. doi:10.1016/S0140-6736(12)61684-5 PMID:23200504

Moutailler, S., Krida, G., Schaffner, F., Vazeille, M., & Failloux, A.-B. (2008). Potential vectors of Rift Valley fever virus in the Mediterranean region. *Vector Borne and Zoonotic Diseases (Larchmont, N.Y.)*, *8*(6), 749–753. doi:10.1089/vbz.2008.0009 PMID:18620510

Murray, K. A., & Daszak, P. (2013). Human ecology in pathogenic landscapes: Two hypotheses on how land use change drives viral emergence. *Curr. Opin. Virol.*, *3*(1), 79–83. doi:10.1016/j.coviro.2013.01.006 PMID:23415415

Murray, K., Rogers, R., Selvey, L., Selleck, P., Hyatt, A., Gould, A., & Westbury, H. et al. (1995). A novel morbillivirus pneumonia of horses and its transmission to humans. *Emerging Infectious Diseases*, *1*(1), 31–33. doi:10.3201/eid0101.950107 PMID:8903153

Murty, U. S., Rao, M. S., & Arunachalam, N. (2010). The effects of climatic factors on the distribution and abundance of Japanese encephalitis vectors in Kurnool district of Andhra Pradesh, India. *Journal of Vector Borne Diseases*, *47*, 26–32. PMID:20231770

Musso, D., Nilles, E. J., & Cao-Lormeau, V. M. (2014). Rapid spread of emerging Zika virus in the Pacific area. *Clinical Microbiology and Infection*, *20*(10), O595–O596. doi:10.1111/1469-0691.12707 PMID:24909208

Muyembe-Tamfum, J. J., Mulangu, S., Masumu, J., Kayembe, J. M., Kemp, A., & Paweska, J. T. (2012). Ebola virus outbreaks in Africa: Past and present. *The Onderstepoort Journal of Veterinary Research*, *79*(2), 451. doi:10.4102/ojvr.v79i2.451 PMID:23327370

Myers, J. E., & Rother, H. (2013). Public health impact of and response to climate change in South Africa. In A. Padarath & R. English (Eds.), *South African Health Review 2012/13*. Durban, South Africa: Health Systems Trust.

Myers, J., Young, T., Galloway, M., Manyike, P., & Tucker, T. (2011). A public health approach to the impact of climate change on health in southern Africa: Identifying priority modifiable risks. *South African Medical Journal*, *101*, 817–819. PMID:22272963

Myers, S. S., Gaffikin, L., Golden, C. D., Ostfeld, R. S., Redford, K. H., Ricketts, T. H., & Osofsky, S. A. et al. (2013). Human health impacts of ecosystem alteration. *Proceedings of the National Academy of Sciences of the United States of America*, *110*(47), 18753–18760. doi:10.1073/pnas.1218656110 PMID:24218556

Myers, S. S., & Patz, J. A. (2009). Emerging Threats to Human Health from Global Environmental Change. *Annual Review of Environment and Resources*, *34*(1), 223–252. doi:10.1146/annurev.environ.033108.102650

Myron, S. (2008). Rudolph Virchow. *Emerging Infectious Diseases*, *14*(9), 1480–1481. doi:10.3201/eid1409.086672

Nabarro, D., & Wannous, C. (2014). The potential contribution of livestock to food and nutrition security: The application of the One Health approach in livestock policy and practice. *Revue Scientifique et Technique (International Office of Epizootics)*, *33*(2), 475–485. Retrieved from http://www.ncbi.nlm.nih.gov/pubmed/25707178 doi:10.20506/rst.33.2.2292 PMID:25707178

Nading, A. M. (2013). Humans, animals, and health: From ecology to entanglement. *Environment and Society: Advances in Research*, *4*(1), 60–78. doi:10.3167/ares.2013.040105

Naeem, S., & Wright, J. P. (2003). Disentangling biodiversity effects on ecosystem functioning: Deriving solutions to a seemingly insurmountable problem. *Ecology Letters*, *6*(6), 567–579. doi:10.1046/j.1461-0248.2003.00471.x

Nakagawa, F., Miners, A., Smith, C. J., Simmons, R., Lodwick, R. K., Cambiano, V., & Phillips, A. N. et al. (2015). Projected Lifetime Healthcare Costs Associated with HIV Infection. *PLoS ONE, 10*(4), e0125018. doi:10.1371/journal.pone.0125018 PMID:25901355

Nanyingi, M. O., Munyua, P., Kiama, S. G., Muchemi, G. M., Thumbi, S. M., Bitek, A. O., … Njenga, M. K. (2015). A systematic review of Rift Valley Fever epidemiology 1931–2014. *Infection Ecology & Epidemiology, 5.*

National Population Commission. (2007). *Federal Republic of Nigeria Official Gadget, 23*(94), 175–198.

National Roundtable for the Economy and Environment (NRTEE). (2011). Paying the price: The economic impacts of climate change for Canada. NRTEE. Retrieved from http://www.fcm.ca/Documents/reports/PCP/paying_the_price_EN.pdf

Naumova, E. N., Jagai, J. S., Matyas, B., DeMaria, A. Jr, MacNeill, I. B., & Griffiths, J. K. (2007). Seasonality in six enterically transmitted diseases and ambient temperature. *Epidemiology and Infection, 135*(2), 281–292. doi:10.1017/S0950268806006698 PMID:17291363

Negredo, A., Palacios, G., Vazquez-Moron, S., Gonzalez, F., Dopazo, H., Molero, F., & Tenorio, A. et al. (2011). Discovery of an ebolavirus-like filovirus in Europe. *PLoS Pathogens, 7*(10), e1002304. doi:10.1371/journal.ppat.1002304 PMID:22039362

Neiderud Carl-Johan. (2015). How urbanization affects the epidemiology of emerging infectious diseases. *Infection Ecology and Epidemiology*, 5. Retrieved from http://www.infectionecologyandepidemiology.net/index.php/iee/article/view/27060 doi:10.3402/iee.v5.27060

Nejjar, R., Lemrani, M., Malki, A., Ibrahimy, S., Amarouch, H., & Benslimane, A. (1998). Canine leishmaniasis due to Leishmania infantum MON-1 in northern Morocco. *Parasite (Paris, France), 5*(4), 325–330. doi:10.1051/parasite/1998054325 PMID:9879555

Nelson, K. E., & Williams, C. F. (2014). Early History of Infectious Disease. In K. E. Nelson & C. F. Williams (Eds.), *Infectious Diseases and Epidemiology* (pp. 3–21). Burlington, MA: Jones & Barlett Learning.

Nerlander, L. (2009). Climate Change and Health, Commission on Climate Change and Development. Retrieved from http://www.pacificdisaster.net/pdnadmin/data/original/CCCD_2009_CC_health.pdf

Nicholas, D. E., Jacobsen, K. H., & Waters, N. M. (2014). Risk factors associated with human Rift Valley fever infection: Systematic review and meta-analysis. *Tropical Medicine & International Health : TM & IH, 19*(12), 1420–1429. doi:10.1111/tmi.12385 PMID:25252137

Nichols, G., Lane, C., Asgari, N., Verlander, N. Q., & Charlett, A. (2009). Rainfall and outbreaks of drinking water related disease and in England and Wales. *Journal of Water and Health, 7*(1), 1–8. doi:10.2166/wh.2009.143 PMID:18957770

Nicky, M., Grigory, N., & Mark, B. (2015). Predicting the effects of climate change on *Schistosoma mansoni* transmission in eastern Africa. *Parasites & Vectors*, *8*(4). doi:10.1186/s13071-014-0617-0 PMID:25558917

NIH - National Institute of Health. (2013). *Severe acute respiratory syndrome (SARS)*. National Institute of Health, US National Library of Medicine, MedlinePlus. Retrieved from http://www.nlm.nih.gov/medlineplus/ency/article/007192.htm

No authors listed. (1890). The Relation of Ground Water to Disease. *Science,* 16(409), 312-313. doi: 10.1126/science.ns-16.409.312

No authors listed. (2008). Global networks for surveillance of rotavirus gastroenteritis, 2001-2008. *Wkly Epidemiol Rec,* 83(47), 421-425.

NRC - National Research Council. (2004). *Indicators for Waterborne Pathogens. Committee on Indicators for Waterborne Pathogens. Washington (DC)*. US: National Academies Press.

Nyokabi, S. (2015). *Biosecurity Measures in Meat and Milk Value Chains: A Study in Bura sub-county, Kenya*. Germany: Hohenheim University.

Oborn, I., Bengtsson, J., Hedenus, F., Rydhmer, L., Stenström, M., Vrede, K., & Magnusson, U. et al. (2013). Scenario development as a basis for formulating a research program on future agriculture: A methodological approach. *Ambio*, *42*(7), 823–839. doi:10.1007/s13280-013-0417-3 PMID:23836311

Odekunle, T. O. (2004). Determining rainfall onset and retreat dates in Nigeria. *Journal of Human Ecology (Delhi, India)*, *16*, 239–247.

O'Donoghue, P. J. (1995). Cryptosporidium and cryptosporidiosis in man and animals. *International Journal for Parasitology*, *25*(2), 139–195. doi:10.1016/0020-7519(94)E0059-V PMID:7622324

OECD/WHO. (2003). *Assessing Microbial Safety of Drinking Water*. OECD Publishing.

Okogun, G. R. A., Anosike, J. C., Okere, A. N., & Nwoke, B. E. B. (2005). Ecology of mosquitoes of Midwestern Nigeria. *Journal of Vector Borne Diseases*, *42*, 1–8. PMID:15999454

Okrah, J., Traore, C., Pale, A., Somerfeld, J., & Muller, O. (2002). Community factors associated with malaria prevention by mosquito nets: An exploratory study in a rural Burkina Faso. *Tropical Medicine & International Health*, *7*(3), 240–248. doi:10.1046/j.1365-3156.2002.00856.x PMID:11903986

Olival, K. J., & Hayman, D. T. (2014). Filoviruses in bats: Current knowledge and future directions. *Viruses*, *6*(4), 1759–1788. doi:10.3390/v6041759 PMID:24747773

Olive, M.-M., Goodman, S. M., & Reynes, J.-M. (2012). The role of wild mammals in the maintenance of Rift Valley fever virus. *Journal of Wildlife Diseases*, *48*(2), 241–266. doi:10.7589/0090-3558-48.2.241 PMID:22493102

Olofin, E.A. (1989) Some Aspect of Physical Geography of Kano and related human responses. Departmental lecture series N01, Department of Geography, Bayero University Kano-Nigeria.

Olson, S. H., Gangnon, R., Elguero, E., Durieux, L., Guegan, J. F., Foley, J. A., & Patz, J. A. (2009). Links between climate, malaria, and wetlands in the Amazon Basin. *Emerging Infectious Diseases*, *15*(4), 659–662. doi:10.3201/eid1504.080822 PMID:19331766

Onda, K., LoBuglio, J., & Bartram, J. (2012). Global access to safe water: Accounting for water quality and the resulting impact on MDG progress. *International Journal of Environmental Research and Public Health*, *9*(3), 880–894. doi:10.3390/ijerph9030880 PMID:22690170

O'Neil, J. M., Davis, T. W., Burford, M. A., & Gobler, C. J. (2012). The rise of harmful cyanobacteria blooms: The potential roles of eutrophication and climate change. *Harmful Algae*, *14*, 313–334. doi:10.1016/j.hal.2011.10.027

ONERC. (2007). *Changements climatiques et risques sanitaires en France*. Rapport de l'Observatoire national sur les effets du réchauffement climatique, Retrieved from http://www.developpement-durable.gouv.fr/IMG/ecologie/pdf/Rapport_ONERC_version_site_27-09-07_-_1.67Mo.pdf

Ontario Public Health Association (OPHA). (2011). Public Health and Planning 101: An Education Module for Public Health and Planning Professionals in Ontario, Survey for Public Health Professionals – Highlights. OPHA. Retrieved from http://opha.on.ca/getmedia/ef89de0c-26ed-4617-b15c-28ba8360d3ac/Summary-of-survey-results-for-public-health-professionals.aspx

Onwuemele, A. (2014). An assessment of the spatial pattern of malaria infection in Nigeria. *International Journal of Medicine and Medical Sciences*, 6(2), 80-86. Retrieved from http://www.academicjournals.org/IJMMS

Oosterbroek, B., de Kraker, J., Huynen, M. M., & Martens, P. (2016). Assessing ecosystem impacts on health: A tool review. *Ecosystem Services*, *17*, 237–254. doi:10.1016/j.ecoser.2015.12.008

Opara, K. N., Usip, L. P., & Akpabio, E. E. (2008). Transmission dynamics of *Simulium damnosum* in rural communities of Akwa Ibom State, Nigeria. *Journal of Vector Borne Diseases*, *45*, 225–230. PMID:18807379

O'Shea, T. J., Cryan, P. M., Cunningham, A. A., Fooks, A. R., Hayman, D. T., Luis, A. D., & Wood, J. L. et al. (2014). Bat flight and zoonotic viruses. *Emerging Infectious Diseases*, *20*(5), 741–745. doi:10.3201/eid2005.130539 PMID:24750692

Ostfeld, R.S. and Brunner, J.L. (2015). Climate change and *Ixodes* tick-borne diseases of humans. *Philosophical Transactions of the Royal Society B*, 370(1665), online ahead of print.

Ostfeld, R. S. (2009). Climate change and the distribution and intensity of infectious diseases. *Ecology*, *90*(4), 903–905. doi:10.1890/08-0659.1 PMID:19449683

Ostry, A., Ogborn, M., Bassil, K. L., Takaro, T. K., & Allen, D. M. (2010). Climate Change and Health in British Columbia: Projected Impacts and a Proposed Agenda for Adaptation Research and Policy. *International Journal of Environmental Research and Public Health*, *7*(3), 1018–1035. doi:10.3390/ijerph7031018 PMID:20617016

Oxford, J. B., Sefton, A., Jackson, R., Innes, W., & Johnson, N. P. A. S. (2002). World War I may have allowed the emergence of "Spanish" influenza. *The Lancet Infectious Diseases*, *2*(2), 111–114. doi:10.1016/S1473-3099(02)00185-8 PMID:11901642

PAHO. (2015, 24 June). Epidemiological Update – *Cholera.* Pan American Health Organization (PAHO)/World Health Organization (WHO). Retrieved from http://www.paho.org/hq/index.php?option=com_docman&task=doc_view&gid=30752+&Itemid=999999&lang=en

Palacios, G., Lowenstine, L. J., Cranfield, M. R., Gilardi, K. V., Spelman, L., Lukasik-Braum, M., & Lipkin, W. I. et al. (2011). Human metapneumovirus infection in wild mountain gorillas, Rwanda. *Emerging Infectious Diseases*, *17*(4), 711–713. doi:10.3201/eid1704.100883 PMID:21470468

Pan American Health Organization. (2014). Health Situation in the Americas: Basic Indicators 2014. Retrieved from http://www.paho.org/chi/index.php?option=com_content&view=article&id=578:situacion-de-salud-en-las-americas-indicadores-basicos-de-salud-2014&catid=681:publicaciones-y-otros-recursos

Panagiotakopulu, E. (2004). Pharaonic Egypt and the origins of plague. *Journal of Biogeography*, *31*(2), 269–275. doi:10.1046/j.0305-0270.2003.01009.x

Panic, M., & Ford, J. D. (2013). A Review of National-Level Adaptation Planning with Regards to the Risks Posed by Climate Change on Infectious Diseases in 14 OECD Nations. *International Journal of Environmental Research and Public Health*, *10*(12), 7083–7109. doi:10.3390/ijerph10127083 PMID:24351735

Parenti, C. (2011). *Tropic of Chaos: Climate change and the new geography of violence*. New York: Nation Books.

Parham, P. E., Waldock, J., Christophides, G. K., Hemming, D., Agusto, F., ... (2015). Climate, environmental and socio-economic change: Weighing up the balance in vector-borne disease transmission. *Philosophical Transactions B, 370*(1665).

Parker, G. (2013). *Global Crisis: War, climate change and catastrophe in the seventeenth century*. New Haven: Yale University Press.

Parry, C. M. (2005). Epidemiological and clinical aspects of human typhoid fever. In P. Mastroeni & D. Maskell (Eds.), *Salmonella Infections: Clinical, Immunological and Molecular Aspects* (pp. 1–24). Cambridge, New York: Cambridge University Press.

Pascual, M., Ahumada, J. A., Chaves, L. F., Rodó, X., & Bouma, M. (2006). Malaria resurgence in the East African highlands: temperature trends revisited. *Proceedings of the National Academy of Sciences of the United States of America*, *103*(15), 5829–5834. doi:10.1073/pnas.0508929103 PMID:16571662

Pascual, M., Bouma, M. J., & Dobson, A. P. (2002). Cholera and climate: Revisiting the quantitative evidence. *Microbes and Infection*, *4*(2), 237–245. doi:10.1016/S1286-4579(01)01533-7 PMID:11880057

Paterson, J. A., Ford, J. D., Ford, L. B., Lesnikowski, A., Berry, P., Henderson, J., & Heymann, J. (2012). Adaptation to climate change in the Ontario public health sector. *BMC Public Health*, *12*(1), 452. doi:10.1186/1471-2458-12-452 PMID:22712716

Paton, N. I., Leo, Y. S., Zaki, S. R., Auchus, A. P., Lee, K. E., Ling, A. E., & Ksiazek, T. G. (1999). Outbreak of Nipah-virus infection among abattoir workers in Singapore. *Lancet*, *354*(9186), 1253–1256. doi:10.1016/S0140-6736(99)04379-2 PMID:10520634

Patton, M. Q. (2011). *Developmental Evaluation: Applying Complexity Concepts to Enhance Innovation and Use*. New York: Guilford Press.

Patz, J. (2005). Satellite remote sensing can improve chances of achieving sustainable health. *Environmental Health Perspectives*, *113*(2), A84–A85. Retrieved from http://www.ncbi.nlm.nih.gov/pubmed/15687039 doi:10.1289/ehp.113-a84 PMID:15687039

Patz, J. A., Campbell-Lendrum, D., Holloway, T., & Foley, J. A. (2005). Impact of regional climate change on human health. *Nature*, *438*(7066), 310–317. doi:10.1038/nature04188 PMID:16292302

Patz, J. A., Daszak, P., Tabor, G. M., Aguirre, A. A., Pearl, M., & Epstein, J. et al. (2004). Unhealthy landscapes: Policy recommendations on land use change and infectious disease emergence. *Environmental Health Perspectives*, *112*(10), 1092–1098. Retrieved from http://www.ncbi.nlm.nih.gov/pubmed/15238283 doi:10.1289/ehp.6877 PMID:15238283

Patz, J. A., Grabow, M. L., & Limaye, V. S. (2014). When It Rains, It Pours: Future Climate Extremes and Health. *Annals of Global Health*, *80*(4), 332–344. doi:10.1016/j.aogh.2014.09.007 PMID:25459335

Patz, J. A., Vavrus, S. J., Uejio, C. K., & McLellan, S. L. (2008). Climate change and waterborne disease risk in the Great Lakes region of the U.S. *American Journal of Preventive Medicine*, *35*(5), 451–458. doi:10.1016/j.amepre.2008.08.026 PMID:18929971

Patz, J., Campbell-Lendrum, D., Gibbs, H., & Woodruff, R. (2008). Health impact assessment of global climate change: Expanding on comparative risk assessment approaches for policy making. *Annual Review of Public Health*, *29*(1), 27–39. doi:10.1146/annurev.publhealth.29.020907.090750 PMID:18173382

Patz, J., & Kovats, R. (2002). Hotspots in climate change and human health. *British Medical Journal*, November, 325. PMID:12424173

Pautasso, M., Döring, T. F., Garbelotto, M., Pellis, L., & Jeger, M. J. (2012). Impacts of climate change on plant diseases—opinions and trends. *European Journal of Plant Pathology*, *133*(1), 295–313. doi:10.1007/s10658-012-9936-1

Paz, S. (2015). Climate change impacts on West Nile virus transmission in a global context. *Philosophical Transactions of the Royal Society B*, 370(1665).

Paz, S. (2006). The West Nile Virus outbreak in Israel (2000) from a new perspective: The regional impact of climate change. *International Journal of Environmental Health Research, 16*(1), 1–13. doi:10.1080/09603120500392400 PMID:16507476

Pelczar, M. J., Chan, E. C. S., & Krieg, N. R. (2010). *Microbiology: An Application Based Approach. New Dehli, India.* Tata: Mac Graw Hill.

Peperzak, L. (2005). Future increase in harmful algal blooms in the North Sea due to climate change. *Water Science and Technology, 51*(5), 31–36. PMID:15918356

Pepin, M., Bouloy, M., Bird, B. H., Kemp, A., & Paweska, J. (2010). Rift Valley fever virus (Bunyaviridae: Phlebovirus): an update on pathogenesis, molecular epidemiology, vectors, diagnostics and prevention. *Veterinary Research, 41*(6), 61. doi:10.1051/vetres/2010033 PMID:21188836

Pereira, C. A. R., & de Lima Barata, M. M. (2014). Organização dos serviços urbanos de saúde frente à mudança do clima e ao risco de desastres na América Latina. *Saúde debate*, 38(102), 624-634.

Perotta, K. (2011). Public Health and Land Use Planning: How Ten Public Health Units Are Working to Create Healthy and Sustainable Communities. Clean Air Partnership. Retrieved from http://www.cleanairpartnership.org/files/CAP%20PHLUP%20Background%20Report%20April%202011.pdf

Perry, B. D., Grace, D., & Sones, K. (2013). Current drivers and future directions of global livestock disease dynamics. *Proceedings of the National Academy of Sciences of the United States of America, 110*(52), 20871–20877. doi:10.1073/pnas.1012953108 PMID:21576468

Pesapane, R., Ponder, M., & Alexander, K. A. (2013). Tracking pathogen transmission at the human-wildlife interface: Banded mongoose and Escherichia coli. *EcoHealth, 10*(2), 115–128. doi:10.1007/s10393-013-0838-2 PMID:23612855

Peterson, A. T. (2006). Ecologic niche modeling and spatial patterns of disease transmission. *Emerging Infectious Diseases, 12*(12), 1822–1826. doi:10.3201/eid1212.060373 PMID:17326931

Peterson, A. T. (2009). Shifting suitability for malaria vectors across Africa with warming climates. *BMC Infectious Diseases, 9*(1), 59–64. doi:10.1186/1471-2334-9-59 PMID:19426558

Pherez, F. M. (2007). Factors affecting the emergence and prevalence of vector borne infections (VBI) and the role of vertical transmission (VT). *Journal of Vector Borne Diseases, 44*, 157–163. PMID:17896618

Phua, K. L. (2015). Meeting the Challenge of Ebola Virus Disease in a Holistic Manner by Taking into Account Socioeconomic and Cultural Factors. *Infectious Diseases, 8*, 39–44. PMID:26604778

Pigott, D. M., Golding, N., Mylne, A., Huang, Z., Henry, A. J., Weiss, D. J., & Hay, S. I. et al. (2014). Mapping the zoonotic niche of Ebola virus disease in Africa. *eLife, 3*, e04395. doi:10.7554/eLife.04395 PMID:25201877

Pike, J., Bogich, T., Elwood, S., Finnoff, D. C., & Daszak, P. (2014). Economic optimization of a global strategy to address the pandemic threat. *Proceedings of the National Academy of Sciences of the United States of America, 111*(52), 18519–18523. doi:10.1073/pnas.1412661112 PMID:25512538

Pinheiro, F. P., Travassos Da Rosa, A. P. A., & Vasconcelos, P. F. C. (1998). An overview of Oropouche fever epidemics in Brazil and neighbour countries. In A. P. A. Travassos da Rosa, P. F. C. Vasconcelos, & J. F. S. Travassos da Rosa (Eds.), *An overview of arbovirology in Brazil and neighbouring countries* (pp. 186–192). Belém, Pará: Instituto Evandro Chagas.

Pinto, C. S., Confalonieri, U. E., & Mascarenhas, B. M. (2009). Ecology of Haemagogus sp. and Sabethes sp.(Diptera: Culicidae) in relation to the microclimates of the Caxiuanã National Forest, Pará, Brazil. *Memorias do Instituto Oswaldo Cruz, 104*(4), 592–598. doi:10.1590/S0074-02762009000400010 PMID:19722082

Pinzon, J. E., Wilson, J. M., Tucker, C. J., Arthur, R., Jahrling, P. B., & Formenty, P. (2004). Trigger events: Enviroclimatic coupling of Ebola hemorrhagic fever outbreaks. *The American Journal of Tropical Medicine and Hygiene, 71*(5), 664–674. Retrieved from http://www.ncbi.nlm.nih.gov/pubmed/15569802 PMID:15569802

Plotkinoff, R. C., Wright, M.-F., & Karunamuni, N. (2004). Knowledge, attitudes and behaviours related to climate change in Alberta, Canada: Implications for public health policy and practice. *International Journal of Environmental Health Research, 14*(3), 223–229. doi:10.1080/0960312042000218633 PMID:15203453

Poland, B. D. (1995). Transcription Quality as an Aspect of Rigor in Qualitative Research. *Qualitative Inquiry, 1*(3), 290–310. doi:10.1177/107780049500100302

Poland, B., Gastaldo, D., Pancham, A., & Ferrence, R. (2009). The interpersonal management of environmental tobacco smoke in the home–a qualitative study. *Critical Public Health, 19*(2), 203–221. doi:10.1080/09581590802395416

Pongsiri, M. J., Roman, J., Ezenwa, V. O., Goldberg, T. L., Koren, H. S., Newbold, S. C., & Salkeld, D. J. et al. (2009). Biodiversity loss affects global disease ecology. *Bioscience, 59*(11), 945–954. doi:10.1525/bio.2009.59.11.6

Portier, C. J., Thigpen Tart, K., Carter, S. R., Dilworth, C. H., Grambsch, A. E., Gohlke, J., . . . Whung, P.-Y. (2010). A Human Health Perspective On Climate Change: A Report Outlining the Research Needs on the Human Health Effects of Climate Change. Research Triangle Park, NC: Environmental Health Perspectives/National Institute of Environmental Health Sciences.; Retrieved from https://www.niehs.nih.gov/health/materials/a_human_health_perspective_on_climate_change_full_report_508.pdf

Porto, S. S. (2012). Saúde em desastres: a experiência de Nova Friburgo. *Nova Friburgo: Cap-Net Brasil*, 30.

Pratlong, F., Rioux, J. A., Dereure, J., Mahjour, J., Gallego, M., Guilvard, E., & Saddiki, A. et al. (1991). *Leishmania tropica* au Maroc. IV. Diversité isozymique intrafocale. *Annales de Parasitologie Humaine et Comparee*, *66*, 100–104. PMID:1776781

PREDICT Consortium. (2014, December). *Reducing Pandemic Risk, Promoting Global Health.*

Press, A., Bergin, A., & Gansey, E. (2013). *Heavy Weather: Climate and the Australian Defence Force*. Barton, Australia: Australian Strategic Policy Institute.

Raghunath, D. (2004) Malaria vaccine: Are we anywhere close? *Journal of postgraduate medicine*, 50(1).

Ramírez, I. J. (2015). Cholera resurgence in Piura, Peru: Examining climate associations during the 1997–1998 El Niño. *GeoJournal*, *80*(1), 129–143.

Ramírez, I. J., Grady, S. C., & Glantz, M. H. (2013). Reexamining El Niño and Cholera in Peru: A climate affairs approach. *Weather, Climate, and Society*, *5*(2), 148–161. doi:10.1175/WCAS-D-12-00032.1

Randolph, S. E., & Dobson, A. D. (2012). Pangloss revisited: A critique of the dilution effect and the biodiversity-buffers-disease paradigm. *Parasitology*, *139*(7), 847–863. doi:10.1017/S0031182012000200 PMID:22336330

Reiter, P. (2001). Climate change and mosquito-borne disease. *Environmental Health Perspectives*, *109*(S1), 141–161. doi:10.1289/ehp.01109s1141 PMID:11250812

Reuveny, R. (2007). Climate change-induced migration and violent conflict. *Political Geography, 26*(6), 656-673. DOI:10.1016/j.polgeo.2007.05.001

Reuveny, R. (2008). Ecomigration and violent conflict: Case studies and public policy implications. *Human Ecology*, *36*(1), 1–13. doi:10.1007/s10745-007-9142-5

Reuveny, R., & Allen, A. P. (2007). On environmental refugees and implications for the future. *Ecología Política*, *33*, 21–36.

Richardson, J. (2012). Deadly African Ebola Virus Linked to Bushmeat. *Food Safety News*, September 17.

Rich, K. M., & Wanyoike, F. (2010). An assessment of the regional and national socio-economic impacts of the 2007 Rift Valley fever outbreak in Kenya. *The American Journal of Tropical Medicine and Hygiene*, *83*(2 Suppl.), 52–57. doi:10.4269/ajtmh.2010.09-0291 PMID:20682906

Rioux, J. A., Lanotte, G., Petter, F., Dereure, J., Akalay, O., Pratlong, F., & Velez, G. … Serres, E. 1986. Les lcishmanioses cutanées du bassin méditerranéen occidental: de l'identification enzymatique à l'analyse éco-épidemiologique, l'exemple de trois 'foyers', tunisien, marocain et français. In Leishmania. Taxonomie et Phylogenése. Applications Écoépidémiologiques, ed. Rioux, J.A. Montpellier, France: Institut Méditerranéen d'Études Épidémiologiques et Écologiques, 365-395.

Rioux, J. A., Petter, F., Akalay, O., Lanotte, G., Ouazani, A., Seguignes, M., & Mohcine, A. (1982). *Meriones shawi* (Duvernoy, 1842) (Rodentia, Gerbillidae), réservoir de *Leishmania major* Yakimoff et Shokhor, 1914 dans le Sud Marocain. *Comptes Rendus de l'Académie des Sciences*, *294*, 515–517.

Robson, B. J., & Hamilton, D. P. (2003). Summer flow event induces a cyanobacterial bloom in a seasonal Western Australian estuary. *Marine & Freshwater Research*, *54*(2), 139–151. doi:10.1071/MF02090

Rønnfeldt, C. F. (1997). Three Generations of Environment and Security Research. *Journal of Peace Research*, *34*(4), 473–482. doi:10.1177/0022343397034004009

Rose, J. B., Epstein, P. R., Lipp, E. K., Sherman, B. H., Bernard, S. M., & Patz, J. A. (2001). Climate variability and change in the United States: Potential impacts on water- and foodborne diseases caused by microbiologic agents. *Environmental Health Perspectives*, *109*, 211–221. doi:10.2307/3435011 PMID:11359688

Rosen, G. E., & Smith, K. F. (2010). Summarizing the evidence on the international trade in illegal wildlife. *EcoHealth*, *7*(1), 24–32. doi:10.1007/s10393-010-0317-y PMID:20524140

Rosen, W. (2007). *Justinian's Flea: The first great plague and the end of Roman Empire*. London, England: Penguin Books.

Roth, A., Mercier, A., Lepers, C., Hoy, D., Duituturaga, S., Benyon, E., & Souares, Y. (2014). Concurrent outbreaks of dengue, chikungunya and Zika virus infections-an unprecedented epidemic wave of mosquito-borne viruses in the Pacific 2012-2014. *Eurosurveillance*, *19*(41), 20929. doi:10.2807/1560-7917.ES2014.19.41.20929 PMID:25345518

Rotureau, B. (2006). Ecology of the Leishmania species in the Guianan ecoregion complex. *The American Journal of Tropical Medicine and Hygiene*, *74*(1), 81–96. PMID:16407350

Rouquet, P., Froment, J. M., Bermejo, M., Kilbourn, A., Karesh, W., Reed, P., & Leroy, E. M. et al. (2005). Wild animal mortality monitoring and human Ebola outbreaks, Gabon and Republic of Congo, 2001-2003. *Emerging Infectious Diseases*, *11*(2), 283–290. doi:10.3201/eid1102.040533 PMID:15752448

Royaume du Maroc. (2010). *Seconde communication nationale à la CCNUCC*. Secrétariat de l'État Chargé de l'Eau et de l'Environnement.

Rubio, A. V., Avila-Flores, R., & Suzan, G. (2014). Responses of small mammals to habitat fragmentation: Epidemiological considerations for rodent-borne hantaviruses in the Americas. *EcoHealth*, *11*(4), 526–533. doi:10.1007/s10393-014-0944-9 PMID:24845575

Russell, R. C. (1998). Mosquito-borne arboviruses in Australia: The current scene and implications of climate change for human health. *International Journal for Parasitology*, *28*(6), 955–969. doi:10.1016/S0020-7519(98)00053-8 PMID:9673874

Sachs, J. (2001). *Macroeconomics and health: Investing in health for economic development. Report of the Commission on Macro Economics and Health*. Geneva: World Health Organization.

Saker, L., Lee, K., Cannito, B., Gilmore, A., & Campbell-Lendrum, D. H. (2004). *Globalization and infectious diseases: a review of the linkages* Retrieved from http://apps.who.int/iris/handle/10665/68726

Salehyan, I. (2008). From climate change to conflict? No consensus yet. *Journal of Peace Research, 45*(3), 315–326. doi:10.1177/0022343308088812

Sall, A., Zanotto, P. de A., Vialat, P., Sene, O., Bouloy, M., Sall, A. A., & Sene, O. K. et al. (1596–1597). M. B. (1998). Origin of 1997-98 Rift Valley fever outbreak in East Africa. *Lancet, 352*(9140).

Sandifer, P. A., & Sutton-Grier, A. E. (2014). Connecting stressors, ocean ecosystem services, and human health. *Natural Resources Forum, 38*(3), 157–167. doi:10.1111/1477-8947.12047

Sang, R., Kioko, E., Lutomiah, J., Warigia, M., Ochieng, C., O'Guinn, M., & Richardson, J. et al. (2010). Rift Valley fever virus epidemic in Kenya, 2006/2007: The entomologic investigations. *The American Journal of Tropical Medicine and Hygiene, 83*(2 Suppl.), 28–37. doi:10.4269/ajtmh.2010.09-0319 PMID:20682903

Satcher, D. (1995). Emerging infections: Getting ahead of the curve. *Perpectives, 1*(1), 1–6. PMID:8903147

Schmid, B. V., Büntgen, U., Easterday, W. R., Ginzler, C., Walløe, L., Bramanti, B., & Stenseth, N. C. (2015). Climate-drive introduction of the Black Death and successive plague reintroductions into Europe. *Proceedings of the National Academy of Sciences of the United States of America, 112*(10), 3020–3025. doi:10.1073/pnas.1412887112 PMID:25713390

Schulte-Herbruggen, B., Cowlishaw, G., Homewood, K., & Rowcliffe, J. M. (2013). The importance of bushmeat in the livelihoods of West African cash-crop farmers living in a faunally-depleted landscape. *PLoS ONE, 8*(8), e72807. doi:10.1371/journal.pone.0072807 PMID:23977355

Schuster, C. J., Ellis, A. G., Robertson, W. J., Charron, D. F., Aramini, J. J., Marshall, B. J., & Medeiros, D. T. (2005). Infectious disease outbreaks related to drinking water in Canada, 1974-2001. *Canadian Journal of Public Health, 96*(4), 254–258. PMID:16625790

Schwartlander, B., Stover, J., Hallett, T., Atun, R., Avila, C., Gouws, E., & Padian, N. et al. (2011). Towards an improved investment approach for an effective response to HIV/AIDS. *Lancet, 377*(9782), 2031–2041. doi:10.1016/S0140-6736(11)60702-2 PMID:21641026

Seas, C., Miranda, J., Gil, A. I., Leon-Barua, R., Patz, J., Huq, A., & Sack, R. B. et al. (2000). New insights on the emergence of cholera in Latin America during 1991: The Peruvian experience. *The American Journal of Tropical Medicine and Hygiene, 62*(4), 513–517. PMID:11220769

Secretariat of the Convention on Biological Diversity. (2014). *An Updated Synthesis of the Impacts of Ocean Acidification on Marine Biodiversity.*

Seguin, J. (Ed.). (2008). Human health in a changing climate: A Canadian assessment of vulnerabilities and adaptive capacity. Minister of Health. Retrieved from http://publications.gc.ca/collections/collection_2008/hc-sc/H128-1-08-528E.pdf

Seifman, R., Kornblet, S., Standley, C., Sorrell, E., Fischer, J., & Katz, R. (2015). Think big, World Bank: Time for a public health safeguard. *Lancet Glob Health*, *3*(4), e186–e187. doi:10.1016/S2214-109X(15)70012-4 PMID:25678307

Sejvar, J. J. (2003). West nile virus: an historical overview. *Ochsner J*, *5*(3), 6-10. Retrieved from http://www.ncbi.nlm.nih.gov/pubmed/21765761

Semenza, J. C., & Menne, B. (2009). Climate change and infectious diseases in Europe. *The Lancet Infectious Diseases*, *9*(6), 365–375. doi:10.1016/S1473-3099(09)70104-5 PMID:19467476

Shin, Y. S., & Ha, J. (2012). Policy Directions Addressing the Public Health Impact of Climate Change in South Korea: The Climate-change Health Adaptation and Mitigation Program. *Environmental Health and Toxicology, 27*.

Shoemaker, T., Boulianne, C., Vincent, M. J., Pezzanite, L., Al-Qahtani, M. M., Al-Mazrou, Y., & Ksiazek, T. G. et al. (2002). Genetic analysis of viruses associated with emergence of Rift Valley fever in Saudi Arabia and Yemen, 2000-01. *Emerging Infectious Diseases*, 8(12). PMID:12498657

Siegert, R., Shu, H. L., Slenczka, W., Peters, D., & Muller, G. (1967). Zur Ätiologie einer unbekannten, von Affen ausgegangenen menschlichen Infektionskrankheit. *Deutsche Medizinische Wochenschrift*, *92*(51), 2341–2343. doi:10.1055/s-0028-1106144 PMID:4294540

Simpson, D. I. H. (1978). Viral haemorrhagic fevers of man. *Bulletin of the World Health Organization*, *56*(6), 819. PMID:310725

Sindato, C., Karimuribo, E. D., Pfeiffer, D. U., Mboera, L. E. G., Kivaria, F., Dautu, G., & Paweska, J. T. et al. (2014). Spatial and temporal pattern of Rift Valley fever outbreaks in Tanzania; 1930 to 2007. *PLoS ONE*, *9*(2), e88897. doi:10.1371/journal.pone.0088897 PMID:24586433

Sindato, C., Karimuribo, E., & Mboera, E. G. (2012). The epidemiology and socio-economic impact of Rift Valley fever epidemics in Tanzania: A review. *The Onderstepoort Journal of Veterinary Research*, *79*(2), E1. doi:10.4102/ojvr.v79i2.467

Smallman-Raynor, M. R., & Cliff, A. D. (2004). *War Epidemics: An Historical Geography of Infectious Diseases in Military Conflict and Civil Strife, 1850-2000*. New York, New York: Oxford University Press.

Smallman-Raynor, M., & Cliff, A. D. (2001). Epidemic diffusion processes in a system of US military camps: Transfer diffusion and the spread of typhoid fever in the Spanish-American War. *Annals of the Association of American Geographers*, *91*(1), 71–91. doi:10.1111/0004-5608.00234

Smallman-Raynor, M., & Cliff, A. D. (2004). The geographical spread of cholera in the Crimean War: Epidemic transmission in the camp systems of the British Army of the East, 1854-55. *Journal of Historical Geography*, *30*(1), 32–69. doi:10.1016/S0305-7488(02)00084-1

Smith, A., Reacher, M., Smerdon, W., Adak, G. K., Nichols, G., & Chalmers, R. M. (2006). Outbreaks of waterborne infectious intestinal disease in England and Wales, 1992-2003. *Epidemiology and Infection*, *134*(6), 1141–1149. doi:10.1017/S0950268806006406 PMID:16690002

Smith, K. M., Anthony, S. J., Switzer, W. M., Epstein, J. H., Seimon, T., Jia, H., & Marano, N. et al. (2012). Zoonotic viruses associated with illegally imported wildlife products. *PLoS ONE, 7*(1), e29505. doi:10.1371/journal.pone.0029505 PMID:22253731

Smith, K. R., Jerrett, M., Anderson, H. R., Burnett, R. T., Stone, V., Derwent, R., & Thurston, G. et al. (2009). Public health benefits of strategies to reduce greenhouse-gas emissions: Health implications of short-lived greenhouse pollutants. *Lancet, 374*(9707), 2091–2103. doi:10.1016/S0140-6736(09)61716-5 PMID:19942276

Smith, K. R., Woodward, A., Campbell-Lendrum, D., Chadee, D. D., Honda, Y., Liu, Q., & Sauerborn, R. (2014). Human health: impacts, adaptation, and co-benefits. In C. B. Field, V. R. Barros, D. J. Dokken, K. J. Mach, M. D. Mastrandrea, T. E. Bilir, & L. L. White et al. (Eds.), *Climate Change 2014: Impacts, Adaptation, and Vulnerability. Part A: Global and Sectoral Aspects. Contribution of Working Group II to the Fifth Assessment Report of the Intergovernmental Panel on Climate Change* (pp. 709–754). Cambridge, UK: Cambridge University Press.

Snow, J. (1855). *On the mode of communication of cholera*. London: John Churchill.

Soumaré, P. O. L., Freire, C. C. M., Faye, O., Diallo, M., de Oliveira, J. V. C., Zanotto, P. M. A., & Sall, A. A. (2012). Phylogeography of Rift Valley Fever virus in Africa reveals multiple introductions in Senegal and Mauritania. *PLoS ONE, 7*(4), e35216. doi:10.1371/journal.pone.0035216 PMID:22539961

Southwood, T. R. E. (1977). Habitat, the templet for ecological strategies? *Journal of Animal Ecology, 46*(2), 337–365. doi:10.2307/3817

Stanke, C., Kerac, M., Prudhomme, C., Medlock, J., & Murray, V. (2013). Health effects of drought: A systematic review of the evidence. *PLoS Currents, 5*. doi:10.1371/currents.dis.7a2cee9e980f91ad7697b570bcc4b004 PMID:23787891

Staples, J. E., & Fischer, M. (2014). Chikungunya virus in the Americas—what a vectorborne pathogen can do. *The New England Journal of Medicine, 371*(10), 887–889. doi:10.1056/NEJMp1407698 PMID:25184860

Stearns, J. K. (2011). *Dancing in the Glory of Monsters: The collapse of the Congo and the great war of Africa*. New York, NY: PublicAffairs.

Steel, J., & Lowen, A. C. (2014). Influenza A virus reassortment. *Current Topics in Microbiology and Immunology, 385*, 377–401. doi:10.1007/82_2014_395 PMID:25007845

Steinmann, P., Keiser, J., Bos, R., Tanner, M., & Utzinger, J. (2006). Schistosomiasis and water resources development: Systematic review, meta-analysis, and estimates of people at risk. *The Lancet Infectious Diseases, 6*(7), 411–425. doi:10.1016/S1473-3099(06)70521-7 PMID:16790382

Sterk, A., Schijven, J., de Nijs, T., & Husman, A. M. D. (2013). Direct and Indirect Effects of Climate Change on the Risk of Infection by Water-Transmitted Pathogens. *Environmental Science & Technology, 47*(22), 12648–12660. doi:10.1021/es403549s PMID:24125400

Stern, N. (2006). The Stern Review on the Economics of Climate Change. HM Treasury. UK: Cambridge University press. Retrieved from http://www.hmtreasury.gov.uk/independent_reviews/stern_review_economics_climate_change/sternreview_index.cfm

Stocker, T. F., Qin, D., Plattner, G.-K., Tignor, M., Allen, S. K., . . . (Eds.), (2013). Climate change 2013: The physical science basis. Contribution of working group I to the 5th assessment report of the IPCC. Cambridge University Press. Retrieved from https://www.ipcc.ch/pdf/assessment-report/ar5/wg1/WG1AR5_Frontmatter_FINAL.pdf

Strange, R. N., & Scott, P. R. (2005). Plant disease: A threat to global food security. *Annual Review of Phytopathology*, *43*(1), 83–116. doi:10.1146/annurev.phyto.43.113004.133839 PMID:16078878

Strassburg, B. B. N., Latawiec, A. E., Barioni, L. G., Nobre, C. A., da Silva, V. P., Valentim, J. F., & Assad, E. D. et al. (2014). When enough should be enough: Improving the use of current agricultural lands could meet production demands and spare natural habitats in Brazil. *Global Environmental Change*, *28*, 84–97. doi:10.1016/j.gloenvcha.2014.06.001

Suarez, E., Morales, M., Cueva, R., Utreras Bucheli, V., Zapata-Ríos, G., Toral, E., & Vargas Olalla, J. et al. (2009). Oil industry, wild meat trade and roads: Indirect effects of oil extraction activities in a protected area in north-eastern Ecuador. *Animal Conservation*, *12*(4), 364–373. doi:10.1111/j.1469-1795.2009.00262.x

Sullivan, G. R., Bowman, F., Farrell, L. P., Gaffney, P. G., Kern, P. J., Lopez, T. J., & Zinni, A. C. et al. (2007). *National security and climate change*. Alexandria, VA: The CNA Corporation.

Susilawati, T. N., & McBride, W. J. (2014). Acute undifferentiated fever in Asia: A review of the literature. *The Southeast Asian Journal of Tropical Medicine and Public Health*, *45*(3), 719–726. Retrieved from http://www.ncbi.nlm.nih.gov/pubmed/24974656 PMID:24974656

Swanepoel, R., & Coetzer, J. A. W. (2004). Rift Valley fever. In J. A. W. Coetzer & R. C. Tustin (Eds.), *Infectious Diseases of Livestock* (2nd ed., pp. 1037–1070). Cape Town: Oxford University Press.

Tabachnick, W. J. (2010). Challenges in predicting climate and environmental effects on vector-borne disease episystems in a changing world. *The Journal of Experimental Biology*, *213*(6), 946–954.

Tabachnick, W. J. (2013). Nature, nurture and evolution of intra-species variation in mosquito arbovirus transmission competence. *International Journal of Environmental Research and Public Health*, *10*(1), 249–277. doi:10.3390/ijerph10010249 PMID:23343982

Talmon, J. L. (1980). European History as the Seedbed of the Holocaust. In J. Sonntag (Ed.), *Jewish Perspectives: 25 years of Jewish writing*. London: Secker & Warburg.

Taubenberger, J. K. (2006). The Origin and Virulence of the 1918 "Spanish" Influenza Virus. *Proceedings of the American Philosophical Society*, *150*(1), 86–112. PMID:17526158

Tauil, P. L. (2002). Aspectos críticos do controle do dengue no Brasil. *Cadernos de Saude Publica*, *18*(3), 867–871. doi:10.1590/S0102-311X2002000300035 PMID:12048612

Tauil, P. L., Santos, J. B., & Moraes, M. A. P. (2005). Febre Amarela. In J. R. Coura (Ed.), *Dinâmica das doenças infecciosas e parasitárias* (pp. 1755–1766). Rio de Janeiro: Guanabara Koogan.

Taylor, L. H., Latham, S. M., & Woolhouse, M. E. (2001). Risk factors for human disease emergence. *Philosophical Transactions of the Royal Society of London. Series B, Biological Sciences*, *356*(1411), 983–989. doi:10.1098/rstb.2001.0888 PMID:11516376

Teeling, E. C., Springer, M. S., Madsen, O., Bates, P., O'Brien, S. J., & Murphy, W. J. (2005). A molecular phylogeny for bats illuminates biogeography and the fossil record. *Science*, *307*(5709), 580–584. doi:10.1126/science.1105113 PMID:15681385

Teixeira, M. G., Paixão, E. S., & Costa, M. C. N. (2013). Recent epidemiological changes in South America. In E. H. Carmo, A. Gemal, Oliveira S. (Orgs), Health Surveillance in South America: epidemiological, sanitary and environmental (125-145). Rio de Janeiro: ISAGS.

Teixeira, A. R., Monteiro, P. S., Rebelo, J. M., Argañaraz, E. R., Vieira, D., Lauria-Pires, L., & Costa, J. M. et al. (2001). Emerging Chagas disease: Trophic network and cycle of transmission of *Trypanosoma cruzi* from palm trees in the Amazon. *Emerging Infectious Diseases*, *7*(1), 100–112. doi:10.3201/eid0701.070100 PMID:11266300

Teklehaimanot, H. D., Schwatrz, J., Teklehaimanot, A., & Lipsitch, M. (2004). Alert Threshold Algorithms and Malaria Epidemic Detection. *Emerging Infectious Diseases*, *10*(7), 1220–1226. doi:10.3201/eid1007.030722 PMID:15324541

Thai, K. T., & Anders, K. L. (2011). The role of climate variability and change in the transmission dynamics and geographic distribution of dengue. *Experimental Biology and Medicine*, *236*(8), 944–954. doi:10.1258/ebm.2011.010402 PMID:21737578

Thonnon, J., Picquet, M., Thiongane, Y., Lo, M., Sylla, R., & Vercruysse, J. (1999). Rift valley fever surveillance in the lower Senegal river basin: Update 10 years after the epidemic. *Tropical Medicine & International Health*, *4*(8), 580–585. doi:10.1046/j.1365-3156.1999.00437.x PMID:10499082

Tilley, S. A. (2003). "Challenging" Research Practices: Turning a Critical Lens on the Work of Transcription. *Qualitative Inquiry*, *9*(5), 750–773. doi:10.1177/1077800403255296

Tirado, M. C., Clarke, R., Jaykus, L. A., McQuatters-Gollop, A., & Frank, J. M. (2010). Climate change and food safety: A review. *Food Research International*, *43*(7), 1745–1765. doi:10.1016/j.foodres.2010.07.003

Toole, M. J., Galson, S., & Brady, W. (1993). Are war and public health compatible? *Lancet*, *341*(8854), 1193–1196. doi:10.1016/0140-6736(93)91013-C PMID:8098086

Tortora, G. J., Funke, B. R., & Case, C. L. (Eds.). (2010). *Microbiology: An Introduction*. San Francisco, CA: Benjamin Cummings.

Toure, Y. T., & Oduola, A. (2004). Focus: Malaria. *Nature Reviews. Microbiology*, *2*(4), 276–277. doi:10.1038/nrmicro870 PMID:15077609

Tourre, Y. M., Lacaux, J.-P., Vignolles, C., & Lafaye, M. (2009). Climate impacts on environmental risks evaluated from space: a conceptual approach to the case of Rift Valley Fever in Senegal. *Global Health Action, 2.*

Towner, J. S., Rollin, P. E., Bausch, D. G., Sanchez, A., Crary, S. M., Vincent, M., & Nichol, S. T. (2004). Rapid diagnosis of Ebola hemorrhagic fever by reverse transcription-PCR in an outbreak setting and assessment of patient viral load as a predictor of outcome. *Journal of Virology*, *78*(8), 4330–4341. Retrieved from http://www.ncbi.nlm.nih.gov/pubmed/15047846 doi:10.1128/JVI.78.8.4330-4341.2004 PMID:15047846

Trari, B., & Carnevale, P. (2011). De la préélimination à l'élimination du paludisme au Maroc. Quels risques pour l'avenir? *Bulletin de la Société de Pathologie Exotique*, *104*(4), 291–295. doi:10.1007/s13149-011-0156-2 PMID:21638201

Travassos Da Rosa, J. F. S., Travassos Da Rosa, A. P. A., Vasconcelos, P. F. C., Pinheiro, F. P., Rodrigues, S. G., Travassos Da Rosa, E. S., . . . Cruz, A. C. R. (1998). Arboviruses isolated in the Evandro Chagas Institute, including some described for the first time in Brazilian Amazon region, their known hosts, and their pathology for man. In A. P. A. Travassos Da Rosa, P. F. C. Vasconcelos And J. F. S. Travassos Da Rosa (Eds.), An overview of arbovirology in Brazil and neighbouring countries (pp. 19-31). Belém, Pará: Instituto Evandro Chagas.

Travassos da Rosa, J. F. S., Pinheiro, F. P., & Vasconcelos, P. F. C. (2005). Febres hemorrágicas viróticas. In *J. R. Coura (Org.), Dinâmica das doenças infecciosas e parasitárias* (Vol. II, pp. 1741–1754). Rio de Janeiro: Guanabara-Koogan.

Tukur, A. I. (2010): Temporal variation of malaria occurrence in Kano municipal local government area. *Bayero Journal of pure and applied science*, 3(1). Retrieved from www.ajol.info/index.php/bajopas/article/viewFile/58745/47071

Turell, M. J., Dohm, D. J., Mores, C. N., Terracina, L., Wallette, D. L. Jr, Hribar, L. J., & Blow, J. A. et al. (2008). Potential for North American Mosquitoes to Transmit Rift Valley Fever Virus 1. *Journal of the American Mosquito Control Association*, *24*(4), 502–507. doi:10.2987/08-5791.1 PMID:19181056

Turell, M. J., Linthicum, K. J., & Beaman, J. R. (1990). Transmission of Rift Valley fever virus by adult mosquitoes after ingestion of virus as larvae. *The American Journal of Tropical Medicine and Hygiene*, *43*(6), 677–680. PMID:2267972

Turell, M. J., Sardelis, M. R., O'Guinn, M. L., & Dohm, D. J. (2002). Potential vectors of West Nile virus in North America. In *Japanese Encephalitis and West Nile Viruses* (Vol. 267, pp. 241–252). Berlin: Springer-Verlag Berlin. doi:10.1007/978-3-642-59403-8_12

Turner, L. R., Alderman, K., Connell, D., & Tong, S. (2013). Motivators and Barriers to Incorporating Climate Change-Related Health Risks in Environmental Health Impact Assessment. *International Journal of Environmental Research and Public Health*, *10*(3), 1139–1151. doi:10.3390/ijerph10031139 PMID:23525029

Turney, C. S. M., Baillie, M., Palmer, J., & Brown, D. (2006). Holocene climatic change and past Irish societal response. *Journal of Archaeological Science, 33*(1), 34-38. doi:10.1016/j.jas.2005.05.014

U.S. Agency for International Development. Pan American Health Organization (2010). The Amazon Malaria Initiative: Goals and Accomplishments October 2001–May 2009 (pp. 121). Washington.

UK Ministry of Defence. (2014). *Global Strategic Trends - Out to 2045*. UK Ministry of Defence.

UNAIDS. (2013). 2013 Global Fact Sheet. Retrieved from http://www.unaids.org/en/resources/campaigns/globalreport2013/factsheet

UNFCC. (1992). United nation framework convention on climate change. Retrieved from: http://unfccc.int/resource/docs/convkp/conveng.pdf

UNHCR. (2015a). Syria regional refugee response. Retrieved from data.unhcr.org/syrianrefugees/regional/php

UNHCR. (2015b). *UNHCR Global Trends Forced Displacement in 2014*. Geneva: UNHCR.

UNICEF and World Health Organization. (2015). Progress on sanitation and drinking water. 2015 update and MDG assessment. USA.

UNICEF. (2012). Pneumonia and diarrhoea: Tackling the deadliest diseases for the world's poorest children. Retrieved from: www.unicef.org/eapro/Pneumonia_and_Diarrhoea_Report_2012.pdf

United Nations Environment Programme. (2014). Illegal Trade in Willife *UNEP Year Book 2014 emerging issues update.*

USAID. (2014). *PREDICT 2*. Retrieved from https://www.usaid.gov/sites/default/files/documents/1864/Predict2-factsheet.pdf

USDA APHIS Veterinary Services. (2015). *Epidemiologic and Other Analyses of HPAI-Affected Poultry Flocks: June 15, 2015 Report.*

USDA APHIS Veterinary Services. (2016). *2016 HPAI Preparedness and Response Plan.* Retrieved from https://www.aphis.usda.gov/animal_health/downloads/animal_diseases/ai/hpai-preparedness-and-response-plan-2015.pdf

USDHEW - U.S. Department of Health, Education, and Welfare. (1963). *Vital Statistics of United States* (Vol. 2). US Department of Health, Education and Welfare. Public Health Service, Washington, DC.

USEPA – U.S. Environmental Protection Agency. (2000). The history of drinking water treatment. United States Environmental Protection Agency, Office of Water, EPA-816-F-00-006.

Valle, D., & Lima, J. M. T. (2014). Large-scale drivers of malaria and priority areas for prevention and control in the Brazilian Amazon region using a novel multi-pathogen geospatial model. *Malaria Journal, 13*(1), 1–13. doi:10.1186/1475-2875-13-443 PMID:25412882

Vasconcelos, P. F., Travassos da Rosa, A., Rodrigues, S. G., Travassos da Rosa, E. S., Dégallier, N., da Rosa, T., & Jorge, F. S. (2001). Inadequate management of natural ecosystem in the Brazilian Amazon region results in the emergence and reemergence of arboviruses. *Cadernos de Saude Publica*, *17*, S155–S164. doi:10.1590/S0102-311X2001000700025 PMID:11426277

Vega-Rúa, A., Zouache, K., Girod, R., Failloux, A. B., & Lourenço-de-Oliveira, R. (2014). High level of vector competence of *Aedes aegypti* and *Aedes albopictus* from ten American countries as a crucial factor in the spread of Chikungunya virus. *Journal of Virology*, *88*(11), 6294–6306. doi:10.1128/JVI.00370-14 PMID:24672026

Vignolles, C., Lacaux, J.-P., Tourre, Y. M., Bigeard, G., Ndione, J.-A., & Lafaye, M. (2009). Rift Valley fever in a zone potentially occupied by Aedes vexans in Senegal: Dynamics and risk mapping. *Geospatial Health*, *3*(2), 211–220. doi:10.4081/gh.2009.221 PMID:19440963

Vignolles, C., Tourre, Y. M., Mora, O., Imanache, L., & Lafaye, M. (2010). TerraSAR-X high-resolution radar remote sensing: An operational warning system for Rift Valley fever risk. *Geospatial Health*, *5*(1), 23–31. doi:10.4081/gh.2010.184 PMID:21080318

Vilaly, A. E. E., Arora, M., Butterworth, M. K., Vilaly, M. A. M. E., Jarnagin, W., & Comrie, A. C. (2013). Climate, environment and disease: The case of Rift Valley fever. *Progress in Physical Geography*, *37*(2), 259–269. doi:10.1177/0309133313478315

Vilela, M. L., Azevedo, C. G., Carvalho, B. M., & Rangel, E. F. (2011). Phlebotomine Fauna (Diptera: Psychodidae) and putative vectors of Leishmaniases in impacted area by hydroelectric plant, state of Tocantins, Brazil. *PLoS ONE*, *6*(12), e27721. doi:10.1371/journal.pone.0027721 PMID:22163271

Vitoria, M., Granich, R., Gilks, C. F., Gunneberg, C., Hosseini, M., Were, W., & De Cock, K. M. et al. (2009). The Global Fight Against HIV/AIDS, Tuberculosis, and Malaria- Current Status and Future Perspectives. *American Journal of Clinical Pathology*, *131*(6), 1844–1848. doi:10.1309/AJCP5XHDB1PNAEYT PMID:19461091

Vittor, A. Y., Pan, W., Gilman, R. H., Tielsch, J., Glass, G., Shields, T., & Patz, J. A. et al. (2009). Linking deforestation to Malaria in the Amazon: Characterization of the breeding habitat of the principal malaria vector, *Anopheles darlingi. The American Journal of Tropical Medicine and Hygiene*, *81*(1), 5–12. PMID:19556558

Vorosmarty, C. J., McIntyre, P. B., Gessner, M. O., Dudgeon, D., Prusevich, A., Green, P., & Davies, P. M. et al. (2010). Global threats to human water security and river biodiversity. *Nature*, *467*(7315), 555–561. doi:10.1038/nature09440 PMID:20882010

Wagenaar, J. A., & Scott, H. M. (2015). Reducing antimicrobial usage in agriculture and aquaculture: Beyond regulatory policy. *Zoonoses and Public Health*, *62*(Suppl. 1), 1–2. doi:10.1111/zph.12197 PMID:25903490

Wagner, D. M., Klunk, J., Harbeck, M., Devault, A., Waglechner, N., Sahl, J. W., & Poinar, H. et al. (2014). *Yersinia pestis* and the Plague of Justinian 541–543 AD: A genomic analysis. *The Lancet Infectious Diseases*, *14*(4), 319–324. doi:10.1016/S1473-3099(13)70323-2 PMID:24480148

Walter, C. T., & Barr, J. N. (2011). Recent advances in the molecular and cellular biology of bunyaviruses. *The Journal of General Virology*, *92*(Pt 11), 2467–2484. doi:10.1099/vir.0.035105-0 PMID:21865443

Wandiga, S. O., Opondo, M., Olago, D., Githeko, A., Githui, F., Marshall, M., & Ouma, G. O. (2010). Vulnerability to epidemic malaria in the highlands of Lake Victoria basin: The role of climate change/variability, hydrology and socio-economic factors. *Climatic Change*, *99*(3), 473–497. doi:10.1007/s10584-009-9670-7

Wardekker, J. A., de Jong, A., van Bree, L., Turkenburg, W. C., & van der Sluijs, J. P. (2012). Health risks of climate change: An assessment of uncertainties and its implications for adaptation policies. *Environmental Health*, *11*(1), 67. doi:10.1186/1476-069X-11-67 PMID:22992311

Watts, N., & al., e. (June 23, 2015). Health and climate change: policy responses to protect human health. *Lancet Commission on Health and Climate Change*.

Watts, N., Adger, W. N., Agnolucci, P., Blackstock, J., Byass, P., Cai, W., et al. (2015). Health and climate change: Policy responses to protect public health. *The Lancet*.

Weaver, S. C., & Reisen, W. K. (2010). Present and future arboviral threats. *Antiviral Research*, *85*(2), 328–345. doi:10.1016/j.antiviral.2009.10.008 PMID:19857523

Welzer, H. (2012). *Climate Wars: Why people will be killed in the twenty-first century* (P. Camiller, Trans.). Cambridge: Polity Press.

Wensman, J. J., Lindahl, J., Wachtmeister, N., Torsson, E., Gwakisa, P., Kasanga, C., & Misinzo, G. (2015). A study of Rift Valley fever virus in Morogoro and Arusha regions of Tanzania - serology and farmers' perceptions. *Infection Ecology & Epidemiology*, *5*(0), 30025. doi:10.3402/iee.v5.30025 PMID:26584830

Whitmee, S., Haines, A., Beyrer, C., Boltz, F., Capon, A., de Souza Dias, B. F., & Yach, D. et al. (2015). Safeguarding Human Health in the Anthropocene Epoch: Report of The Rockefeller Foundation-Lancet Commission on Planetary Health. *Lancet*, *386*(10007), 1973–2028. doi:10.1016/S0140-6736(15)60901-1 PMID:26188744

WHO - World Health Organization. (2004). *WHO Guidelines for drinking Water* (3rd ed.). Geneva, Switzerland: WHO press.

WHO - World Health Organization. (2015). *World Health Statistics 2015*. Geneva, Switzerland: WHO press.

WHO. (2015). S., Lau, S., Woo, P., & Yuen, K. Y. (2007). Bats as a continuing source of emerging infections in humans. *Reviews in Medical Virology*, *17*(2), 67–91. Retrieved from http://www.who.int/csr/disease/ebola/enWong doi:10.1002/rmv.520

Wilcox, B. A., & Gubler, D. J. (2005). Disease ecology and the global emergence of zoonotic pathogens. *Environmental Health and Preventive Medicine*, *10*(5), 263–272. doi:10.1007/BF02897701 PMID:21432130

Wiley, L.F. (2010). Mitigation/ adaptation and health: Health policymaking in the global response to climate change and implications for other upstream determinants. *Journal of Law, Medicine and Ethics*, 38(3), 629.

Wilkie, D. S., & Carpenter, S. F. (1999). Bushmeat hunting in the Congo Basin: An assessment of impacts and options for mitigation. *Biodiversity and Conservation*, *8*(7), 927–955. doi:10.1023/A:1008877309871

William, B. (1997). *Infectious Diseases in History: a guide to causes and effects*. Prepared by Brian Williams in 1997. For distribution within the Department of Economic and Social History at the University of Hull, England. http://urbanrim.org.uk/diseases.htm

Winkler, M. S., Krieger, G. R., Divall, M. J., Cisse, G., Wielga, M., Singer, B. H., & Utzinger, J. et al. (2013). Untapped potential of health impact assessment. *Bulletin of the World Health Organization*, *91*(4), 298–305. doi:10.2471/BLT.12.112318 PMID:23599554

Wolfe, N. D., Dunavan, C. P., & Diamond, J. (2007). Origins of major human infectious diseases. *Nature*, *447*(7142), 279–283. doi:10.1038/nature05775 PMID:17507975

Wong, G., Liu, W., Liu, Y., Zhou, B., Bi, Y., & Gao, G. F. (2015). MERS, SARS, and Ebola: The Role of Super-Spreaders in Infectious Disease. *Cell Host & Microbe*, *18*(4), 398–401. PMID:26468744

Woods, C. W., Karpati, A. M., Grein, T., McCarthy, N., Gaturuku, P., Muchiri, E., & Swanepoel, R. et al. (2002). An outbreak of Rift Valley fever in northeastern Kenya, 1997-98. *Emerging Infectious Diseases*, 8(12). PMID:11897064

World Bank. (2012). *People, Pathogens and Our Planet: The Economics of One Health*. Retrieved from https://openknowledge.worldbank.org/handle/10986/11892

World Bank. (2012). *People, pathogens, and our planet. Volume 2: the economics of one health*. Washington, DC. Retrieved from http://documents.worldbank.org/curated/en/2012/06/16360943/people-pathogens-planet-economics-one-health

World Bank. (2014). *The economic impact of the 2014 Ebola epidemic: short and medium term estimates for West Africa.*

World Economic Forum. (2015). *The Global Risks report 2015*. Retrieved from http://reports.weforum.org/global-risks-2015/

World Health Organisation. (2011). Guidelines for drinking-water quality, fourth edition. Retrieved from: http://www.who.int/water_sanitation_health/publications/dwq_guidelines/en/

World health organization (WHO) year book 2002. World Health Organization (WHO)/ World Meteorological Organization (WMO)/ United Nations.

World Health Organization (WHO), & Pan-American Health Organization (PAHO). (2013). *Protecting health from climate change: vulnerability and adaptation assessment*. World Health Organization. Retrieved from http://apps.who.int//iris/handle/10665/104200

World Health Organization (WHO). (2008). *Protecting Health from Climate Change - World Health Day 2008*. Geneva, Switzerland: WHO Press.

World Health Organization and Convention on Biological Diversity. (2015). *Connecting Global Priorities: Biodiversity and Human Health, a State of Knowledge Review*.

World Health Organization and UNICEF. (2014). Progress on drinking water and sanitation. Joint Monitoring Programme update 2014.

World Health Organization. (1976b). Ebola haemorrhagic fever in Zaire. *Bulletin of the World Health Organization*, *56*, 271–293. PMID:307456

World Health Organization. (1990). Health education in the control of schistosomiasis (Ch. 8). Freshwater snails. Intermediate hosts of schistosomiasis and foodborne trematode infections, 337p.

World Health Organization. (2006) The African malaria Report, Regional office for Africa and East Mediterranean.

World Health Organization. (2009). Para salvar vidas: hagamos que los hospitales sean seguros en las situaciones de emergencia.

World Health Organization. (2012a). *Accelerating work to overcome the global impact of neglected tropical diseases: a roadmap for implementation*. Geneva, Switzerland: WHO.

World Health Organization. (2012b). *Élimination de la schistosomiase au Maroc, une réalité et un succès après trois décennies de lutte*.

World Health Organization. (2013). Diarrhoeal disease. Retrieved from: http://www.who.int/mediacentre/factsheets/fs330/en/

World Health Organization. (2013). Impact of Dengue World Health Organization, Geneva. IPCC WGII AR5, 2013.

World Health Organization. (2014). *Information to States Parties regarding determination of fulfilment of IHR national Core Capacity requirements and potential extensions*. Retrieved from http://apps.who.int/iris/bitstream/10665/112789/1/WHO_HSE_GCR_2014.9_eng.pdf?ua=1

World Health Organization. (2014). *WHO Guidance for Countries on Combining Indoor Residual Spraying and Long-Lasting Insecticidal Nets*. Geneva, Switzerland: World Health Organization.

World Health Organization. (2016). *Ebola Situation Report - 17 February 2016*. Retrieved from http://apps.who.int/ebola/current-situation/ebola-situation-report-17-february-2016

World Heath Organization. (1976a). Ebola haemorrhagic fever in Sudan. Report of a WHO/International Study Team. *Bulletin of the World Health Organization*, *56*, 247–270. PMID:307455

Worrall, E., Connor, S. J., & Thomson, M. C. (2007). A model to simulate the impact of timing, coverage and transmission intensity on the effectiveness of indoor residual spraying (IRS) for malaria control. *Tropical Medicine & International Health*, *12*(1), 75–88. doi:10.1111/j.1365-3156.2006.01772.x PMID:17207151

Wu, X., Lu, Y., Zhou, S., Chen, L., & Xu, B. (2016). Impact of climate change on human infectious diseases: Empirical evidence and human adaptation. *Environment International*, *86*, 14–23. doi:10.1016/j.envint.2015.09.007 PMID:26479830

WWF / Dalberg. (2012). *Fighting illicit wildlife trafficking: A consultation with governments.*

Xie, H., You, L., Wielgosz, B., & Ringler, C. (2014). Estimating the potential for expanding smallholder irrigation in Sub-Saharan Africa. *Agricultural Water Management*, *131*, 183–193. doi:10.1016/j.agwat.2013.08.011

Yahaya, A., Tyav, Y. B., & Nura, M. S. (2014). Seroprevalence of Paediatric Malaria Infections in Two Hospitals in Kano State, Nigeria. *International Journal of Sciences: Basic and Applied Research*, *14*(1). Retrieved from http://gssrr.org/index.php?journal=JournalOfBasicAndApplied&page=issue&op=view&path%5B%5D=82

Yardley, J., Sigal, R. J., & Kenny, G. P. (2011). Heat health planning: The importance of social and community factors. *Global Environmental Change*, *21*(2), 670–679. doi:10.1016/j.gloenvcha.2010.11.010

Yassi, A., Kjellstrom, T., de Kok, T., & Guidotti, T. L. (2001). *Basic Environmental Health*. New York, New York: Oxford University Press. doi:10.1093/acprof:oso/9780195135589.001.0001

Yates, T. L., Mills, J. N., Parmenter, C. A., Ksiazek, T. G., Parmenter, R. R., Vande Castle, J. R., & Peters, C. R. (2002). The ecology and evolutionary history of an emergent disease: Hantavirus pulmonary syndrome. *Bioscience*, *52*(11), 989–998. doi:10.1641/0006-3568(2002)052[0989:TEAEHO]2.0.CO;2

You, L. Z. (2008). Africa: Irrigation investment Needs in Sub-Saharan Africa. Retrieved from https://openknowledge.worldbank.org/handle/10986/7870

You, L., Xie, H., Wood-Sichra, U., Guo, Z., & Wang, L. (2014). Irrigation potential and investment return in Kenya. *Food Policy*, *47*, 34–45. doi:10.1016/j.foodpol.2014.04.006

Young, C. C., & Olival, K. J. (2016). Optimizing Viral Discovery in Bats. *PLoS ONE*, *11*(2), e0149237. doi:10.1371/journal.pone.0149237 PMID:26867024

Younger, M., Morrow-Almeida, H. R., Vindigni, S. M., & Dannenberg, A. L. (2008). The built environment, climate change, and health: Opportunities for co-benefits. *American Journal of Preventive Medicine*, *35*(5), 517–526. doi:10.1016/j.amepre.2008.08.017 PMID:18929978

Yuen, K. Y. (2015). From SARS to MERS and Ebola. *Journal of Microbiology, Immunology, and Infection*, *48*(2), S3. doi:10.1016/j.jmii.2015.02.002

Yusuf, D. M. (2008, August 25-28). An assessment of malaria prevalence in Bichi local government area Kano state. *Proceedings of the 50th Annual national conference The Association Nigerian Geographers (ANG)*, University of Calabar Cross River State.

Zaki, A. M., van Boheemen, S., Bestebroer, T. M., Osterhaus, A. D., & Fouchier, R. A. (2012). Isolation of a novel coronavirus from a man with pneumonia in Saudi Arabia. *The New England Journal of Medicine*, *367*(19), 1814–1820. doi:10.1056/NEJMoa1211721 PMID:23075143

Zhang, D. D., Brecke, P., Lee, H. F., He, Y.-Q., & Zhang, J. (2007). Global climate change, war, and population decline in recent human history. *Proceedings of the National Academy of Sciences of the United States of America*, *104*(49), 19214–19219. doi:10.1073/pnas.0703073104 PMID:18048343

Zhang, D. D., Lee, H. F., Wang, C., Li, B., Pei, Q., Zhang, J., & An, Y. (2011). The causality analysis of climate change and large-scale human crisis. *Proceedings of the National Academy of Sciences of the United States of America*, *108*(42), 17296–17301. doi:10.1073/pnas.1104268108 PMID:21969578

Zhang, D. D., Zhang, J., Lee, H. F., & He, Y.-Q. (2007). Climate change and war frequency in eastern China over the last millennium. *Human Ecology*, *35*(4), 403–414. doi:10.1007/s10745-007-9115-8

Zhang, D.D., Jim, C.Y., Lin, G.C.-S., He, Y.-Q., Wang, J.J., & Lee, H.F. (2006). Climatic change, wars and dynastic cycles in China over the last millennium. *Climatic Change*, *76*(3-4), 459–477. doi:10.1007/s10584-005-9024-z

Zhang, G., Cowled, C., Shi, Z., Huang, Z., Bishop-Lilly, K. A., Fang, X., & Wang, J. (2013). Comparative analysis of bat genomes provides insight into the evolution of flight and immunity. *Science*, *339*(6118), 456–460. doi:10.1126/science.1230835 PMID:23258410

Zhou, J., Lau, K. M., Laughlin, L. W., Masuoka, M., Chamberlin, J., & Lawyer, P. (2001). The effect of regional climate variability on outbreak of Bartonellosis epidemics in Peru. *Proceedings of the third Symposium on Environmental Applications: Facilitating the Use of Environmental Information*. Orlando, Florida: American Meteorological Society.

Zhou, G., Minakawa, N., Githeko, A. K., & Yan, G. (2004). Association between climate variability and malaria epidemics in the East African highlands. *Proceedings of the National Academy of Sciences of the United States of America*, *101*(8), 2375–2380. doi:10.1073/pnas.0308714100 PMID:14983017

Zubair, L., Galappaththy, G. N., Yang, H., Chandimala, J., Yahiya, Z., Amerasinghe, P., & Connor, S. J. et al. (2008). Epochal changes in the association between malaria epidemics and El Niño in Sri Lanka. *Malaria Journal*, *7*(1), 140. doi:10.1186/1475-2875-7-140 PMID:18652697

About the Contributors

Maha Bouzid is currently a Senior Research Associate at the Norwich Medical School, University of East Anglia. Her broad research interests are emerging infectious diseases and Global Health. Her primary interests are the epidemiology and pathogenesis of waterborne and vector-borne diseases. She has a Master's in Microbiology, Molecular Biology and Biotechnology and a PhD in Medical Microbiology awarded from the University of East Anglia, investigating mechanisms of virulence in *Cryptosporidium* species using genomic data. Her research interests have recently expanded to the field of climate change and human health as well as evidence based decision making in the public health sector.

* * *

Mohamed Behnassi, specialist in Environment and Human Security Politics, is Associate Professor, Director of the Research Laboratory for Territorial Governance, Human Security and Sustainability (LAGOS), and ex-Head of Public Law Department (2014-2015) at the Faculty of Law, Economics and Social Sciences, Ibn Zohr University of Agadir, Morocco. He was the Founder and Director of the North-South Center for Social Sciences (NRCS) from 2008 to 2015, and currently Director of The Center for Research on Environment, Human Security and Governance (CERES). Behnassi is Member of the Moroccan Association for Regional Sciences (AMSR), and Active member of the National Council for Climate Change, Sustainable Development and Leadership (NCCSD), India. He is also Senior Social Compliance Auditor (practicing consultancy by monitoring human rights at work and the sustainability of the global supply chain). Behnassi obtained his Ph.D in 2003 from the Faculty of Law, Economics and Social Sciences, Hassan II University of Casablanca for a Thesis titled: Multilateral Environmetal Negotiations: Towards a Global Governance for Environment. In 2011, he completed a U.S. State Department-sponsored Civic Education and Leadership Fellowship (CELF) at the Maxwell School of Citizenship and Public Affairs, Syracuse University, USA and in 2014 he obtained a Diploma in Diplomacy and International Environmental Law from the University of Eastern Finland and the United Nations Environment Programme (UNEP), Finland. Behnassi has pursued several post-doc trainings since the completion of his Ph.d. His core teaching and expertise areas cover: Environmental change, human security, sustainability, climate change politics and governance, human rights, CSR, etc. He has published numerous books with international publishers such as: Vulnerability of Agriculture, Water and Fisheries to Climate Change (Springer 2014); Science, Policy and Politics of Modern Agricultural System (Springer 2014); Sustainable Food Security in the Era of Local and Global Environmental Change (Springer 2013), Global Food Insecurity (Springer, 2011); Sustainable Agricultural Development (Springer, 2011); Health, Environment and Development (European University Editions, 2011), and Climate Change, Energy Crisis and Food Security (Ottawa University Press, 2011). He has also published numerous research papers and communicated several presentations in relevant international conferences. Behnassi has also organized many international conferences covering the above research areas in collaboration with national and international organizations and managed many research and expertise projects on behalf of various institutions such as the Royal Institute of Strategic Studies (IRES) in Morocco (Food and health security face to climate change in Morocco), GIZ (Mainstreaming climate risk in local development planning), Carrefour Associatif and UNICEF (University Social Responsibility).

Bernard Bett is a veterinary epidemiologist based at the International Livestock Research Institute. He studied epidemiology at the University of Nairobi, Kenya and the University of Guelph, Canada. He completed his PhD research, which focused on tsetse and tryapnosomosis control in Kenya, in 2008. Since then, he has implemented multiple studies on emerging infectious diseases in Africa and Asia which investigate the effects of climate, land use and biodiversity changes on the emergence and transmission of these diseases.

Samia Boussaa is currently Professor in ISPITS (Institut Supérieur des Professions Infirmières et des Techniques de Santé) of Ministry of Health in Marrakesh (Morocco). She obtained her PhD, in Ecology and Epidemiology, at Cadi Ayyad University (Morocco) and at Strasbourg University (France) in 2008. Her research is focused on Medical Entomology, Vector Borne Disease and especially on Epidemiology of leishmaniasis in Morocco.

Devin C Bowles is a PhD candidate at the National Centre for Epidemiology and Population Health at the Australian National University. His current research focuses on the socially-mediated health effects of climate change, including through conflict, migration and health inequity. Previous publications cover a range of topics, including climate change, religion and prosopagnosia (face blindness).

Chris G Buse received his PhD from the Dalla Lana School of Public Health at the University of Toronto. He is now leading a new initiative at the University of Northern British Columbia analyzing the cumulative and community impacts of resource development, climate change, and environmental degradation.

Ulisses Confalonieri is a Veterinarian and Medical Doctor, and has a Ph.D. in Sciences and is a full professor at the Oswaldo Cruz Foundation and the Federal University in Rio de Janeiro, Brazil. He has developed studies on epidemiology; ecology of infectious diseases, especially emerging diseases; medical entomology; health of natural ecosystems and impacts of global environmental changes on health, especially climate change. He has also coordinated the health committee at the Intergovernmental Panel on Climate Change (IPCC) and at the Millennium Ecosystem Assessment.

Samir Dervisevic works as a Consultant Medical Virologist and Head of the Department of Microbiology, the Eastern Pathology Alliance and as an Honorary Senior Lecturer at the University of East Anglia in Norwich.

Delia Grace, Program Leader Food Safety and Zoonoses, International Livestock Institute, Nairobi and Theme Leader, Agriculture Associated Disease, CGIAR Research Program Agriculture for Nutrition and Health Delia is an epidemiologist and veterinarian with 20 years' experience working in developing countries in Africa and Asia. She leads research on zoonoses and foodborne disease at the International Livestock Research Institute in Kenya and the CGIAR Research Program on Agriculture for Human Nutrition and Health. Her research interests include emerging diseases, participatory epidemiology, gender and animal welfare. Her career has spanned the private sector, field-level community development and aid management, as well as research.

Kahime Kholoud is a PhD student at the University of Marrakesh – Morocco. She holds a Master of Science in Environmental-Health Engineering. Currently, Kholoud is a PhD student working on leishmaniasis, vector ecology, parasite trend, epidemiology, and vulnerability and adaptations capacities. Ms Kahime is focalized to identify the impacts of surface climate changes on the occurrence of leishmaniasis in Morocco with a specific intent to model the relation between changes in surface parameters, and the cycle of the leishmaniasis and its incidence.

Salisu Lawal Halliru is a Lecturer in the Department of Geography, Federal College of Education Kano, Nigeria. He had his Bachelor degree of Science in Geography and postgraduate Diploma on Environmental Management at Bayero University Kano. He had Master's Degree in Geography at Kampala International University, Uganda. He undertakes few projects on climate change issues, sustainable development and poverty indices were he presented his research work on both national and international conferences; he is a member Association of Nigerian Geographers, African Geopark network, and member research and development network in Nigeria among others. He authored and co-authored Book chapter contributions, numerous journal articles, including International Journal of Sustainable Development, Volume 2, Number 3, 2009; Climate Change and rural water supply in Nigeria, An Appraisal of Climate Change: Global risks challenges and community Based Adaptation Approaches for Sustainable Development in Northern Nigeria. He is also an associate editor International Journal of Climate Change: Impacts and Responses Volume 2, number 1 and 2, in 2010. His research interest focus in the areas of Environmental Management and Sustainable development, Population Dynamics, Water resources management, rural development, Poverty and Education sustainability, with special interest in Geopark, Geoheritage and Goe-Education, Climate Change and food security. He is currently working on Climate change, Adaptation, and Human migration in northwestern Nigeria.

Johanna Frida Lindahl is a veterinary epidemiologist working at Swedish University of Agricultural Sciences and International Livestock Research Institute (ILRI). Johanna graduated from Swedish University of Agricultural Sciences after doing her PhD working on Japanese encephalitis virus in Vietnam. Since her PhD she has been focusing her research on food safety, mycotoxins, and vector-borne, zoonotic and emerging infectious diseases in developing countries, mainly East Africa and South and Southeast Asia.

Catherine Machalaba is the Program Coordinator for Health and Policy at Eco-Health Alliance. Her research interests focus on One Health approaches to human, animal, and environmental health challenges and strengthening public health systems. She is a researcher on the USAID Emerging Pandemic Threats PREDICT project. Ms. Machalaba serves as the Science Officer for the Future Earth oneHEALTH project, which is aimed at generating solutions-oriented research to address the health impacts of global environmental change. She holds an undergraduate degree in Biology and a Master's in Public Health, and is currently a doctoral student in environmental and occupational health at the City University of New York School of Public Health.

Carina Margonari is a Biologist, with a Ph.D. in parasite biology, and is a researcher at the René Rachou Research Center and works with molecular biology, parasitology, entomology and epidemiology, especially with topics related to tropical diseases, phlebotomines and leishmaniasis.

Júlia Alves Menezes is a Biologist, MSc in Sciences by the René Rachou Research Center, Oswaldo Cruz Foundation. Her work themes are related to public health, eco-epidemiology of infectious diseases, impacts of climate change on human health and socio and environmental vulnerability to environmental change.

Timothy Robinson is a Principal Scientist within the Livestock Systems and Environment (LSE) research theme of the International Livestock Research Institute (ILRI). He has more than 20 years of experience working in the field of spatial analysis in relation to agriculture, food security and poverty alleviation, during which he has worked within the United Nations, the CGIAR system, UK universities and government departments. He has published extensively on a variety of topics; most recently relating to the global livestock sector. His research interests include the application of spatial analytical techniques to understanding and predicting current and future livestock species and production systems distributions – particularly in the context of social, environmental and epidemiological risks and opportunities associated with a changing livestock sector.

Cristina Romanelli works at the nexus of biodiversity conservation, ecosystem and human health with the United Nations Secretariat for the Convention on Biological Diversity where she coordinates the joint work programme with the World Health Organization. Her work focuses on global policy development, building international cross-sectoral partnerships, conducting interdisciplinary research at the science-policy interface, and building national capacity to support the implementation of the Strategic Plan for Biodiversity 2011-2020, the Sustainable Development Goals and post-2015 development agenda. She has organized several regional capacity-building workshops including in the Americas and Africa regions. She is a research fellow with the United Nations University - International Institute for Global Health, Malaysia. She has also co-authored various book chapters, UN reports, and peer-reviewed publications, including with The Rockefeller Foundation-Lancet Commission on Planetary Health. She recently led the development of the state of knowledge review, Connecting Global Priorities: Biodiversity and Human Health co-published by WHO and SCBD. She holds an MA in international relations, an MSc in environmental impact assessment and is pursuing Doctoral studies at University College London, UK.

Maria Ines Zanoli Sato graduated in Biomedicine at São Paulo Federal University, obtained the PhD in environmental microbiology from São Paulo University in 1995. Director of the Department of Environmental Analysis at CETESB, the Environmental Company of São Paulo State, where leads a team of biologists, pharmaceutical-biochemists, chemists, biomedical scientists and technicians responsible for the analytical and scientific support of environmental quality monitoring programmes and enforcement activities carried out by the Company. Has strong scientific knowledge and experience in the environmental sciences and the management of analytical laboratories and scientific projects.

Peter Stoett is Director of the Loyola Sustainability Research Centre and Professor in the Department of Political Science at Concordia University in Montreal. His main areas of expertise include international relations and law, global environmental politics, and human rights. His latest single-authored book is *Global Ecopolitics: Crisis, Governance, and Justice* (University of Toronto Press). He is currently working on several issues including invasive species policy, transnational environmental crime, and oceans governance.

Index

E

F

G

H

I

K

L

M

N

O

P

R

S

T

U

V

W

Z

www.ingramcontent.com/pod-product-compliance
Lightning Source LLC
LaVergne TN
LVHW061549290926
849162LV00004B/950